Praise for

REVOLUTIONARIES

"*Revolutionaries* is much more than a convenient inventory of truths. While Rakove provides a cogent summary of what scholars know about the political history of the late-eighteenth century, what gives his book real distinction is the skill with which he delivers this knowledge through a series of interlocking biographical narratives."

—*Miami Herald*

"Rakove offers a consolation to modern liberals: that no matter how serious the crises, we will somehow find what we need to make it through. [*Revolutionaries*] is a bedtime story for grown-ups."

—*Washington Post*

"A brilliant account of the Revolution that focuses on the men who led it. Rakove offers the reader a new and fascinating method of combining analytic and narrative history. By using some marvelous vignettes and beautifully crafted portraits of the principal characters as a means of exploring the crucial issues of the Revolutionary era, including state constitution-making, diplomacy, anti-slavery, and the framing of the federal government, he has written a remarkable work of history."

—Gordon S. Wood, author of *Empire of Liberty* and *Revolutionary Characters*

"Rakove works these familiar individual lives and stories together into a seamless and authoritative narrative of the Revolution, which shows against all the odds that there is something new still to be said about even these deadest and whitest of dead white men."

—TheNewRepublic.com

"A mighty book . . . indispensable to understanding the whys and hows of the creation of the United States."

—Raymond Seitz, *Literary Review*

"In *Revolutionaries*, Jack Rakove delivers what he promises: a 'new history of the invention of America.' Under his deft touch, historical subjects come alive in a series of richly drawn portraits. Deeply researched and elegantly written, this book marks a milestone in the study of America's revolutionary period. It should not be missed."

—Annette Gordon-Reed, author of *The Hemingses of Monticello*

"Essential reading for both scholars and fans of Joseph Ellis (*American Creation*) and David McCullough (*1776*) as well as a poignant and engaging glimpse into the world of the revolutionary generation."

—*Library Journal*

"Distilling an academic career's worth of understanding about the Revolution, Rakove vibrantly restores it as the lived experience of those who led it." —*Booklist*

"An ambitious, intelligent exploration into the intellectual underpinnings of the Revolution."

—*Kirkus Reviews*

"A sparkling, authoritative work . . . Everyone interested in the founding of the U.S. will want to read this book."

—*Publishers Weekly*

REVOLUTIONARIES

Books by Jack Rakove

The Beginnings of National Politics: An Interpretive History of the Continental Congress (1979)

James Madison and the Creation of the American Republic (1990)

Original Meanings: Politics and Ideas in the Making of the Constitution (1996)

Declaring Rights: A Brief History with Documents (1997)

The Annotated U.S. Constitution and Declaration of Independence (2009)

Revolutionaries

A
New History
of the
Invention
of
America

JACK RAKOVE

MARINER BOOKS
HOUGHTON MIFFLIN HARCOURT
BOSTON NEW YORK

First Mariner Books edition 2011

www.hmhbooks.com

Library of Congress Cataloging-in-Publication Data
Rakove, Jack N.
Revolutionaries : a new history of the invention of America / by Jack Rakove
p. cm.
Includes bibliographical references and index.
ISBN 978-0-618-26746-0
ISBN 978-0-547-52187-9 (pbk.)
1. United States—History—Revolution, 1775–1783. 2. Revolutionaries—United States—History—18th century. 3. Statesmen—United States—History—18th century. 4. United States—Intellectual life—18th century. 5. United States—Politics and government—1775–1783. I. Title.
E208.R167 2010
973.3—dc22

Book design by Brian Moore

Printed in the United States of America

DOM 10 9 8 7 6 5 4 3 2

For Bernard Bailyn,
the creative historian

Contents

PROLOGUE

The World Beyond Worcester

On March 15, 1756, a young schoolteacher in Worcester, Massachusetts, sat in his "great Chair at School," surveyed his pupils at work around him, and fancied himself "as some Dictator at the head of a commonwealth." That evening, in his diary, John Adams mused about his charges. In "my little school" he could find "all the great Genius's, all the surprizing actions and revolutions of the great World in miniature." Numbered among his students were "several renowned Generalls but 3 feet high, and several deep-projecting Politicians in peticoats." The whole variety of humankind was present, he boasted: "Kings, Politicians, Divines, L.D. [LL.D.'s?], Fops, Buffoons, Fidlers, Sychophants, Fools, Coxcombs, chimney sweepers, and every other Character drawn in History or seen in the world." It took a lively imagination to detect such diversity in the children of Worcester's farmers, all destined to follow in their parents' stolid ways. The teacher seemed content with his lot. "Let others waste the bloom of Life, at the Card or billiard Table." He would "rather sit in school, and consider which of my pupils will turn out in his future Life, a Hero, and which a rake, which a phylosopher, and which a parasite." It could not have occurred to him to ask which might become a revolutionary.

Still, the learning Adams had acquired at Harvard College in Cambridge was outrunning its uses in Worcester, and other professions already beckoned. Most Harvard men still entered the ministry, preaching the familiar biblical passages that generations of Puritan preachers had expounded to generations of New England farmers—though without quite the fervor of their forerunners. Adams dutifully attended Sabbath worship, where the same minister who had lured him to Worcester held the pulpit. Only the day before the minister had spoken "all Day upon Abrahams Faith, in offering up Isaac." Two months later the teacher recorded some of the minister's favored expressions: "Carnal, ungodly persons. Sensuality and voluptuousness. Walking with God. Unregeneracy. Rebellion against God." Often Adams copied sermons into his commonplace book of instructive quotations. Yet his Christian beliefs were also lodged within the universe of Sir Isaac Newton, the great genius who had devoted obsessive years to searching scriptural study. "When we consider that Space is absolutely infinite and boundless, that the Power of the Deity is strictly omnipotent, and his Goodness without Limitation," Adams asked after one sermon, "who can come to a Stop in his Thoughts, and say hither does the Universe extend and no farther?"

Public events sparked similar reflections. "The last year is rendered conspicuous by a Series of very remarkable Events," Adams noted on July 22, 1756. Britain and France were warring for control of the Ohio Valley. Across the Atlantic "the british Nation has been making very expensive and very formidable Preparations" against a French invasion. A catastrophic minutes-long earthquake, followed by an equally massive tsunami, had leveled Lisbon the previous November, killing tens of thousands and inspiring thinkers as diverse as the French philosophes Rousseau and Voltaire and the founding preacher of Methodism, John Wesley, to rush into print their reflections on the meaning of the tragedy. The conventional religious moral Adams drew from these events lay closer to Wesley's repent-now spiritual lessons than to Rousseau's rational observations on the danger of packing population into large cities. "Is it not then the highest Frensy and Distraction," Adams asked, "to neglect these Expostulations of Providence and continue a Rebellion against that Potentate who alone has Wisdom enough to perceive and Power enough to procure for us the only certain means of Happiness"?

Although his religious views were more earnest than fervent, Adams wondered whether a career in the ministry would enable him "to do more good to his fellow men and make better provision for his own future Happiness." But that summer he committed his future to the law. "I am not without Apprehensions, but I am much less troubled with them than I was before I was determined what Profession to follow," he wrote to a Harvard friend. To his future brother-in-law, he confessed that "The Engines of frightful Ecclesiastical Councils, of diabolical Malice and Calvinistical good nature never failed to terrify me exceedingly whenever I thought of Preaching." A life in the law was one in which "I shall have Liberty to think for myself without molesting others or being molested myself."

Nine years later, newly wed to Abigail Smith, a prominent minister's very bright daughter, and with his legal practice prospering, Adams recorded a fresh threat to his serenity. His entry for December 18, 1765, only the fourth for the whole year, lamented "my Loss, in neglecting to keep a regular Journal" since spring. "The Year 1765 has been the most remarkable Year of my Life," made so by the spirit with which the colonists were opposing "That enormous Engine, fabricated by the british Parliament, for battering down the Rights and Liberties of America, I mean the Stamp Act"—the controversial legislation for raising revenue by imposing a tax on every form of paper, from legal documents and newspapers to playing cards and calendars. Adams actively supported the resulting American protests, drafting resolutions for his Braintree townsmen and publishing several essays in the *Boston Gazette.* Rather than permit the stamped paper to be used in legal proceedings, as the act required, the protesting Sons of Liberty insisted that the courts simply close, effectively suspending much of the daily business of governance. Adams viewed the threat to American liberty in strikingly personal terms. "I was but just getting into my Geers, just getting under Sail, and an Embargo is laid upon the Ship," he wrote; "just become known, and gained a small degree of Reputation, when this execrable Project was set on foot for my Ruin as well as that of America in General, and of Great Britain."

The next day the town of Boston summoned Adams to join two more eminent attorneys, Jeremiah Gridley and James Otis, in an effort to persuade Governor Francis Bernard to ignore the Stamp Act and allow the courts to reopen. He was "wholly at a loss to conjecture" why he had been called on this errand. In Boston on December 20, he learned that Otis and Gridley expected their young associate "without one Moments Opportunity to consult any Authorities, to open an Argument, upon a Question that was never made before."

Like many a provincial intellectual, the young John Adams harbored ambitions and dreams that outran anything his society could promise to satisfy. Had the American Revolution not intervened, the most life could have offered him was a reasonably prosperous career as an attorney. Courtroom success would have brought its rewards. He could then avoid the endless suits over debts and trespasses that were the lot of most lawyers—the eighteenth-century equivalent of personal injury cases. Opportunities for public service would have arisen, as they did when he represented the colony in boundary disputes with New Hampshire and New York. From time to time his Braintree neighbors would have sent him to the General Court, the provincial legislature that met in the handsome Georgian red-brick building that now cowers below the high-rise office buildings of downtown Boston. Though he probably would never cross the Atlantic, the dispute with New York could have brought him to its capital city at the lower tip of Manhattan island. From there he could have ventured farther to Philadelphia, the largest town in British North America, though its population amounted to a scant fraction of the residents of the imperial metropolis, London.

The Revolution did intervene, however. Or rather, a series of political crises broke upon the colonies after 1765, eventually becoming a revolution that transformed the private lives and provincial identities of most Americans. Few of these lives were more enlarged than that of John Adams, the deacon's son who would soon be received at the courts of Versailles and St. James, the plain-speaking advocate who rarely allowed discretion to get the better part of polemical valor. Yet for all the restless ambition that had festered within him since his teaching days in Worcester, Adams remained uncertain whether he was meant for a life in politics. His support for the cause of colonial rights never flagged, but after the Stamp Act was repealed in 1766, he balked at committing himself fully to political life. That conversion came only when the British Parliament responded to the Boston Tea Party of 1773 with a vindictive program of legislation directed against his native province and its capital. These events swept British North America into a frenzy of political agitation that soon carried Adams to the First Continental Congress in Philadelphia, where the colonies rallied to support Massachusetts, "now suffering in the common cause." Twenty-five years later, with barely a respite from public life, John and Abigail Adams became the first occupants of the presidential mansion erected in Washington, District of Columbia, the newly built capital of a national republic.

At the moment in 1756 when John Adams was choosing the law, another colonist was entering his third year of legal studies at the Inns of Court in London. John Dickinson was almost exactly three years Adams's senior. Sailing from Philadelphia in October 1753, he spent five weeks of the eight-week voyage seasick in his cabin—not the happiest way to mark his twenty-first birthday. Such a journey "home" lay beyond the means, if not the imagination, of an Adams. But Dickinson hailed from a prominent family. His father, Samuel, had amassed a Maryland estate of nine thousand acres before moving to Delaware in 1741. There he acquired another three thousand acres to provide for John and his younger brother, Philemon, the offspring of a second marriage. Like other members of the colonial gentry, John's parents envisioned his London sojourn as a mark of the gentility they already enjoyed and a source of the greater refinement they sought. Legal study in London *might* make Dickinson a better lawyer in Pennsylvania. More likely it would aid in the search for patronage that was so essential for a budding politician in the British Empire.

While in London, the young Dickinson wrote regularly to his "Honoured Father" and "Honoured Mother." His letters are respectful, affectionate, earnest, and richly detailed, and they reveal the tangle of attitudes that colored colonists' views of the mother country. Dickinson knew what he was supposed to feel when encountering sites and persons familiar from reading or by reputation. Sometimes he found himself too well primed: "as I had heard or seen particular descriptions of everything," he dryly noted after six weeks in London, "nothing excited my admiration, but only confirmed or lessend . . . my former notions." A visit to the House of Lords proved downright disappointing. Their chamber was "much inferior" to the meeting room of the Pennsylvania assembly. The nobility "drest in their common cloths" and looked to be "the most ordinary men I have ever facd." To judge from "the heaviness & foppery of their looks & behaviour," Dickinson observed, "many of them are more indebted to fortune than their worth for a seat in that August place."

Soon enough, however, Dickinson discovered that London had delights unequaled in America. Before dinner he joined the "gay assembly" who paraded on the Mall between St. James's Palace and the park beyond. At the Vauxhall Gardens on the south bank of the Thames, an orchestra played to welcome the evening crowds, who strolled past a Chinese temple and down "beautiful walks" lit by "a thousand lamps" hung on trees. At the theater performers such as David Garrick and Hannah Pritchard "easily perswade you they are the very persons they represent," though lesser talents appeared so "stiff & affected" as to be "disgustful." By May 1754 Dickinson was ready to ask his parents to extend his stay for another two years. Without additional study, he could never fulfill the high expectations that Americans often formed "of young fellow's coming to England." There was a further advantage to an extended stay: "it cannot be disputed that more is learnt of mankind here in a month than can be in a year in any other part of the world." London offered knowledge of character and status that could never be gained in the American hinterland. "Here a person sees & converses with people of all ranks, of all tempers," Dickinson wrote. "He acquires an ease & freedom of behaviour with his superiors, complaisance and civility to his inferiors."

Though hardly of the rank to hobnob with aristocrats, Dickinson enjoyed good connections in London. Thomas Penn, the proprietor of Pennsylvania and Delaware, took the law student in his own coach to St. James's Palace for the royal birthday reception on November 10, 1755. Dickinson had earlier seen George II at the theater, where he struck the young man as having "the most cheerful face in the whole house." A closer view at the palace left a different impression. The king exchanged only banal pleasantries with those around him; when not engaged, "he constantly cast his eyes on the ground. In short, this seemed so painful a tax upon majesty that I *pitied* him." Nor did the pleasure of seeing the royal family at close quarters obscure the fact that Dickinson left the reception with "an empty stomach, & satisfied in a little chophouse the hunger which I had procured in a palace." Even so, the occasion set his mind racing as he imagined what it would be like to be a regular at court. Then a suitably American reaction set in as he thought about "the difficulty of rising" within the ranks of an aristocratic society. And for what purpose—"to see a king at a little nearer distance or to wear a blue ribbond"?

Dickinson knew some young Americans, however, whose upbringing prevented them from grasping the lessons London could teach. In perhaps his most interesting letter, Dickinson reflected on the damage that youthful exposure to slavery had inflicted on his colonial acquaintances. The passions of "pride, selfishness, peevishness, violence, anger, meanness, revenge & cruelty" that young Americans learned from owning and commanding slaves unfitted them to deal with their social equals, much less their superiors, in England, where "the first lesson a person learns is that he is nothing." This discovery threw many of his countrymen into futile attempts to imitate what they could never acquire. It was better to avoid these pretensions, Dickinson thought, and let the great theater of London reveal "human nature in all shapes" to the careful observer.

There was one realm of British life, however, that Dickinson felt confident to judge critically: politics. He arrived in time for the parliamentary elections of 1754, brilliantly satirized in William Hogarth's *An Election*, a series of four paintings based on the sharply fought contest in Oxfordshire (they now hang in the wonderful Sir John Soane Museum, hard by the Inns of Court where Dickinson studied). The series begins with *An Election Entertainment*, in which the Whig candidates feast their voters at an inn while a Tory crowd pounds at the door and hurls bricks through the window. The central figure in the next painting, *Canvassing for Voters*, is an ostensibly independent voter, simultaneously accepting bribes from agents of both parties. Meanwhile a Tory gentleman, eyeing two women on a balcony above, is buying handkerchiefs from a Jewish peddler—a poke at the Tories' violent opposition to an innocuous bill for naturalizing Jewish immigrants. In the third painting, *The Polling*, a procession of the lame and the blind ascends the stand where the sheriff is recording the votes. The series ends with the riotous *Chairing the Candidate*, in which the victor is carried precariously through the streets amid violence and mayhem while his wife swoons in shock.

Hogarth was immensely popular, and Dickinson would surely have seen popular engravings of his paintings. But he already knew that British politics was rife with corruption. For thirty years and more, opposition writers had inveighed against the rot sapping the "boasted constitution" of Britain, and Americans had consumed these writings with as much fervor as Britons displayed in snapping up Hogarth's prints. The recent elections only confirmed what Dickinson already knew. In boroughs where members of Parliament were chosen by handfuls of electors, as much as two hundred guineas, he heard, was paid for a single vote. "I think the character of Rome will equally suit this nation," he concluded: "'Easy to be bought, if there was but a purchaser.'" When Parliament met later that year, scores of "controverted elections" remained to be resolved because of allegations of bribery at the polls. This too was evidence that the decline of "old fashioned religion" in Georgian Britain had led to "the most unbounded licentiousness & utter disregard of virtue which is the unfailing cause of the destruction of all empires."

Twenty years later Edward Gibbon would trace the decline and fall of the Roman Empire to the opposite cause: the spread of the original old-fashioned religion, primitive Christianity. Yet the Britain where young Dickinson studied in the mid-1750s was the center of an expanding empire, not a contracting one. Coming as he did from a devoutly Quaker family, he sometimes recoiled from the worldly corruptions that London conspicuously displayed. Still, his sojourn in London enabled Dickinson to experience the allure and the benefits of empire with an immediacy that eluded the vivid imagination of John Adams. There were lessons to learn everywhere, whether listening to "some of the greatest men in England, perhaps the world," as they argued cases in the Cockpit at Whitehall Palace or at Westminster Hall; meditating on fame and greatness at Westminster Abbey or the solitary tomb of Sir Francis Bacon; or contrasting the respectful attention of Thomas Penn with "that brutal, mean, false pride of some peoples [*sic*] who, because they are independent, think no oblidgingness, no honour, due to any person."

Empire, for Dickinson, was a matter of culture as well as law and power. He relished the remarkable access he enjoyed in London and the kindness showered upon him. Yet the ease he felt at the seat of empire never erased the knowledge that he was a visitor from its distant periphery. The allure of the metropolis was always tempered by the recognition that Kent County, Delaware, was home, and nearby Philadelphia a poor substitute for London. When he and a Maryland friend took new lodgings near Hampton Court, the two housemates "resolved to remember we are AMERICANS, to live soberly & prosecute our business."

This ambivalence shaped the distinctive role that Dickinson took in the politics of resistance after 1765. During the Stamp Act crisis, he wrote directly to William Pitt, the "great commoner" who led the British government to its victories in the Seven Years' War, to warn this friend of America that Britain could easily alienate colonial affections. Yet soon afterward, his *Letters from a Farmer in Pennsylvania* sharpened the colonial argument against parliamentary taxation, giving its author a celebrity few Americans attained. That celebrity proved a political burden. When the crisis of empire broke in 1774, Dickinson supported the radical measures that the Continental Congress was driven to adopt, while longing for reconciliation with the mother country he still loved. Thus it was that he found himself first courted and then mocked by John Adams, drafting the Articles of Confederation that formed the nation's first constitution, while resisting the Declaration of Independence that announced that nation's existence.

While Dickinson was sailing to England in 1753, another young colonist, nine months his senior, undertook a different journey, traveling northwest from Williamsburg, Virginia, toward the general destination of Lake Erie. Compared to the hardships this young man met on his return and the greater perils he faced on two succeeding missions, Dickinson's seasickness was a trifling discomfort. Dickinson went to London to seek the benefits of empire; George Washington trekked west in its service. Dickinson reflected on the examples of fame he encountered in London; Washington's missions made him famous there in his own right, when a postscript to a letter he wrote to his brother John in 1754 found its way into the London press: "I can with truth assure you, I heard Bulletts whistle and believe me there was something charming in the sound." This brief stab at bravado elicited a critical comment from the same king whom Dickinson had twice glimpsed. "He would not say so," snorted George II, a veteran campaigner himself, "if he had been used to hear many." Even so, there were few American names, other than Benjamin Franklin, that any British king had to know; Washington quickly became one of them.

Like Dickinson, Washington was the first offspring of a second marriage. Their fathers occupied roughly comparable positions, owning estates that placed them well above ordinary farmers but a notch or two below the great families who dominated the lands surrounding the Chesapeake. Like the law student, Washington might well have gained the benefits of an English education, as had his two older half-brothers. But the death of his father, Augustine, in 1743, when George was only eleven, foreclosed that possibility. Washington inherited a small farm, other parcels of land, and ten slaves, and he obtained a modicum of schooling. Most of his learning, however, was self-attained, hard-won, and never wholly worn with the confidence, deserved or not, that English schooling would have bestowed. Washington thought about joining the Royal Navy, but when he was sixteen, he fixed on a different vocation: surveyor. In this he gained the aid of Virginia's greatest landowner, Thomas, Lord Fairfax, proprietor of five million acres in the colony's Northern Neck. Fairfax hired Washington to survey his extensive holdings in the Shenandoah Valley, and the youth used this opportunity to acquire fifteen hundred acres for himself. He also became an officer in the Virginia militia. In 1751 he sailed to Barbados with his ailing half-brother, Lawrence, who hoped the voyage would improve his health. It did not: Lawrence died the next summer of the tuberculosis that no voyage could cure. George was lucky to survive the smallpox he contracted in Barbados on what proved to be his only trip beyond the North American mainland.

It was not eastward across the Atlantic that Washington looked, but westward into its interior. Between the autumn of 1753 and summer of 1755, he led two missions to the Forks of the Ohio River and then served in a third under Major General Edward Braddock, the commander of British forces in North America. The first mission ended in polite rebuff from local French commanders, the second in a military debacle, and the third in the disaster of Braddock's defeat and death. The conflict that began with frontier skirmishes in the woods below Lake Erie escalated into the global Seven Years' War, fought in North America, the Caribbean, West Africa, central Europe, Gibraltar, the Indian subcontinent, and the Philippine archipelago. Washington's missions were not the *cause* of this escalation, but they did mark its beginnings.

His 1753 mission was sparked by reports that French forces were erecting forts below Lake Erie, perhaps as far south as the point where the Allegheny and Monongahela rivers join to form the Ohio. Both Britain and France based their claims to that territory on the usual array of legal arguments that European powers deployed to justify their domination of American lands. By linking their main settlements in Quebec and Louisiana through a network of forts and trading posts, the French hoped to confine the more populous English colonies to the Atlantic seaboard.

The mission that Washington received from Robert Dinwiddie, Virginia's newly arrived lieutenant governor, was to verify these reports and to request the French to withdraw from the Ohio Valley. Whether their forts fell within the claimed boundaries of Pennsylvania or Virginia was a question no one could yet answer. But in this affair of real estate, precise location did not wholly matter. A dominant French presence between the Great Lakes and the Ohio would prevent American expansion into the interior and thwart the schemes of land speculation that Dinwiddie and other prominent Virginians were already forming.

On his first expedition, Washington encountered a handful of French officers occupying a cabin along the Allegheny. After informing the young officer that he was trespassing on land belonging to King Louis XVI, they treated him politely and allowed him to proceed to Fort LeBoeuf, on Lake Erie. Once arrived, he was again greeted hospitably while the French commanders pondered their reply to the message he carried from Dinwiddie. With winter setting in, Washington left the fort in mid-December. A month's arduous journey brought him back to Williamsburg on January 16, 1754.

Dinwiddie asked Washington to draft a report of his mission, which was soon published as a pamphlet, reprinted in colonial newspapers, and then excerpted in the London press. He then instructed Washington to take a company of Virginians back to the Forks of the Ohio River and there erect a fort to contest the French claims. Should they meet resistance, they were authorized to capture or "kill and destroy" the French soldiers. An advance guard of Virginians reached the Forks in early spring and built a primitive fort. Its low palisade could obviously not withstand the five hundred French soldiers, armed with eighteen cannon, who appeared on April 17, 1754. Once again, good manners prevailed. When the French commander offered the Virginians the chance to abandon the fort and withdraw, Ensign Edward Ward gratefully accepted.

Washington, meanwhile, was still plodding westward, with his larger contingent of Virginians. Their slow progress was well known to the French. On May 27, nearing the Monongahela, Washington learned that a French reconnaissance party was heading their way. That evening he found out the location of the French bivouac from Tanghrisson, a Seneca chief known as the Half-King, who was encamped nearby.

In "heavy rain" through a pitch-black night, Washington marched forty men, "frequently tumbling one over another," to meet the Half-King, and together they approached the French camp. When the skirmish was over, ten of the French lay dead, including their commander, Ensign Jumonville. Despite being wounded, Jumonville had tried to read Washington the letter he was carrying, restating his nation's claim to the Ohio country. Before his message could be translated, the Half-King smashed his ax into the ensign's skull and reached inside to fondle the brains. Other wounded French prisoners were killed and scalped by the Seneca warriors.

Washington remained in the field another month. On June 28, however, he learned that a large French force was approaching. His resolve wisely crumbled. The next day he started his men on an exhausting retreat to Fort Necessity, a small palisade his troops had hastily built. They reached it on July 1. The French force, led by the vengeful brother of the butchered Jumonville, arrived the next day and attacked on July 3. Pinned down in their trenches, drenched by heavy rain, their powder soaked and their muskets useless, the Virginians suffered a hundred casualties (including thirty killed) before the French commander generously allowed Washington to surrender and withdraw without further reprisal. So began another retreat, with the wounded in agony, soldiers constantly deserting, and the remnant trudging along, hungry and tattered.

The original charm that Washington found in whistling bullets faded quickly under the French fusillade. But the reaction that mattered most took place in distant London. Within weeks of learning of the debacle, the British government adopted an ambitious plan to send additional regiments to North America, appoint Braddock to command royal forces there, and launch multiple attacks on French positions, not only in the Ohio Valley but also from northern New York to Nova Scotia.

In late April 1755, after declining an invitation to command Virginia's provincial forces, Washington secured an appointment as an aide-de-camp to General Braddock, who valued his knowledge of the difficult terrain he would have to traverse to besiege Fort Duquesne. In early summer the British troops slogged their way across the mountains and forests of Pennsylvania, hauling the usual train of artillery and supplies. On July 9, they forded the Monongahela a scant ten miles from their objective. But the French commander was tracking their progress, and rather than wait to be besieged, he dispatched nine hundred men—mostly Indian allies—to intercept the British. Contact was made in the early afternoon. The Indians streamed down both flanks of the advancing column, directing withering volleys against the British from the surrounding woods. Braddock struggled valiantly to maintain discipline, but the impressive skills he brought to the battlefield were irrelevant to his last combat. He was completely disdainful of Indian warriors, both as allies and as foes, and once they flanked his troops, he had no idea of how to regain the initiative. Only after Braddock himself was gravely wounded did the British will to fight finally break, and a headlong retreat began.

Washington was there when the attack started. For days he had suffered "from a violent illness that had confin'd me to my Bed, and a Waggon." That he survived the battle and retreat was miraculous. "I had four Bullets through my Coat," he wrote to his mother, "and two Horses shot under me." Washington did not fault Braddock or his officers for the defeat. Rather, it was the "cowardice" and "the dastardly behaviour of thos they call regular's" that brought death to so many "other's that were inclined to do their duty." The surviving officers' efforts to halt the panicked retreat met with as much "success as if we had attempted to have stopd the wild Bears of the Mountain or rivulets with our feet."

Once back in Virginia, Washington secured appointment as colonel commanding the Virginia regiment and was given the task of defending frontier settlers against expected Indian raids. But his aspirations remained imperial more than provincial: to secure a commission in the Royal Army. In February 1756, he journeyed to Boston, where Governor William Shirley had succeeded Braddock as commander for North America, only to return to Virginia disappointed. A year later he renewed his bid with Shirley's successor, Lord Loudoun. Again the desired commission was withheld. His final military service in the war came in the fall of 1758, leading one of two Virginia regiments in a new expedition that finally occupied Fort Duquesne after its outmanned garrison fled. Suffering severely from dysentery, Washington resigned his commission in December.

Prior to the Duquesne expedition, he had proposed marriage to a young and very wealthy widow, Martha Parke Custis, and then secured election to the Virginia legislature. The marriage took place on January 6, 1759, giving Washington control of hundreds of slaves and thousands of acres. Two years later the death of his brother Lawrence's widow made him owner of the Mount Vernon estate that he had leased since 1754. Washington now pursued the life of a Virginia squire. No longer seeking to rise within the empire, he became a consumer of its culture, importing luxury goods from England as he tried to convert the house, on its bluff above the Potomac, into a replica of an English country seat.

Nothing in Washington's life after 1759 suggests that he regretted the thwarting of his military ambitions, though perhaps it rankled in his memory. With the decisive British victory over France and the end of French rule in Canada, the idea of a North American military career lost its allure. Just past thirty when the war ended, Washington found opportunities enough as a planter: expanding his holdings, pursuing the land speculations for which his youthful experience as surveyor prepared him, and cultivating wheat in place of the tobacco that robbed the land of its fertility. In the 1760s he was well on his way to attaining everything someone of his social station desired, with one exception—offspring of his own to add to the stepchildren he gained through marriage.

Yet as a member of the Virginia political elite, he could not escape the new political questions that arose in the wake of victory. Lacking the education of the colony's best-known leaders, he was not in the vanguard of opposition to the new imperial policies of the mid-1760s. It was for lawyers such as Adams and Dickinson, or his own well-read neighbor, George Mason, to rebut Parliament's claim to jurisdiction over the colonies. When he met Adams and Dickinson at the First Continental Congress in 1774, the impression he left may have been more physical than political. Yet his political views and loyalties were the same as theirs, while the military episodes of his twenties gave him qualifications few colonists could claim. Within months his stature as the commander of the untested forces of a new nation gave him a reputation that far eclipsed the early fame he had gained when bullets first whistled over him.

Adams, Dickinson, and Washington were still in their early forties when the crisis that led to independence erupted in the summer of 1774. They were, in that sense, the generation that made the American Revolution and provided its leadership. Yet none was a revolutionary in the modern sense of the term. None joined the protests of the 1760s and early 1770s out of a secret hankering for independence, or from calculations of ambition and power, or in resentment over the cruel hand that life and fate had dealt him. Each was ambivalent about engaging too deeply in public life. At any point before 1774, each would have preferred to see the Anglo-American dispute resolved peacefully. For Dickinson, that sentiment lasted even longer, down to the moment in July 1776 when he gave the last major speech in Congress opposing independence.

The notion that some generations are marked by distinctive experiences is, of course, hardly a great historical discovery. The journalist Tom Brokaw popularized the term "the greatest generation" as a badge of honor for Americans who came of age with the Great Depression and then fought the good war against fascism and Japanese militarism. Historians write of the Spanish "generation of 1898," who tasted the final defeat of a once-global empire, and the European "generation of 1914," who marched off to the guns of August, visions of military glory dancing in their heads, only to confront slaughter on an industrial scale in Flanders fields, on the Somme, at Verdun. American historians have long written about the Revolutionary generation and tried to grasp its distinctive characteristics. Gordon Wood, for example, gave his recent collection of essays, *Revolutionary Characters*, the subtitle *What Made the Founders Different*, and then located their defining generational trait in their attachment to a set of Enlightenment and gentlemanly ideals that America's democratic future would soon eclipse. Joseph Ellis's fabulously successful book *Founding Brothers* is subtitled *The Revolutionary Generation*, but Ellis curiously devotes his actual narrative to the post-Revolutionary period stretching from 1790 to 1804, with only the odd glance back to the Revolution proper. There may not be quite as many notions of what constitutes a generation as there are historians. Yet clearly anyone who tackles this theme needs to be as precise as possible about the experiences, attributes, and events that give a generation its defining character.

At the start, we need to recognize that there were at least two generations of 1776: an older cohort who led the colonies into independence (such as the Adamses, Washington, Mason, Dickinson) and another that came of age with it, "young men of the Revolution" (such as John Jay, James Madison, and Alexander Hamilton, who, coincidentally, co-authored *The Federalist* essays of 1787–1788). Scholars often blur or collapse this distinction by speaking simply of the "founding generation" or "the founders"—terms that elide the distinct political movements that culminated in the Declaration of Independence of 1776 and the federal Constitution of 1787 into one larger process of forming an independent national republic. Yet the experience of the older cohort, who had to decide upon independence, could not have been wholly identical with that of the younger group, who were only reaching maturity as independence was being declared for them.

There were other traits, however, that many of these men shared. Each grew up in provincial societies that were imperial outposts on the far shore of the Atlantic world, and their provincial status may have been their most profound shared experience. Not that it took the same form for each. In Massachusetts, John Adams was only one of many talented men who resented the monopoly of high offices held by the extended family of Thomas Hutchinson and their allies. Like colonists everywhere, Adams gloried in the triumphs of the Seven Years' War, but the closed circle of imperial favor in the Bay Colony grated on his own ambitions. John Dickinson labored under no such disadvantages. He repaid the hospitality that Thomas Penn showed him in London by opposing Benjamin Franklin's futile campaign to have the Crown strip the Penn family of its proprietary right to the government of Pennsylvania and Delaware. Yet Dickinson acted not out of personal loyalty to the Penns but rather from a principled desire to protect the distinctive political culture that had flourished in Pennsylvania ever since William Penn launched his "holy experiment" in the 1680s. Though not formally a Quaker himself, he was steeped in Quaker political philosophy and its pacific virtues of moderation and respect. In Virginia, George Washington had abandoned his quest for a military career, but marriage and his reputation for bravery brought him into the highest stratum of his colony's planter elite. The ambitions of all three men were bounded and shaped by the distinctly provincial worlds they knew.

The leaders of the colonial protests against Britain were thus all provincials before they became revolutionaries, revolutionaries before they became American nationalists, and nationalists who were always mindful of their provincial roots. Understanding how these traits and experiences fit together and played upon one another is essential to explaining the puzzle that keeps drawing Americans back to the founding era—and which this book has been written to explore. Whether we call them "Revolutionary Characters" or "Founding Brothers," or whenever we ponder *The Genius and Ambiguities of the American Founders* (the subtitle of another recent set of essays by Bernard Bailyn), we find ourselves asking one recurring question: how do we explain the appearance of the remarkable group of leaders who carried the American colonies from resistance to revolution, held their own against the premier imperial power of the day, and then capped their visionary experiment by framing a Constitution whose origins and interpretation still preoccupy us over two centuries later?

Like many simple questions, it is not easily answered. The sense of historical destiny that surrounds the Revolution challenges our capacity to think our way back into the contingencies of the past and to appreciate how improbable an event it was. Part of that improbability lies in the record of misguided decisions that led the British government to fulfill its worst fears by driving the colonists down the road to independence. It took a peculiarly flawed process of framing bad policies and reacting to the resulting failures to convince the government of George III and Lord North that the best way to maintain the loyalty of their North American subjects was to make war on them. But that improbability immediately leads us to another that is a central theme of this book. The men who took commanding roles in the American Revolution were as unlikely a group of revolutionaries as one can imagine. Indeed to call them revolutionaries at all is almost ironic. With the possible (and doubtful) exception of Samuel Adams, none of those who took leading roles in the struggle actively set out to foment rebellion or found a republic. They became revolutionaries despite themselves. Or rather, they became revolutionaries because a crisis in a single colony spiraled out of control in 1773–1774, and the empire's harsh response to the challenge to its authority persuaded colonists everywhere that the British government really was bent on abridging their basic rights and liberties. Until then, the men who soon occupied critical positions in the struggle for independence were preoccupied with private affairs and hopeful that the troubles that had roiled the empire in the 1760s would soon be forgotten. To catch them (as we repeatedly shall) at those moments when they individually realized that would not be the case is to understand that the Revolution made them as much as they made the Revolution.

Yet if they were not revolutionaries by design, they became so in action. Their origins may have been provincial, and their involvement in the Revolution may have been forced upon them. But once engaged, they quickly grasped what an opportunity it was. None of them captured this discovery better than John Adams, our schoolteacher turned lawyer turned statesman, when he exulted at his good fortune at being "sent into life, at a time when the greatest law-givers of antiquity would have wished to have lived." That thought occurs repeatedly in the writings of the revolutionaries, sometimes in works meant for publication, as with Adams, and sometimes in private letters when, amid the hurly-burly of public business, they snatched an odd moment to reflect on what they were doing. That the business was a hurly-burly of an extraordinary kind also helps explain the commitments they were forging. As the political philosopher Hannah Arendt shrewdly observed of them, it was only in "the speech-making and decision-taking, the oratory and the business, the thinking and the persuading, and the actual doing" that the revolutionaries discovered "the rather obvious fact that they were enjoying what they were doing far beyond the call of duty."

Down to 1774, the structure of the imperial controversy reinforced the colonists' provincialism even as it exposed its limits. Americans repeatedly insisted that they sought only the restoration of their traditional political rights. Before 1763, each colony was largely free to frame its own laws, subject only to review by the king's Privy Council. Occasional parliamentary acts regulating American commerce were easily evaded within a trading system that benefited both countries regardless of how strictly its rules were enforced. But this era of "salutary neglect" abruptly ended in 1763. The treaty of peace recognizing Britain's decisive victory over France became the occasion for a host of reforms and initiatives that profoundly challenged the colonists' understanding of their place within the empire.

Well before Washington's missions to the Ohio triggered the great conflict, some imperial officials were already fretting that the colonies enjoyed too much autonomy. These concerns had begun to cohere into a tentative agenda of reform. The war itself offered ample proof of the need to replace "salutary neglect" with efficient administration. The colonists' propensity to use prisoner exchanges with the French West Indies to obtain the molasses needed to distill rum was a disturbing reminder of how often they had flouted the navigation laws that Parliament had been enacting since the previous century. Britain's courtship of Indian allies spurred recognition that native territorial rights should be secured, even if that meant limiting the movement of European settlers into the interior. Securing the king's conquests in Canada and his territories west of the Appalachians required stationing additional forces above the St. Lawrence River and in the Ohio Valley—an expensive proposition in itself. The king's ministers in Whitehall and backbenchers in the House of Commons naturally thought that Americans should bear more of the costs of the expanded empire from which they would directly benefit.

The year 1763, then, was a moment when concerns once confined to the middle ranks of the British bureaucracy gained real force. The government of George III, the young king who came to the throne on the death of his grandfather in 1760, responded in several ways. It ordered absentee holders of imperial posts in America to either take up their positions or resign their commissions. (Lieutenant Governor Dinwiddie, for example, came to Virginia in 1752 as deputy for its true governor, Lord Albemarle, who enjoyed the fruits of office as a matter of patronage without having to visit his dominion.) In October 1763 the new ministry of Sir George Grenville issued a royal proclamation establishing a line separating areas of European settlement from lands reserved for their Indian inhabitants. In March 1764 Parliament enacted its first systematic overhaul of the imperial trade system since 1733. Americans called it the Sugar Act because it halved the duty on foreign molasses in the hope of converting a tax the colonists had evaded into one they would pay, in the process generating substantial revenue to defray the costs of empire.

The Sugar Act, though unpopular, did not threaten Americans' understanding of their rights. They had long used their numerous harbors and estuaries to evade full compliance with the Navigation Acts. Often they bribed or bullied customs officials into looking the other way. But they had never denied that Parliament could regulate their commerce in the interest of imperial harmony. It was altogether different with the Stamp Act that Parliament enacted in March 1765. Unlike import duties, which fell on goods that many Americans rarely purchased, the various forms of paper that would need stamps—from playing cards to legal documents—affected most households. In the language of the time, the stamp tax was akin to a "direct tax," which would fall on everyone, not an "indirect tax," which individuals could avoid by not purchasing dutied items.

Perhaps the broad sweep of the Stamp Act explains the violence of the colonists' initial protests. In August 1765 riots broke out in several northern ports. Community leaders quickly acted to channel opposition into more orderly forms. Following tacit rules of political protest familiar to Britons and Americans, colonists intimidated the hapless stamp distributors into resigning their commissions. Rather than allow legal proceedings using stamped paper to go forward, colonial leaders insisted that courts cease operations until Parliament reconsidered its measure. Once it became clear that the Stamp Act would not take effect, the courts reopened, proceeding without the accursed paper. Colonists also boycotted British goods, in the hope that British manufacturers and merchants would petition Parliament to repeal the Stamp Act. In October, the Stamp Act Congress, representing eight colonies, met in New York City to adopt resolutions justifying resistance on the broad claim that Parliament could not tax the king's unrepresented subjects in North America.

Issues of taxation and representation were the starting point for the great controversy that ultimately rent the empire. The Stamp Act violated the fundamental constitutional principle that taxes were the "free gift" of the people, to be granted only with the explicit consent of their elected representatives. No colonial delegates sat in the House of Commons, nor did Americans believe they could ever be properly represented in a body that met an ocean away. The prevailing British theory of virtual representation, however, saw things differently. Though every county in the realm sent two "knights of the shire" to the Commons, numerous communities, including populous towns in the Midlands, returned no members. The fiction of virtual representation nevertheless treated the whole people as somehow present in Parliament, whose members were duty-bound to act not for the parochial interests of their constituents but for the good of the whole. Only a modest leap across the Atlantic was required to cover the colonists. As subjects of the king, they too were part of his people.

Americans rejected this theory. Their practice of representation was "actual," not virtual. The right to send delegates to the colonial assemblies was routinely extended to communities as they were organized, not treated as a privilege doled out at royal pleasure. In America most free adult males could vote because access to land was relatively easy. In the colonies too, norms of actual representation emphasized that lawmakers should be held accountable to their constituents. Representatives were commonly viewed as the agents of their constituencies, not a separate governing class only incidentally tied to the voters who elected them.

Colonial arguments on these points enjoyed a sympathetic reception among those Britons who regarded the inequities of representation in the "unreformed" Parliament as evidence that the vaunted British constitution was indeed rotting. In response, defenders of Parliament's right to tax America abandoned the idea of virtual representation for the higher ground of sovereignty. The Glorious Revolution of 1688 had conclusively recognized that Parliament was the sovereign, highest source of law within the British polity. And sovereignty, as it was defined, was an absolute, final, unlimited power. Sir William Blackstone stated the orthodox view in his first volume of *Commentaries on the Laws of England*, coincidentally published in the year of the Stamp Act. In every government "there is and must be," Blackstone wrote, "a supreme, irresistible, absolute, uncontrolled authority, in which the *jura summi imperii*, or the rights of sovereignty, reside." In Britain, this authority was manifestly lodged in Parliament. If the colonies were part of the realm that Parliament governed, it was nigh impossible to see how they could be exempt from its jurisdiction, however strong their arguments on representation might be.

As the colonial protests took effect, a new ministry led by the young Marquess of Rockingham moved to repeal the Stamp Act. (Grenville had been dismissed, not because the Americans were defying his policies, but because he had run afoul of the king's mother.) As part of this repeal effort, Benjamin Franklin was called to the Commons to testify about American attitudes and positions. Franklin had spent nearly a decade in London, pursuing his campaign to make Pennsylvania a royal colony while relishing the intellectual and social pleasures of the metropolis. He was also out of touch with developments at home—so much so that he had worked hard to obtain the post of stamp distributor for his friend, John Hughes. Yet Franklin was the shrewdest American in London, and the most respected. In four hours of testimony on February 13, 1766, he insisted that Americans would never pay the stamp duty or any similar tax levied by Parliament. But he also muddied the question in one key respect. If "internal" taxes, like the stamp tax, were objectionable to Americans, the authority of Parliament to levy "external" taxes—duties on imported goods—was not.

As soothing and persuasive as Franklin might have been, his testimony alone could not carry the case for repeal. Few members of Parliament were willing to relent on both taxation and American claims to exemption from parliamentary authority. To ensure the repeal of the Stamp Act, the Rockingham ministry consented to the adoption of the Declaratory Act, baldly stating that Parliament retained the power to legislate for the colonies "in all cases whatsoever." Whether that power would soon be exercised was not at issue. Tactical retreat on the Stamp Act could not be construed as a confession that the colonists were right as a matter of principle. While grudgingly yielding to colonial protests, neither the Crown nor Parliament was prepared to exempt Americans from the highest legal authority within the British Empire.

News of the repeal took two months to reach America. When it did, John Adams recalled, it "hushed into silence almost every popular Clamour, and composed every Wave of Popular Disorder into a smooth and peaceful Calm." The next threat to imperial harmony arose a year later. The Stamp Act's repeal had done nothing to alleviate the empire's need for revenue. In January 1767, Charles Townshend, the chancellor of the exchequer, proposed levying duties on miscellaneous goods imported into the colonies—glass, lead, paint, paper, pasteboard, all items that Americans could not easily manufacture. Townshend was an experienced official, well versed in colonial affairs. Although these goods would produce only small revenues at first, he clearly conceived his scheme as a way of habituating Americans to the payment of new taxes. He also hoped to exploit Franklin's distinction between internal and external taxes, the former objectionable on constitutional grounds, the latter presumably acceptable under Parliament's general authority over trade. Parliament adopted Townshend's scheme in June, adding tea as another taxable item, and on July 2 the king signed the act into law.

It took the colonists nearly two years to mount another effective boycott of British goods as an incentive for the repeal of the Townshend duties. But one of the most noteworthy American responses came almost immediately. In early December 1767 Dickinson published the first of twelve *Letters from a Farmer in Pennsylvania*, denouncing the new duties and other recent measures of government. Like all eighteenth-century polemicists, he wrote under a pseudonym, but his identity was soon disclosed and Dickinson found himself widely hailed as "the Farmer." Dickinson's letters closed the ambiguity in the colonial position that Franklin's testimony had opened. It was specious to distinguish between different kinds of taxes, Dickinson argued. Duties levied for protectionist purposes might be legitimate. But if their avowed aim was revenue, as Townshend freely conceded, they were subject to the same constitutional objections as the Stamp Act.

The clarifying impact of Dickinson's letters was not fully matched by the colonies' patchwork efforts to organize a new boycott. In 1765 the Stamp Act Congress had helped establish a framework for intercolonial unity. That precedent was not followed again. Instead, the initiative for a boycott came from individual assemblies, beginning with a circular letter from the Massachusetts General Court, and from associations of merchants in particular ports. Their appeals were generally but not universally supported, and as a result, resistance leaders applied intimidating tactics against those suspected of violating the boycott. Shops selling British goods were smeared with the mixture of mud and feces called "Hillsborough paint" to mock the British minister for colonial affairs. Boston proved particularly turbulent, especially after the first of five regiments of British troops began debarking on October 1, 1768. Their presence had been requested by commissioners of the Board of Customs—another of Grenville's reforms—whom a Boston mob had assaulted when they attempted to seize John Hancock's sloop *Liberty* for its alleged smuggling of Madeira wine. In Boston's narrow streets, nasty incidents erupted between townspeople and troops, especially when poorly paid soldiers undercut town workers for casual labor. On March 5, 1770, a squad of soldiers guarding the customs house on King Street opened fire on a harassing crowd, killing five and wounding another six, an event that came to be known as the Boston Massacre.

On the same day, an ocean away, Parliament began debating whether to lift the Townshend duties. Repeal was supported by the king's new chief minister, Lord North, the first leader with whom George III was able to form a comfortable, enduring relationship. Though sympathetic to British manufacturers who complained that the duties had interrupted their colonial trade, North supported Townshend's goal of securing American revenues. He was also unwilling to abandon Parliament's claim of authority over the colonies. Retaining a duty on tea was consistent with both goals. Like the Declaratory Act of 1766, it affirmed that Parliament's legal authority remained intact. And as an object of mass consumption, tea might eventually generate a significant revenue—if Americans relaxed their constitutional scruples and purchased it. In April, Lord North and his parliamentary whips beat back an amendment to repeal the tea duty, and it remained in place when the other duties were removed. When news of this repeal reached America, some leaders hoped the boycott could continue until the tea duty was also lifted. But despite revulsion over the Boston Massacre, enthusiasm for political action was already waning, and the non-importation agreement collapsed.

Twice in five years colonists had defied an act of Parliament, and the government "at home" had backed down—but without abandoning its crucial constitutional claim. These two episodes had exposed a potentially dangerous fissure between Britain and America. That fissure was not really about the costs of empire. The true fault line was constitutional. Americans had mustered powerful arguments explaining why Parliament lacked any authority to impose taxes or laws on royal subjects whose only lawful representatives gathered in quiet provincial capitals from Savannah to Portsmouth. Some exception might have to be made, out of respect for precedent and reasons of convenience, to acts regulating the flow of commerce. But that would not countermand the general principle that a free people could be governed only by laws to which their own politically accountable representatives had assented.

That belief, deeply rooted in the traditions of Anglo-American constitutionalism, had collided with another equally powerful principle: the legal supremacy of a sovereign Parliament. An impartial observer might well conclude that both principles were well grounded in a common tradition. Each represented a legitimate strand in English constitutional thinking. Each could be upheld by its proponents as a decisive solution to the imperial controversy. Each seemed persuasive to its adherents, even as they struggled to blunt the force of the rival argument. In the absence of an authoritative written constitution and a neutral arbiter of its meaning, neither could decisively vanquish the other. That was what made this seemingly abstract controversy so dangerous. If pushed to their logical conclusion, the two positions were irreconcilable. American norms of representation and government by consent could not tolerate the claim that Parliament could give law to the colonies "in all cases whatsoever." Nor could orthodox British conceptions of sovereignty and limited monarchy allow whole areas of imperial governance to escape the oversight of a supreme Parliament.

Yet to say that these positions were irreconcilable in theory does not mean that they had to reach the point of confrontation in practice. Should such a crisis ever come, it would result not from the logic of political theory but from the dictates of political calculation. And if it did arrive, each side would face serious weaknesses. For the British, the challenge would be to overcome a lack of political assets in the colonies. Repeated assertions of parliamentary sovereignty were more a confession of imperial weakness than a description of how the empire actually worked. The reality was that British authority did not penetrate deeply into the American countryside. It was merely a thin template resting on distant societies where governance had always been highly decentralized. Had the empire and Parliament really exercised the authority they claimed, Americans would need no reminders of where sovereignty resided.

The colonists faced a different challenge. Their refusal to submit to Parliament was directed more to what it might do in the future than to the burdens it wanted to impose in the present. American political thinking emphasized the aggrandizing nature of power. Those who wielded power always wanted more and plotted constantly to attain it. But many, and probably most, colonists sensibly wondered why they should sacrifice the current benefits of empire to ward off an indefinite future danger. True, the removal of the French threat north of the St. Lawrence River made Americans less dependent on the mantle of British power. But if that strategic consideration was felt at all, it did not outweigh the rich sources of affection and attachment that united the colonies and Britain: religion, language, commerce, culture, history. There were real benefits to remaining within the empire, real costs to challenging the Atlantic world's greatest power, and real reasons to think that both countries had more to gain from restoring harmony than from pushing the constitutional controversy to the brink of confrontation.

Thus when the Townshend duties were lifted in the spring of 1770, the apparatus of colonial resistance again dissolved, just as it had four years earlier. Most colonists could plausibly hope that these two episodes had been just that: not chapters in a drama whose conclusion was yet to be written, but aberrant disputes that sounder heads could have averted and would strive to avoid in the future. A few colonial radicals still harbored darker views of British aims, and some imperial officials still believed that the colonies had to be brought to obedience. Yet a visitor to the American colonies in the early 1770s would not have discovered a society seething with revolutionary discontent or writhing with resentment at its callous mistreatment by cruel imperial masters. He would have instead observed a prosperous countryside and bustling towns still enticing immigrants from Britain and Germany with ready access to land (sometimes as tenants, but with freeholds to follow) and the promise of higher wages than they could earn at home. True, closer scrutiny would reveal sources of anxiety, misery, and conflict. From Maryland to Georgia, hundreds of thousands of slaves toiled from dawn to dusk on southern plantations. In New England, fourth- and fifth-generation descendants of the original Puritan immigrants worried that they could not bequeath a "decent competence" of land to their sons. In the middle colonies lying in between, a polyglot mixture of peoples and sects were pioneering a strangely pluralistic society. And in particular colonies, political violence occasionally erupted. In North Carolina, for example, tidewater militia mobilized in 1771 to put down frontier Regulators protesting misrule by the provincial government. In Rhode Island, a crowd happily sacked and burned the HMS *Gaspée*, a royal schooner whose zealous enforcement of imperial regulations had been making life miserable for local merchants after it ran aground in June 1772.

But in most of the colonies, the early 1770s were a period of political torpor that gave no hint of the explosion to follow. In only one colony did the issues agitated in the 1760s remain in the forefront of politics. That colony was Massachusetts.

PART I

THE CRISIS

1

Advocates for the Cause

IN THE GATHERING DUSK of December 16, 1773, a mass meeting of "the Body of the People" of Boston waited restlessly for Francis Rotch to return to Old South Church. Only twenty-three years of age, Rotch was a Quaker merchant from Nantucket and a co-owner of the *Dartmouth*, the first-arrived of the three ships now in harbor bearing East India Company tea. Rotch was more worried about his ship than its cargo, which he did not own. If Boston's protesting citizens forced the ships to sail with the tea unloaded and its duty unpaid, the *Dartmouth* might be subject to seizure, either by the Royal Navy patrolling just beyond the harbor or by customs officials in England; Rotch might also be liable for the value of his ship's cargo. At the town's order, Rotch had ridden the seven miles from Boston to Governor Thomas Hutchinson's country house at Milton in a final bid to persuade the empire's loyal servant to grant the necessary clearances. On another occasion the two men might have ambled to the shore to admire the peaceful view north to Boston Bay. But Rotch had time only to make one last plea and then return to the capital. As expected, Hutchinson refused to permit the *Dartmouth*, *Eleanor*, and *Beaver* to sail. Once they legally entered the harbor and customs officials registered their cargo, the law required the goods to be offloaded and duties paid within three weeks, or else confiscated.

When Rotch returned to Old South, he told the waiting crowd of the governor's refusal. Within minutes, Samuel Adams, the driving force on the town's Committee of Correspondence, arose to declare that "they had now done all they could for the Salvation of their Country."

Soon shouts erupted from the gallery and door, and many of the five or six thousand attending headed outside. To the sound of mock war whoops and cries of "Boston harbor a tea-pot tonight," the crowd, some fancifully dressed as Mohawk warriors, descended to Griffin's Wharf, where the tea ships lay docked. A merchant drawn outdoors by the clamor recalled thinking "that the inhabitants of the infernal regions had been let loose," before returning to his house to finish his own pot of tea. But once the "Mohawks," numbering fifteen or twenty a vessel, boarded the ships, the crowd watched silently as 340 massive chests of East India Company tea were hauled on deck and whacked open with axes; then the contents were dumped overboard. By 9 P.M. a cargo valued at a hefty nine thousand pounds sterling was weakly brewing in the low-tide waters.

Had the value of the tea not been so dear, the Boston Tea Party might be remembered, if at all, as a minor piece of political theater, with critics hailing the players' costumes as its most noteworthy feature. Americans were heirs to a rich tradition of extralegal political protest—effigy burnings and the like—which communities mounted when acts of government threatened their basic rights and interests. Some of these popular actions combined symbolic protest with dollops of violence, like the rare tarring and feathering, which left victims painfully burnt. With its gross assault on private property, however, the Tea Party crossed the line between *extralegal* and *illegal*, defying the authority of the British government in ways that smearing "Hillsborough paint" on merchants' houses and shops did not.

The ministry of Lord North answered this challenge with a punitive program of parliamentary legislation that made Boston a garrisoned city and Massachusetts a tinderbox of rebellion. But had Boston's protests taken a milder form, or had Hutchinson let the tea ships go, the crisis might have been averted and the Revolution itself delayed, or perhaps even avoided. Just as we speculate whether the guns of August 1914 might never have fired had Archduke Franz Ferdinand's driver not made the wrong turn in Sarajevo on June 28, the Boston Tea Party is one of those events that leaves us to wonder whether history—even History—might easily have turned out differently.

Sixteen months after Rotch's futile trip, another rider, far better remembered, also set out on horseback from Boston, headed not south to Milton but west toward Lexington and Concord. The purpose of

Paul Revere's mission on the fateful night of April 18–19, 1775, was to warn Samuel Adams and John Hancock that British troops were sortieing from Boston, intent on capturing the two men and seizing provincial munitions. Revere reached Lexington but was snatched by a British patrol before he could continue to Concord. But the alarm was already spreading by word of mouth, and warning shots fired in the night air. "You know the rest," wrote the poet Henry Wadsworth Longfellow four score and five years later. First at Lexington, then at Concord, British regulars and colonial militiamen exchanged fire.

For "the fate of a nation" to have been "riding that night" required something more than the bravery and ingenuity Revere and William Dawes showed in slipping out of occupied Boston. In the sixteen months between the Rotch and Revere missions, two developments had altered the underlying structure of American politics, laying a foundation for revolutionary upheaval. First, the British program to punish Boston had produced exactly the opposite of its intended result. Instead of making Massachusetts an object lesson in the costs of defying imperial policy, the British response unified colonial opinion in support of that defiance. Just as important, that unity was no longer a matter of mere opinion or sentiment. On their own, Americans had created a new central political authority in the Continental Congress, which first met at Philadelphia in September 1774 and was set to reconvene in May 1775. Already observers were marveling that its decisions would be like the laws "of the Medes and the Persians, which must not be altered."* As yet that Congress was something less than a national government. But it was already becoming something more than the grand diplomatic assembly that the delegations to the First Continental Congress imagined they were attending.

Beleaguered Massachusetts sent four delegates to that First Continental Congress, and the two best remembered were the distant cousins Samuel and John Adams. The challenge they faced on their dip-

* This phrase appears in two places in the Hebrew Bible, in stories well known to American readers. Both of them describe traps that evil counselors may set for their unsuspecting royal sovereigns. The primary reference is in Daniel 6, where the phrase introduces the treacherous royal decree that causes Daniel to be cast into the lions' den. It also appears in the opening chapter of Esther, the book whose villain is that most sinister of all royal counselors, Haman. At some level, this phrase reminded colonists that their own king should reject the evil plots of his ministers and render to his loyal American subjects the justice they deserved.

lomatic mission to Philadelphia illustrated a deep uncertainty in the character of the colonial resistance movement. British strategy in Massachusetts presumed that Americans did not constitute a nation and that a decisive show of force in this single irksome province would prevent their becoming one. The Adamses and their colleagues faced a different challenge. Creating an American nation was not their avowed goal. But once the British government responded to the Boston Tea Party as it did, it became essential to ensure that the other colonies would support Massachusetts, "now suffering in the common cause," and that they do so fully recognizing that war might indeed be the outcome. Their cause and America's, they thought, were one. But in critical respects, Massachusetts *was* different, and the fact that the British government had selected it for retribution only reinforced its people's notion that history and providence had singled them out for a special role. Was the purpose of the "common cause" to support Massachusetts in its time of peril, or to transcend the explosive situation in that single province in the name of forging a new, larger, and avowedly American community? Massachusetts was where the Revolution began, and to explain why that was so, we have to begin our story there.

For months before the Tea Party, colonists elsewhere had indeed watched events in Massachusetts with a mixture of admiration and alarm. In nearly every other province, the imperial controversies of the late 1760s were a fading memory. That was where most colonists were happy to consign them. It was only natural to hope that a government mindful of the protests of the 1760s would find other ways to persuade Americans to help defray the costs of the empire from which they drew so many benefits. True, every colony had a few radicals who suspected that the new ministry of Lord North harbored more sinister designs. But even these men hardly constituted a cadre of revolutionary agitators anxious to provoke a crisis in order to gain influence or seize power.

Almost everywhere in America, then, the political fevers of the 1760s had broken and subsided. The exception was Massachusetts. There, different conditions prevailed to make politics more volatile and less manageable.

For starters, Massachusetts had an unusually cohesive cluster of political leaders, centered in the capital of Boston but with reliable contacts in outlying towns, who remained suspicious of the secret designs of the British government. The most active and disciplined was Samuel Adams, a man whose only true vocation was politics. None could match Adams for energy, but he had collaborators who shared his views: other activists such as William Cooper, the town clerk, and Thomas Young, the radical physician; and prominent townsmen such as John Hancock, New England's wealthiest merchant; Thomas Cushing, speaker of the Massachusetts house; William Molineux; and Dr. Joseph Warren.

Like the other New England colonies, Massachusetts also had a remarkably homogeneous population. From New York south, the colonial population was an ever-changing blend of the descendants of early settlers and a steady and swelling flow of immigrants—free, half-free, and unfree—from the British Isles, Germany, and Africa. But most inhabitants of Massachusetts descended from the great Puritan migration of 1630–1642. They retained a deep sense of their colony's history, its foundation in the religious idealism of that first formidable generation of visible saints, and the defiance their forebears had shown toward the despised Stuart kings of the previous century. The religious enthusiasms of that era had cooled by several degrees. But currents of religious revivalism preserved the idea—improbable to outsiders—that Massachusetts could still play a role in the unfolding of grand providential designs.

In their fabled town meetings, the people of Massachusetts also had a ready mechanism for forging their political views. Elsewhere in America a population dispersed across the landscape had to trek miles to attend county court days or the revelry of an election. In the densely settled communities of Massachusetts, townspeople knew each other all too well. Petty disputes festered for decades, feeding upon an intimacy that nurtured bitter grudges as easily as lasting friendships. (Since the Salem horrors of 1692, though, neighbors no longer accused one another of witchcraft.) Resolving these disputes was one task of the town meeting. In ordinary times its business revolved around electing a small platoon of officials to manage routine affairs, dickering over how much firewood to allot the town minister, or identifying whose pigs were doing the most damage to fields and crops. (In New England, good fences did make better neighbors.) But in extraordinary times the town meeting offered a forum that enabled communities to make up their minds quickly—a political infrastructure waiting to be energized.

Massachusetts was different in one other, accidental respect. It had in its governor an exceptionally capable servant of the British Empire, a man who naturally assumed that Britain and Massachusetts shared a common welfare. But Thomas Hutchinson's policies, and his family's monopoly on high offices, made him the object of a vitriolic jealousy unique in the annals of colonial politics, which had seen repeated conflicts between royal governors and their provincial opponents. Hutchinson too descended from the founding generation of the 1630s. One ancestor was the controversial Anne Hutchinson, banished to Rhode Island after claiming divine revelation as a direct source for her radical theology. The governor included an account of her prosecution in the ambitious history of Massachusetts he was writing as an avocation. But the Hutchinsons now worshiped with the Church of England, a natural choice for a family that linked its interests to the empire's. To his detractors, proud of their Puritan heritage, that was another token of betrayal.

These factors began to converge in the fall of 1772, when reports circulated that Hutchinson and other high officials would receive salaries from the civil list of the British Crown. Royal governors were customarily paid by the colonial legislatures, which thereby gained significant leverage over their behavior. Granting Hutchinson a royal salary would make him even less amenable to the influence of the General Court (the legislature) and its constituents in the towns. In protest, Samuel Adams and his circle launched a provocative initiative. With the approval of the town meeting, they organized the Committee of Correspondence and promptly transmitted to other towns the inflammatory resolutions that Boston had adopted condemning the Crown salaries. The towns quickly replied, conveying new denunciatory resolutions approved by their own meetings.

The governor answered this barrage with a salvo of his own. When the General Court met in Boston on January 6, 1773, he used the customary speech opening the session to review the essential principles of the imperial relationship. Hutchinson hoped to bypass his opponents and talk sense to the community at large. Instead he found himself sucked into a contentious debate with the legislature. Newspapers in other colonies reprinted these exchanges, which quickly revived the basic question at the heart of the imperial controversy: were Americans bound to obey the acts of a Parliament to which they sent no members?

Hutchinson's rash decision to reopen the issues that had been agitated during the 1760s dismayed his superiors back in London. They had no wish to foment a fresh crisis in American affairs.

Hutchinson had another critic in London: Benjamin Franklin, the most eminent American of the day, Boston's best-known (if long departed) son, retired printer, true scientist, ingenious inventor, and prominent citizen of the international republic of letters. He had lived in London since 1758, acting as agent for the legislature of his adopted colony of Pennsylvania and later for Massachusetts, New Jersey (whose governor was his illegitimate son, William), and Georgia. Franklin loved cosmopolitan London far more than provincial America. He also believed that the continued association of both countries would bring great advantages to each. But the frustration of dealing with British officials, great and small, also left him exasperated with their attitudes toward America and Americans. By June 1773, he thought that the colonies should call a congress to discuss common concerns, including their future relationship to the empire.

Franklin's critical intervention in Massachusetts politics had begun some months earlier. In December 1772 he sent Speaker Cushing a packet containing copies of letters that Hutchinson had written to an unidentified London correspondent. In the slow season of winter crossings, the packet took four months to reach Boston. But once arrived, its explosive contents plunged the province into new turmoil. The letters were full of biting comments about provincial politics and Hutchinson's critics. They also suggested that Americans should gladly accept some reduction in the full range of English liberties in order to enjoy the real benefits of empire. Franklin instructed Cushing to show the letters only to select associates. But inevitably rumors of their contents became common gossip, rendering the restriction pointless. The letters were published in Boston newspapers in mid-June and reprinted elsewhere. The Massachusetts assembly promptly petitioned the Crown to remove Hutchinson and his brother-in-law, Lieutenant Governor Andrew Oliver, from office. Franklin, as the assembly's agent, submitted the petition to William Legge, Lord Dartmouth, the new secretary of state for America.

Why had Franklin sent these letters? Even before reading Hutchinson's ill-advised speech, he had concluded that the governor's presence in volatile Massachusetts threatened the stability of the empire that

Franklin, no less than Hutchinson, wished to preserve. To Franklin it was the governor who was the real provocateur. Destroying his influence and reputation seemed a small price to pay for restoring political calm in Massachusetts and harmony to the empire.

The torrents of invective that followed publication of the letters did not surprise Franklin, the veteran of many a political combat and polemical skirmish of his own. But the consequences were not those he had imagined. The frenzy in Massachusetts restored the ministry's support for the hapless Hutchinson. In the end Franklin fell victim to his own stratagem. In December 1773 he was forced to acknowledge his role in the affair after a duel was fought between the son of the letters' original recipient and another man whom he had wrongly accused of purloining his father's correspondence. A month later Franklin suffered a public humiliation as vicious as the one he had brought on Hutchinson. On January 29, 1774, the day assigned for the Privy Council to consider the petition urging Hutchinson's dismissal, Franklin stood mutely as Solicitor General Alexander Wedderburn subjected him to a harsh dressing-down in the hearing room known as the Cockpit. British onlookers chattered and snickered as Wedderburn laid the blame for the disorder in Massachusetts not on Hutchinson, as Franklin once intended, but on Franklin himself. Americans present watched in disgust but marveled at Franklin's stoic demeanor. For the moment Hutchinson retained his governorship, but Franklin lost his lucrative position as deputy postmaster general for North America.

One other consequence of the affair of the letters dwarfed all the others. It steeled Hutchinson to turn enforcement of the Tea Act, passed by Parliament in May 1773, into a confrontation with his opponents. The governor and his family had a personal stake in seeing the act enforced. He owned stock in the East India Company, the giant enterprise that the act was designed to benefit. And his sons, Elisha and Thomas Jr., were among the small group of consignees who would market the tea once it reached America. Once again, the interests of the Hutchinsons and the empire converged.

The Tea Act attracted little attention as it made its way through Parliament in the spring of 1773. Nearing bankruptcy, its warehouses bulging with surplus tea, the East India Company sought help from the government. Lord North responded by granting it a monopoly over the sale of tea legally imported into America. The colonists were still boycotting legal tea because it carried the duty levied by the Townshend Act of 1767. To induce them to comply with the act, the government proposed to lower the duty, making East India Company tea more competitive with the illegal tea that was easily smuggled into the colonies. Skeptical members of Parliament urged North to drop the duty altogether, the better to help the company, but he refused.

For colonial radicals the act offered an unexpected occasion to revive the apparatus of resistance. The old tactics of the 1760s were put back into action, successfully. Crowds marched, burned effigies, and intimidated officials, ship captains, and tea consignees into doing the right thing. The tea ships returned to England, their cargoes still on board.

Only in Boston did the ships' arrival spark an actual crisis. Hutchinson claimed that he initially opposed allowing the tea ships to enter the harbor and thereby trigger the legal requirements of having their cargo registered by the customs commissioners and duties paid. But whatever his misgivings, the fact remained that the ships did moor at Griffin's Wharf, allowing the apparatus of imperial law to swing into play. From this point on, Hutchinson welcomed the test of wills that ensued. With British naval vessels anchored outside Boston, the three ships could depart only if he issued the necessary clearances. This was exactly what he refused to do. Believing he had law on his side and having a personal stake in seeing the tea landed and the duties paid, Hutchinson refused to let the mounting protests sway his judgment. When Francis Rotch begged him to release the ships, Hutchinson curtly replied that "he could not think it his Duty in this Case" to grant the request. In response to the governor's intransigence, the patriot leaders formed their Mohawk war party on the night of December 16.

Hutchinson accepted the confrontation for two reasons, one political, the other personal. For years he had urged correspondents in London to promote a firm and consistent policy toward America. It was not that he wanted Americans to learn to kiss the whip of imperial rule. There was no whip to kiss, and Hutchinson knew that American loyalty could never be coerced. But he did believe that popular respect for the empire would decline if the government did not act consistently or if it forever allowed demagogues to mock its authority.

These were prudent calculations, drawn from years of thoughtful analysis. But by December 1773, prudence and calculation alone no longer controlled Hutchinson's thinking. How could they, after months of vilification had made him the object of suspicion up and down the seaboard? With the law offering an opportunity to force his opponents either to back down or to commit some desperate folly, Hutchinson welcomed confrontation, not only to regain political mastery but for personal vindication and perhaps even revenge. Having served the empire faithfully over the years, Hutchinson now summoned the empire to his own cause.

So the ships lay at anchor, the Mohawks swung their axes, the crowd watched silently, the harbor fish encountered a foreign substance in the dockside waters—and the British Empire in North America awaited the crisis on which it would founder. Five months passed before the government's response reached American shores.

Of Hutchinson's many opponents, history remembers three best. One is John Hancock, who placed his beautiful bold signature on the Declaration of Independence barely hours after the University of Oxford awarded the exiled Hutchinson an honorary degree on July 4, 1776. But Hancock's impact on events is difficult to measure, in part because he left few political papers. It is perhaps fitting that he is most remembered not for anything he wrote but for the epochal document he signed.

No such uncertainty colors the historical memory of Hutchinson's two other great foes, Samuel and John Adams. Much of Samuel's correspondence has been lost, and much of what survives blandly masks more than it reveals. Yet his place in the politics of resistance is well established. By the early 1770s he had emerged as the one radical leader most suspicious of British motives and most likely to play an active part in whatever form of resistance proved necessary. John's papers, by contrast, run to volumes yet unpublished and scores of reels of microfilm. Far from revealing too little, they almost divulge too much. For John Adams was a writer whose heart and mind flowed through the quill of his pen, who never used a single well-chosen word when six impetuous synonyms would do just as well. Where Samuel Adams was ultimately eclipsed by the independence movement to which he once seemed indispensable, John Adams was liberated by it, released into that wider world of activity and thought and the very stream of history that the Harvard graduate and Worcester schoolteacher had only imagined he might one day join.

Their descent from a common ancestor formed little part of their connection. Samuel, born in 1722, was older by thirteen years. The son of a prominent Boston brewer, he entered Harvard at the usual age of fourteen. In an era when class rank was assigned upon entrance, not at graduation, and was used to mark social status, not intellectual achievement, Samuel placed sixth among the twenty-three-member class of 1740. Like most young men of that era, he followed the occupation of his father. But in the son's case that notion of vocation had a dual meaning. For the elder Samuel was an active ally of Elisha Cooke, an early pioneer in the line of urban political bosses that later included such legendary figures as Samuel Tweed, James Curley, Erastus Corning, and two Richard Daleys. Though colonial Boston, a town of fewer than twenty thousand, differed from the great nineteenth- and twentieth-century metropolises, a leader of Boston's South End would no doubt recognize, say, a precinct captain from Chicago's South Side as a kindred practitioner of the politician's art.

The elder Adams started his son in commerce, first placing him in a merchant's countinghouse, then staking him a thousand pounds to trade on his own. But the world of trade never engaged Samuel's heart or talents. Temperament and the times conspired to steer Samuel away from commerce and into the political activity he really loved. Nor did the depressed local economy of the 1740s and 1750s make Boston the most promising place to launch into business. With a relatively poor agricultural hinterland and rival ports all along the New England shore, Boston was already lagging behind New York and Philadelphia as a commercial entrepôt. In such a risky environment, the family business might seem a safe port, but after his father's death in 1748, Samuel managed to run the brewery into bankruptcy. A quarter-millennium would pass before the name Samuel Adams again brought a smile to the lips of American tipplers.

Other legacies mattered more to him. From his Calvinist upbringing, Harvard education, and early entrance into politics, Samuel Adams absorbed a set of attitudes and idioms that placed him squarely within the tradition of opposition politics that flourished in colonial America. Adherents of this tradition, which was nurtured in the religious and revolutionary turmoil of seventeenth-century England, were ever alert to the danger of tyranny that lurked whenever the concentrated power of monarchy went unchecked. The Puritan revolutionaries who beheaded Charles I in 1649 acted on the radical Protestant conviction that submission to tyranny was *not* a Christian duty. The English Whigs who opposed the reigns of Charles II (1660–1685) and his younger brother, James II (1685–1688), bequeathed a similar legacy of political suspicion. After James II was deposed in the Glorious Revolution of 1688, the Convention Parliament that met during the interregnum adopted the Declaration of Rights, which made acceptance of the principle of parliamentary supremacy a condition of bestowing the throne on James's son-in-law, William of Orange. The principle was confirmed in the Act of Settlement of 1701, which established a line of succession from King William, a childless widower, and his sister-in-law and successor, Queen Anne—whose children had predeceased her—to the elector of Hanover, a Protestant state in what is now northern Germany.

Hanover remained the true home of the first two Georges who ruled Britain from 1714 to 1760. While it did, the British constitution underwent a major transformation. Beginning with the government of Sir Robert Walpole, the king's ministers of state developed new means and tactics for controlling Parliament. Some relied on the enormous political influence still wielded by the aristocracy, with their networks of dependents and retainers. Others used the revenues and resources that a booming economy placed at the government's disposal to purchase the loyalty of members of Parliament and many of their electors. The use of patronage and influence turned Parliament into a docile body that often debated but rarely challenged ministerial decisions. Though supreme in theory, Parliament was easily managed in practice.

Alarm over these developments was commonly sounded in the coffeehouses and popular journals where public opinion in Georgian Britain was formed. Writers such as John Trenchard and Thomas Gordon, authors of the popular *Cato's Letters*, warned that avaricious ministers were striving to subvert the principles of 1688 and deprive Britons of their precious birthright of liberty. These charges had little political impact in Britain. But in America this literature of opposition attained surprising popularity. Some colonists were attracted to the image of a mother country sinking into corruption and the seductive allure of "luxury," a loaded word that implied that the refined manners that commercial society was supposed to bring would turn liberty-loving Britons into "effeminate" hedonists. Many of William Hogarth's most popular prints—from *Gin Lane* to *A Rake's Progress*—were morality tales satirizing the vices into which different classes were likely to fall as they pursued these pleasures. But these images and writings also fit colonial politics peculiarly well. For in America, unlike Britain, the constitutional quarrels of the seventeenth century still resonated. There, governors and legislatures continued to skirmish over their respective powers and rights, privileges and duties. The more conscientiously governors tried to enforce their instructions from London, the more easily they could be tarred as agents of the corrupt ministers who were sapping the principles of 1688.

Samuel Adams was one of countless colonial politicians who absorbed these writings and viewed the doings of government through the lens they provided. Like many others, he used this idiom to advance his own political ambitions. With little chance of cracking the circle of imperial patronage, an aspiring provincial politician could better succeed by casting himself as a tribune of popular liberties. But this conclusion suggests that men adopt and adapt views that will best help them pursue their ambitions—and it is far from clear that Adams possessed ambition as we define that term, or that if he did, he could ever admit it to himself. Like more modern revolutionaries—but without understanding that he was one—Adams inhabited his ideology. His identity and his politics fused so completely that he probably did not know where one left off and the other began.

For his younger distant cousin John, by contrast, anxieties about ambition and identity seemed to fester daily. If any of his Puritan ancestors kept diaries, they would have recorded their spiritual travails while they struggled with the Calvinist doctrine of predestination and the signs of their own depravity. John's diary and letters contain the odd reflection on religion. But it is life and work, not soul and faith, that animate these writings. John could acknowledge his ambitions all too easily. The hard part was determining which to pursue.

His origins were modest. His father, though respected in Braintree as a church deacon and town selectman, worked a typical fifty-acre farm. Most New England farmers aspired to nothing more than acquiring enough land to ensure that their sons' status would equal their own, enough to form a decent "competence." But the elder John Adams wanted more for his firstborn namesake. At his insistence John entered Harvard at age fifteen, ranked fourteen out of that year's twenty-four entrants. John's indecision about which profession to pursue was resolved by 1756, when he began reading law. The professional success he began to enjoy in the 1760s relieved him of the fear that his practice might never escape the tedium of debts and trespasses. Marriage to Abigail Smith, daughter of a respected minister, in 1764 provided another source of stability. Well read, with an independent streak of her own, she proved remarkably resourceful in running the family farm as John's legal practice called him away from home.

There were, however, facets of the famous Adams personality that prevented him from ever attaining equanimity. He had a keen eye for observation, which he often cast on colleagues at the bar. Not only did he regularly itemize their strengths and failings; he also constantly measured himself against them. He did this because reputation mattered deeply to him, and not only professional standing, but all the other qualities by which one person judges another. He knew himself too well to think that others always appreciated his ability. He was aware that he was opinionated, that he wore his feelings on his sleeve—both sleeves, really—and that he spoke too directly and candidly. Above all, he knew that somehow, on some occasion, he aspired to do great things, though what they might be and how he might accomplish them were a mystery. As well read as any American of his era, a compulsive seeker after knowledge, Adams always sensed there must be some larger stage on which he could test and show his abilities.

Samuel Adams was always on the lookout for political recruits, and as John became known in Boston and began to write for the press, Samuel recognized his cousin's talents. There was no need for the veteran politician to convert the ambitious attorney to the cause of colonial rights. In their views of the issues dividing Britain and America, there was nothing to distinguish one Adams from the other. But John was argumentative and academic in a way that Samuel was not. When the older Adams wrote for the press, he restated familiar themes and arch warnings that generations of radical Whig polemicists on either side of the Atlantic had long pronounced. John was a more original and dogged controversialist, eager to prove points, muster evidence, and run legal arguments back to their sources and forward to their conclusions. Samuel wrote occasional pieces that any of his collaborators could have drafted. John's works were the product of an assertive thinker finding his voice as an advocate of colonial rights and the special cause of Massachusetts.

Occasional writing for the press hardly constituted a complete commitment to politics, however. Before 1774 Adams measured his political engagements carefully. He had a legal career to advance and a growing family to provide for, and his practice frequently carried him on horseback from one county to another. A diary entry for June 22, 1771, finds Adams in Ipswich "in the usual Labours and Drudgery of Attendance upon Court." While there he took tea with Justice Edmund Trowbridge, who cautioned him against political enthusiasm much as he might warn another man against strong drink. You'll ruin your health, the judge let Adams know, "if you tire yourself with Business, but especially with Politics." But "I don't meddle with Politicks, nor think about em," Adams protested. "'Except, says he, by Writing in the Papers.'—I'le be sworn, says I, I have not wrote one Line in a Newspaper these two Years." In a later age Adams could have posed as a recovering drinker who had been dry for years.

Adams punctuated his diary with ambivalent confessions of his anxieties and ambitions. "What is the End and Purpose of my Studies, Journeys, Labours of all Kinds of Body and Mind, of Tongue and Pen?" he asked on a cold January night in 1768. Whatever plan he settled on, he gloomily concluded, "will neither lead me to Fame, Fortune, Power nor to the Service of my Friends, Clients or Country." How could they, when the endless cycle of circuit courts forced him into "a rambling, roving, vagrant, vagabond life"? Four years later, with his legal practice flourishing, Adams bought a brick house near the Suffolk County courthouse in Boston and planned to move his family to town. "If I do, I shall come with a fixed Resolution, to meddle not with public Affairs of Town or Province." He had forgone opportunities for "Money, and Preferment," Adams complained, lest they tempt him "to forsake the Sentiments of the Friends of this Country." Yet those same friends "are such Politicians, as to bestow all their Favours upon their professed and declared Enemies." His disillusionment with politics owed much to his decision to defend the British soldiers charged with perpetrating the Boston Massacre of 1770. When Samuel Adams and Samuel Pemberton asked him to deliver the oration commemorating its third anniversary, John begged off, claiming that he "was desirous to avoid even thinking upon public Affairs." By accepting, Adams replied, "I should only expose myself to the Lash of ignorant and malignant Tongues on both sides of the Question. Beside that," he lamely added, "I was too old to make Declamations."

Yet Adams privately gloried in what he himself described as the "gallant, generous, manly and disinterested" role he played in defending the soldiers, and he regarded himself as one of the few reliable watchmen intent on preserving the colony's traditional liberties. When he and Samuel formed half of "the Boston seat" in the General Court, he was delighted when a former royal governor was heard to scoff, "where the Devil this Brace of Adams's came from, I cant conceive." Was "it not a pity," Adams wrote in his diary, "that a Brace of so obscure a Breed, should be the only ones to defend the Household, when the generous Mastiffs, and best blooded Hounds are all hushed to silence by the Bones and Crumbs that are thrown to them?" The master of this better-bred pack was Hutchinson, and the "Sentiment" that bound the brace of Adamses to their lonely path was that Massachusetts had "more to fear" from Hutchinson's evil designs "than from any other Man, nay than from all Men in the World."

A year later, with that conviction unshaken, John did allow himself to be ensnared in politics, helping to draft the assembly's replies to Hutchinson's ill-advised speech opening the 1773 legislative session. The arrival of the purloined Hutchinson letters agitated him further. "What shall I write?—say?—do?" Adams asked himself. "Bone of our Bone, born and educated among us!" By the May elections, he was reconciled to his duty, if summoned "to take a Part in public Life," to "Act a fearless, intrepid, undaunted Part, at all Hazards." For a time, Adams was spared. The General Court nominated him to a seat on the provincial council, but Hutchinson vetoed his appointment.

There is no evidence that John had any role in organizing the opposition to the Tea Act. But he and the Tea Party's leading planner, Samuel Adams, understood this event identically. John had the final silence of the crowd in mind when he marveled that there was "a Dignity, a Majesty, a Sublimity in this last effort of the Patriots." They had been "so daring, so firm, intrepid and inflexible" in this action, "that I cant but consider it as an Epocha in History." By contrast, "The malicious pleasure with which Hutchinson" and those around him "have stood and looked upon the distresses of the People, and their Struggles to get the Tea back to London, and at last the destruction of it, is amazing," he wrote. It was "hard to believe Persons so hardened and abandoned" could feel so assured of their own rectitude when their countrymen were risking so much. John even wondered whether Hutchinson and his crowd wanted to see as many "dead Carcasses" as there had been chests of tea "floating in the Harbour." Then he briefly indulged a lethal fantasy of his own, speculating that "a much less Number of Lives however would remove the Causes of all our Calamities."

Months had to pass before the colonists would know whether the government had learned its lesson. One could hope for the best and fear the worst. Consistent with his dark view of the sinister forces at work in Britain, Samuel most likely expected the government to pursue a policy of repression. Writing to John Dickinson in early April 1774, he noted that "*We* [the Bostonians] have borne a double share of ministerial Resentment, in every Period of the Struggle for American Freedom. I hope this is not to be attributed to our having, in general, imprudently acted our Part," when the real blame lay with Hutchinson and others "whose Importance depended solely upon their blowing up the flame of Contention." Samuel was already wondering whether the other colonies would rally to Boston's support should the government indeed opt for vengeance. John favored a different conclusion. As late as April 1774 he clung to his long-held opinion "that there is not Spirit enough on Either side to bring the Question to a compleat Decision—and that We shall oscilate like a Pendulum and fluctuate like the Ocean, for many Years to come, and never obtain a compleat Redress of American Grievances, nor submit to an absolute Establishment of Parliamentary Authority." Perhaps, he wrote to his friend James Warren, "Our Children, may see Revolutions, and be concerned and active in effecting them of which we can form no conception."

Rather than anticipate the crisis that was about to break, John wondered whether he had "patience, and Industry enough to write an History of the Contest between Britain and America." Like any working historian, he wondered where such a narrative should begin: with the accession to the throne of George III in 1760? Or perhaps the peace treaty of 1763, which had brought "the cession of Canada, Louisiana, and Florida to the English"? He even jotted down a cast of characters whose actions would have to be recounted. With two exceptions—Benjamin Franklin and William Lee, the younger son of a prominent Virginia family and now sheriff of London—the list consisted entirely of prominent leaders in Britain and Massachusetts. In his mind, Britain and Massachusetts, or even London and Boston, were the protagonists of this history. Had Adams pursued this project, he would have had to bring the other colonies into the story. But in April 1774 the lawyer soon to turn revolutionary was little less provincial than the Worcester schoolteacher of 1756. There was an America out there, its political destiny awaiting discovery, but John Adams was still a stranger to it.

Remarkably for an eighteenth-century government, the ministry of Lord North did have a partial contingency plan in reserve as it pondered its response to the Boston Tea Party. Plans to revise the Massachusetts charter of government had been broached in 1768 and again in 1770 but were shelved both times. Still, a belief persisted that Massachusetts was the great source of the mischief emanating from America. At some point something had to be done to strengthen imperial authority in this quarrelsome province. The chief suggestion was to allow the Crown, rather than the lower house of the assembly, to appoint the council, which operated both as an advisory body to the governor and as an upper legislative chamber. News of the Tea Party justified a more radical approach.

The government responded in a succession of acts that Parliament adopted in the spring of 1774. The first closed the port of Boston until restitution was made for the tea. Next, the Massachusetts Government Act altered the colony's 1691 royal charter by allowing the king to appoint members of the council. It further limited town meetings to an annual session for the sole purpose of electing officers. Then the Administration of Justice Act placed British officials and soldiers accused of murder and other capital offenses beyond the reach of colonial courts, implying that they could engage in violent acts against the colonists with impunity. Last, the Quartering Act provided legal authority to billet royal soldiers, typically drawn from the dregs of British society, in unoccupied houses and other spare buildings. The ministry also recalled Governor Hutchinson to England, there to brief his superiors on American affairs. His acting replacement was to be General Thomas Gage, commander in chief of His Majesty's forces in North America and an early advocate of the policies the government was about to enforce.

These Coercive Acts turned the episodic political controversies of the previous decade into a revolutionary crisis. They did so for three reasons. One was the severity of the collective punishment imposed on all of Boston, not specifically the unknown perpetrators who dumped the tea into the harbor. So harsh a penalty might be legitimate for Ireland or for the Highlands of Scotland, which was subject to the "clearances" that followed the failed Jacobite rising of 1745, when the clans rallied to the standard of Bonnie Prince Charlie, grandson of the deposed James II. But Americans were not a conquered nation like the Irish or a suppressed people like the Highlanders; they claimed all the rights of freeborn Englishmen. Second, altering the colony's charter and denying the right of its juries to protect injured citizens threatened the equally fundamental principle that legislatures and juries both existed to prevent the arbitrary exercise of executive power. Third, and most important, in making legislative acts the vehicle for punishing Massachusetts, the government offered its definitive assertion of the reach of parliamentary sovereignty. When Parliament adopted the Declaratory Act back in 1766, affirming its power to govern the colonies "in all cases whatsoever," that ominous phrase stated only a broad principle, not a plan of action. Now it described a legislative program that evidently knew no limit. What could Parliament not do, Americans asked, not only to Massachusetts, but to any other colony?

British officials hoped that the leaders of other colonies would ask a different question: why should they risk the grave costs of supporting a colony whose provocative politics exceeded the proper bounds of opposition? This was not a foolish calculation. Massachusetts was regarded as an unruly place to govern, the Tea Party had been a wanton destruction of property, and the colony's Puritan character did not sit well with residents of other colonies where religious matters were taken less seriously. By making the costs of defiance so evident, British officials reasonably assumed that other provinces would draw the obvious lessons.

The moral that Americans did draw was not the one the government wanted to teach. Resolutions adopted in local meetings throughout the colonies in the summer of 1774 agreed that the people of Boston and Massachusetts were "now suffering in the common cause" of American liberty. If excesses had occurred there, that was because they had been singled out for special attention, perhaps because Hutchinson had indeed convinced his royal masters that they merited it. This was why Hutchinson's troubles of 1773—his debate with the General Court and the affair of his letters—mattered so much. They neutralized whatever doubts and criticisms other colonists might voice about the tenor of opposition politics in Massachusetts.

One question remained: what was to be done? Even before news of the Port Act arrived on May 12, Samuel Adams fretted that colonists elsewhere might view Boston as a town ripe for comeuppance. Once word of the act was received, Adams and his collaborators decided to propose an immediate suspension of trade with Britain and the West Indies, where the slaves who produced the lucrative sugar crop depended on American foodstuffs for their sustenance. With the support of a hastily called town meeting, the Committee of Correspondence began writing to other communities and colonies, proposing an embargo to be known as the Solemn League and Covenant. The name carried a significance that many colonists instantly grasped. It was borrowed from a famous agreement of 1643, at the outset of the English Revolution, between the English Puritan opponents of King Charles I and their Scottish Presbyterian allies. At a moment when the Bay Colony desperately needed aid from other provinces, this was an astonishingly bad choice. In New England the beheading of Charles I in 1649 and the ensuing rule of Oliver Cromwell were still recalled with some favor, but elsewhere the turmoil of the 1640s and 1650s remained an object lesson in the dangers of revolution and political excess. If the goal of the Massachusetts radicals was to secure support from moderates in other colonies, the Solemn League of 1643 was the last symbol they should have invoked.

That was not the only miscalculation in Boston's early response to "the vengeful Stroke of the hand of Tyranny." Boston wanted immediate action, but other communities and colonies preferred consultation first. As the replies of neighboring towns and other colonies revealed, Americans believed that any opposition to London's punitive measures demanded careful consideration and broad support. The stakes were too high for impetuous action, however sorely pressed the Bostonians might be. Within weeks, support rapidly coalesced for another measure, first proposed in Virginia. There the legislature was meeting at Williamsburg when news of the Port Act arrived. Lord Dunmore, the royal governor, dissolved the assembly before it could act, but a rump caucus of members reconvened in the Long Room of Raleigh Tavern and issued a call for an intercolonial Congress.

By early June Samuel Adams had to concede that an immediate boycott had no chance of success. It faced opposition even within Boston and mustered little support elsewhere. When Hutchinson sailed for

London on June 1—unknowingly into a permanent and sad exile from the native province for which he ever after longed—he carried an address of thanks signed by seventy-five prominent citizens. A week later the General Court met at Salem, summoned there by Governor Gage in order to insulate the legislators from the intimidating presence of the Boston crowd. The assembly immediately appointed a committee to consider what the colony should do. Samuel Adams was its chair, but it also included Daniel Leonard, a known loyalist. The patriots on the committee gulled Leonard into thinking they were contemplating moderate measures, and then Robert Treat Paine—one of John Adams's great rivals at the bar—persuaded Leonard to accompany him to a meeting of the county court at Taunton. In his absence the committee drafted a resolution calling for an intercolonial Congress. When it was reported to the assembly on June 17, another loyalist feigned illness and left to inform Gage of what was happening. The governor quickly sent the provincial secretary to dissolve the assembly. But Samuel Adams had taken the precaution of barring the door, and Thomas Flucker could read his proclamation only "out-of-doors," while inside, the assembly approved the resolution and elected a delegation to attend the Congress proposed for Philadelphia in early September.

The delegation numbered five: Speaker Thomas Cushing, Paine, the wealthy merchant James Bowdoin, and both Adamses. John had been in Boston when the Port Act arrived. His first reaction was that the town "must suffer Martyrdom: It must expire." He was quickly caught up in helping to organize relief measures for a bustling port whose economy ground to a halt when the act took effect on June 1. But soon his thoughts moved on to Philadelphia.

For both Adamses, the imminent Congress would fulfill deep aspirations. Samuel's were primarily political. He had been troubled by the ineffective coordination of colonial opposition to British policies since at least 1770, when repeal of the Townshend duties largely unraveled the network of communications that the colonists had forged in the 1760s. Disappointed as he was by the response to the Solemn League and Covenant, Adams was too experienced an organizer to ignore the obvious point: in New England and elsewhere, there was a strong consensus that any response to the assault on Boston had to be reached by common agreement.

John's aspirations were personal as well as political. For twenty years he had yearned for a more prominent theater of activity than the circuit of county courts and greater causes than the routine pleadings he had to argue. Now "a new, and a grand Scene open before me," he wrote in his diary days after his election. Once again the synonymous images poured out as the old tension between private concerns and public ambitions made itself felt. Congress would be like "the Court of Ariopagus, the Council of the Amphyctions, a Conclave, a Sanhedrim, A Divan, I know not what," he mused. "A School of Political Prophets I Suppose—a Nursery of American Statesmen." Marooned at a county court at York, he longed to return to Boston to converse with his fellow delegates and be "furbishing up my old Reading in Law and History, that I might appear with less Indecency before a Variety of Gentlemen, whose Educations, Travel, Experience, Family, Fortune, and every Thing will give them a vast Superiority to me." He professed to be "unequal to this Business," deficient in his "Knowledge of the realm, the Colonies and of Commerce, as well as of Law and Policy." He knew nothing "of the Characters which compose the Court of great Britain" nor "of the people who compose the Nation." An "American Senator" or an "American Statesman" needed just as much knowledge "as was ever necessary for a British, or a Roman Senator, or a British or Roman General." Yet, as he wrote to his friend James Warren in mid-July, "Our New England Educations, are quite unequal to the Production of Such great Characters."

Adams protested too much. No American education was adequate to the challenge the Congress would face. No knowledge of Britain offered an obvious solution to the questions and concerns that had brought the empire to this impasse. The Congress was an occasion for decision more than discovery. The knowledge the delegates needed most was exactly what they would have to acquire from one another during their deliberations. How far were they and their constituents prepared to go in support of Massachusetts, in fashioning a definitive statement of colonial rights and grievances, in agreeing upon the tactics to be used to persuade Britain to retreat? It was not lack of knowledge on these points that made Adams feel deficient, but rather the gravity of the decisions to be taken and the risks incurred.

Still, the ease with which Adams did imagine himself as an American statesman or senator, and not merely an envoy from Massachusetts, suggests how far ahead his ambitions were already racing. They sped faster too than the leisurely but politic pace of the delegates' journey to Philadelphia. On a very warm August 10, the Massachusetts members (save for Bowdoin, who chose not to attend) gathered at Cushing's house, boarded a coach-and-four (John had wondered what form of conveyance they would take), and set out for nearby Watertown. There they dined with several score gentlemen who rode out from Boston "and prepared an entertainment for them" at Coolidge's Inn. At 4 P.M., with the "fervent prayers of every man in the company," the delegation said farewell. "The scene was truly affecting, beyond all description affecting," John wrote. Similar scenes followed over the next three weeks as they crossed Connecticut, entered New York, were feted in New York City, and then traversed New Jersey before finally reaching Philadelphia on August 29.

Everywhere they received lavish hospitality and expressions of sympathy and support. They were also objects of profound curiosity. As they approached New Haven, an escort party greeted them a good seven miles out. Then upon arriving "all the Bells in Town were sett to ringing, and the People Men, Women and Children, were crouding at the Doors and Windows as if it was to see a Coronation."

The delegation played the part of curious tourists in turn. In New York their guide was Alexander McDougall, the city's closest counterpart to Samuel Adams. He took them to Battery Point, where the governor's "magnificent" house had burnt only days after the Tea Party (a portent of the 1776 fire that devastated the city). They admired the statue of George III on horseback, "solid Lead, gilded with Gold," on its high marble pedestal—to be pulled down in July 1776, the lead melted for shot. "We then walked up the broad Way," visited "several Marketts," and read newspapers at a coffeehouse. John Adams could not avoid noting that "The Streets of this Town are vastly more regular and elegant than Boston," with its twisting lanes never set straight, "and the Houses are more grand as well as neat." Local manners left him less impressed. "They talk very loud, very fast, and alltogether," he complained. "If they ask you a Question, before you can utter 3 Words of your Answer, they will break out upon you, again—and talk away." They spent six days there, then crossed New Jersey—"This whole Colony," Adams wrote, "is a Champaign"—visiting the college at Princeton and the "pretty village" of pre-industrial Trenton before arriving at Philadelphia. Once again a large party of dignitaries greeted them. "Dirty, dusty, and fatigued as we were, we could not resist the invitation" to dine at the new City Tavern on Second Street, "the most genteel one in America."

Beneath these shows of support and convivial bonhomie, there was political work to do. Whenever possible, the Massachusetts delegates sounded out the dignitaries who received them. Some they would see again in Philadelphia: Silas Deane at Hartford, Roger Sherman at New Haven, two New Hampshire delegates at New York hastening on because neither was inoculated against smallpox, four of the New York delegates. They were repeatedly pleased to learn that everyone agreed that their colony was indeed suffering in the common cause. But they were also anxious to correct untoward suspicions, such as the fear that the "Levelling Spirit" of New England might infect the other colonies and the belief that Massachusetts remained a bastion of religious intolerance, as it had been during "our hanging [of] the Quakers" a century ago.

During the three weeks of their leisurely progress, the situation back home deteriorated. Thomas Gage had come to Boston to rule as a civilian governor, not by martial law, but for all practical purposes he was fast becoming a military ruler whose authority extended as far as the reach of his troops—but no farther. A regiment of redcoats was now encamped on the Boston Common, reviving memories of the massacre of 1770, and more troops were expected. To assert authority outside the capital was virtually impossible. Eighteenth-century Boston was still a peninsula, a bulging top-shaped town resting on the thin Boston Neck, its link to Roxbury and Dorchester, and surrounded by the Atlantic and the estuaries of the Charles and Mystic rivers. Eastern Massachusetts was the most densely settled region of North America. Gage's troops could not move undetected, and the colonists could concentrate large numbers of militia on Boston at short notice.

Nor was Gage's political position secure. In August he delivered royal commissions to the new councilors whose appointments were supposed to fortify the empire's hand within the provincial government. But they were quickly subjected to a tactic the colonists had first deployed in 1765. The easiest way to thwart imperial measures, Americans knew, was to bring the pressure of public opinion to bear on those unwise enough to enlist in the cause of tyranny. One would have to be an especially hardy soul to resist the entreaties of hundreds of townspeople gathered at your front door, urging you to renounce the tendered office under pain of being burnt in effigy, tarred and feathered, and shunned—not to mention fielding the occasional death threat. These crowds made the councilors an offer they dared not refuse—unless they sought refuge among Gage's troops.

Gage began his governorship hopeful that order and reason would prevail. The illusion was short-lived. By late August his discouragement was measured by the descriptive terms that crept into his dispatches: "Phrenzy," "popular Fury," "further Extravaganzies," "the pitch of Enthusiasm," loaded expressions all. "Civil Government is near it's End," he wrote to Lord Dartmouth, secretary of state for America, on September 2. His immediate hope was "to avoid any bloody Crisis as long as possible," he added. "Nothing that is said at present can palliate, Conciliating, Moderation, Reasoning is over, Nothing can be done but by forceable Means."

In Philadelphia, meanwhile, the Massachusetts delegates were thrown into a round of continuous meetings with other delegations, over dinner at the City Tavern, at coffee, in strolls around town, and at receptions in gentlemen's parlors. Glasses were lifted in numerous patriotic toasts, from calls for colonial union and congressional unanimity to "'a constitutional death to the Lords Bute [once tutor to the king], Mansfield [the chief justice], and North.'" The delegates were naturally anxious to reassure one another of their mutual seriousness. In nearly every case, they were personally meeting their colleagues from other colonies for the first time. Avowing their common political sentiments offered an easy mode of introduction.

Monday, September 5, was the day appointed for the First Continental Congress to convene. The delegates met at City Tavern, strolled a few blocks to the newly constructed Carpenters' Hall, meeting place for one of the city's most numerous class of artisans, and promptly agreed "that this was a good room." (Modern visitors see a spacious room around which the delegates could easily have spread. But originally the main floor was split into "an excellent Library" and "a convenient Chamber opposite" to it, and the delegates would have found themselves in a more confined and intimate space.) Without further ado, they elected Peyton Randolph, speaker of the Virginia House of Burgesses, as their president. The delegates then read their commissions, and that done, elected Charles Thomson of Philadelphia as their secretary. John Adams had already heard Thomson described as "the Saml. Adams of Phyladelphia—the life of the Cause of Liberty, they say." And in fact, back in December 1773, Thomson had taken the initiative of writing an extraordinary letter to the Boston Committee of Correspondence carefully assaying the limits and prospects for American resistance—exactly the sort of letter that Samuel himself might have written.

The next obvious step was to set the rules of deliberation. Many of the delegates were veteran legislators, but their experience offered no ready solution for one obvious issue: the rule of voting. Should it be by colony, with each delegation casting a single vote; by "poll," with each delegate voting as an individual; or by "interests," which meant proportioning the vote of each colony to its population, wealth, or trade? The last idea was raised by Patrick Henry, the legendary Virginia orator with the common touch, and was quickly challenged by John Adams. To adopt such a rule and then find the information needed to make it work, Adams warned, "will lead us into such a Field of Controversy as will greatly perplex us." Henry was unconvinced. The stirring speech he gave the next day would still be recalled at another great meeting in Philadelphia thirteen years later. "Fleets and Armies and the present State of Things shew that Government is Dissolved," Henry declaimed. "We are in a State of Nature" wherein provincial loyalties no longer applied. "I am not a Virginian, but an American."

This effusion of nationalist sentiment was headily patriotic in one sense and transparently political in another. Henry's obvious motive was to give Virginia and other populous colonies a hefty, if not dominant, say in the Congress's decisions. But the whole point of the Congress was to enable the colonists to speak as Americans, to prove that Massachusetts could not be isolated. The delegates needed to reach decisions less by counting heads than by consensus, which made the rules for voting largely irrelevant. As delegates from the small states liked to say, "their all" was as much at stake as that of their populous neighbors. Every colony had an equal stake in the right decisions, regardless of disparities in population and wealth. Even could they have agreed on the justice of "equal Representation" (what we call "one person, one vote"), the delegates had come "unprepared with Materials to settle that Equality" since they lacked adequate data about population and wealth.

That unanswerable objection led Congress to approve a different norm of equality, one that gave each colony one vote.

With that troubling question resolved, Congress proceeded to adopt other rules. But then at 2 P.M. an express message from Boston arrived and plunged the entire city into despair, leading Congress to adjourn hastily while Philadelphia's churches tolled ominously muffled bells. The occasion was a report that British troops and ships "had fired on the People & Town at Boston," with unknown casualties. On Wednesday morning, another rider confirmed the news. In the Quaker City, Silas Deane wrote to his wife, Elizabeth, "All gather indignation, & every Tongue pronounces Revenge." Congress had asked the Reverend Jacob Duché to open its next session with a prayer, and the lesson for the day proved "accidently extremely Applicable": Psalm 35, which begins, "Plead my cause, O Lord, with them that strive with me: fight against them that fight against me." Duché was a minister of the same Church of England whose persecutions had sent the Massachusetts delegates' Puritan ancestors to New England in the 1630s. But that did not stop Samuel Adams, a Calvinist by conviction but ecumenical in politics, from proposing the clergyman. The delegates were deeply moved by Duché's reading and the extempore prayer that followed. (The ability to preach spontaneously was much admired by congregants of all denominations, who had heard too many ministers mumble through sermons the way an Oxford don might lecture on medieval Byzantine scholarship without noticing the audience before him.)

In fact the reports were wrong. There had been no cannonade, no spilling of innocent blood. Gage had merely sent troops to nearby Cambridge and Charlestown to seize arms that the colonists were storing there. The raid was a tactical success and a strategic shock. In response to confused reports, thousands of militia from Massachusetts and even Connecticut mobilized and descended on Boston. Gage now knew that it would be difficult to conduct any military operations outside the capital unless he obtained reinforcements well beyond his current force of three thousand.

In Philadelphia, relief that civil war had *not* erupted was tempered by the realization that Massachusetts was a tinderbox; a misstep there, by either side, could end any chance for peaceful resolution of the great dispute. The Powder Alarm (as it is known) had a profound effect on Congress. It made all the difference that the provocation had come from

Gage. Everyone expressed admiration for the restraint shown by the people of Massachusetts. "By g-d, says one I dont believe there is such a People in the World!" John Adams reported one member exclaiming as he read the reports from Boston, to be echoed by another who marveled that they could be "So cool, So cautious, so prudent, and yet So unalterably determined." This perception explains why Congress readily endorsed the strongly worded resolutions that the Suffolk County convention had adopted on September 9 and hastily shuttled to Philadelphia. The resolutions were replete with overwrought references to "the vengeance but not the wisdom of Great-Britain," "the arbitrary will of a licentious minister," "the parricide which points the dagger to our bosoms," and the "military executioners" who "thronged" the streets of Boston. The resolves also laid out a program of noncompliance with the existing legal institutions if they attempted to operate under the discredited authority of the Massachusetts Government Act. But careful readers—and the delegates read everything carefully—would note the language of the twelfth resolution: "we are determined to act merely upon the defensive, so long as such conduct may be vindicated by reason and the principles of self-preservation, but no longer." The threat was there, but so was the promise of restraint. The idea of limiting militant action to "defensive" measures, as opposed to authorizing offensive acts against British soldiers, was the promise that Congress seized upon by making its hasty approval of the resolutions its first published act.

Yet this invocation of "the principles of self-preservation" carried an ominous connotation. When Thomas Hobbes published his controversial book *Leviathan* in 1651, he made the right to self-preservation the first law of nature. In his *Two Treatises of Government*, published four decades later, John Locke extended the idea of self-preservation to embrace a right to revolution—"the appeal to heaven"—against tyrannical misrule. With each passing week the people of Massachusetts increasingly thought that they were falling into something like the state of nature that Hobbes and Locke had described. The more precise condition was "a dissolution of government" in which legal institutions had lost either the capacity or moral authority to rule. Though the General Court had been summoned to meet at Salem, it was evident that the people would never acknowledge its authority. Juries were refusing to serve, courts were closed, the royal councilors who accepted their commissions had fled to Gage's protection—and over all there now lay the grim specter of civil war. In such conditions, a people could claim a natural right to constitute a new government, which was exactly what many in Massachusetts thought they were entitled to do. Perhaps they could resume legal government under their original charter of 1629, which allowed the colony to govern itself without royal interference. Or more boldly still, they might simply create a new government of their own devising.

The members of Congress universally opposed both ideas. There were limits beyond which it was too dangerous to allow Massachusetts to pass. To establish legal government under either of these schemes would be tantamount to renouncing the authority of the Crown as well as Parliament. Once that link was severed, no tie would remain binding the colony to the empire. Massachusetts would somehow have to soldier on, maintaining some improvised semblance of law and order as best it could.

For John and Samuel Adams, these were difficult weeks. Not that they doubted that Congress would do the right thing. Once it endorsed the Suffolk Resolves—"one of the happiest days in my life," John wrote—it seemed evident "that America will support the Massachusetts or perish with her." Their letters home reassured correspondents that the other delegates were united in admiration for the Bay Colony, expressed in meeting after meeting and dinner after dinner. Yet privately the Adamses worried whether some delegates were still suspicious about the true character of their native province. After weeks of hearing "the most figurative Panegyricks upon our Wisdom Fortitude and Temperance," they still felt they were engaged in an awkward courtship with fifty "Strangers," all unacquainted "with Each others Language, Ideas, Views, Designs. They are therefore jealous, of each other—fearfull, timid, skittish."

John jotted these sentiments down in a brief note to Abigail on September 25, then evidently thought better of trusting these remarks to an unreliable post. Samuel Adams recorded similar reflections the same day. The colony's old reputation for being "intemperate and rash" had been replaced by a new "character" for being "cool and judicious as well as Spirited and brave." But there remained "a certain Degree of Jealousy in the Minds of some that we aim at total Independency not only of the Mother Country but of the Colonies too: and that as we are a hardy and brave People we shall in time over run them all." Baseless as this fear was, "it ought to be attended to." Unless directly attacked, Massachusetts had to avoid hostilities involving Gage's occupying force. If it was feasible for the colony "to live wholly without a Legislature and Courts of Justice as long as will be necessary to obtain Relief," John wrote the next day, "the general Opinion is, that We ought to bear it."

The situation in Massachusetts might have impelled Congress to act with greater urgency. But as one Rhode Island delegate complained, "The Southern Gentlemen have been used to do no Business in afternoon so that We rise about 2 or 3 o'Clock & set no more that Day." Northern merchants and lawyers were accustomed to long hours in their stores, offices, and studies. Southern planters were more used to spending hours on horseback, trying to coax an honest day's labor from their slaves while setting a good example for the overseers whose own feckless habits could be as aggravating as the slaves' uncanny knack for breaking tools, ignoring instructions, and defying common-sense notions of efficiency.

The easiest decision was to agree upon a plan for a commercial boycott, the favored tactic used against the Stamp Act and the Townshend duties. This new scheme was more ambitious than its predecessors. It would begin on December 1 with a ban on the importation of all goods from Britain and Ireland. Should the British government not offer redress, the next step would be to ban exports to Britain, Ireland, and the West Indies after September 1775. The delegates knew that American commerce was essential to the prosperity of the home islands. Their great hope was that the threat of losing that commerce would outweigh the other calculations (or miscalculations) underlying British policy. They believed they had "friends" to plead their cause, and by jeopardizing the vital interests of merchants and manufacturers and the very lives of Irish linen weavers and the slave gangs of the West Indies, Congress could mobilize the support of influential constituencies. Under such pressure, the government might well relent.

The boycott met with only one noteworthy dissent, from Joseph Galloway, speaker of the Pennsylvania assembly and once a close ally of Benjamin Franklin. Rather than risk the misery a prolonged boycott would bring, Galloway offered a more radical idea. Congress should propose the establishment of an intercolonial assembly to act as "an inferior and distinct branch" of Parliament. This "grand council" would regulate affairs of general concern and, in wartime, pass "bills for granting aid to the crown." Matters of merely "internal policy" would remain the business of the colonial assemblies.

Although Galloway had not prepared the groundwork for this radical scheme, Congress debated it seriously before tabling it for later discussion. That discussion never took place; instead, Congress expunged any mention of the resolution from its journals. Later, as a loyalist refugee in London, Galloway traced the rejection of his bold initiative to the wiles of Samuel Adams, the leader of a "republican" (meaning antimonarchical) faction already bent on independence. In Galloway's view, Adams was "a man, who though by no means remarkable for brilliant abilities, yet is equal to most men in popular intrigue, and the management of a faction. He eats little, drinks little, sleeps little, thinks much, and is most decisive and indefatigable in pursuit of his objects." It was Adams, Galloway alleged, who kept Paul Revere shuttling between Boston and Philadelphia, coordinating events in both places to thwart the prospects for reconciliation.

Samuel Adams might have relished the tribute, but he would have resisted taking credit for the rejection of Galloway's plan. Most delegates had already concluded that Congress should insist that Britain restore the *status quo ante*—that is, the condition under which colonial affairs had been administered in 1763, before the empire launched its first reforms. That strategy was problematic enough, but far less so than the introduction of a wholly new scheme. Congress could have no assurance that such a plan would be favorably received either in the provincial capitals of North America or in London. Nor would it exert any pressure on the British government to relieve the suffering in Boston and restore legal government in Massachusetts.

In the first days of October, thoughts about redress and reconciliation took other forms. One conciliatory gesture came from New York: a proposal that Congress offer to pay for the tea destroyed at Boston while defending the town for its action and insisting that its residents "be instantly relieved" of their suffering. After a number of lengthy speeches, this idea was unanimously rejected. Congress then considered a second proposal, introduced by James Duane of New York, to remind the king that the colonies "have always cheerfully complied with the royal Requisitions for raising Supplies of Men and Money" for "their common defense," and remained prepared to do so still "in any plan consistent with constitutional Liberty." When the delegates discussed this proposal on Monday, October 3, Richard Henry Lee of Virginia offered an amendment that gave Duane's motion a radically different thrust. Lee's language baldly stated that there was no longer any need for Britain to station permanent forces in North America, which is "able, willing, and under Providence determined to Protect Defend and Secure itself." Congress should urge the colonies to take immediate steps to invigorate their militias and make sure that they were "well provided with Ammunition and Proper Arms." There being no imminent threat from any potential enemy other than Britain, the meaning of this recommendation was transparent. As one South Carolina delegate promptly rejoined, this was tantamount to "a Declaration of Warr, which if intended, no other Measure ought to be taken up." Patrick Henry replied with an outburst so stirring that Silas Deane scrambled to get his exact language. "Arms are a Resource to which We shall be forced, a Resource afforded Us by God & Nature," Henry argued, "& why in the Name of both are We to hesitate providing them Now whilst in Our power?"

Lee almost certainly concocted his resolution with the active cooperation of the Adamses. John had previously drafted an even more militant proposal. Without seeking to force or provoke events in Massachusetts, they had concluded that a military confrontation there was likely, and perhaps sooner rather than later. Lee even composed a resolution urging the Bostonians to flee their city and "find a safe asylum among their hospitable Countrymen."

Congress was unwilling to go that far. It amended Duane's original resolution to affirm that the colonies were prepared to defray "all the necessary expenses of supporting government, and the due administration of Justice"; that the militias, if properly provided for, were adequate to protect the colonies in peace; and that the colonies would happily vote the necessary funds to support additional troops in wartime. It refused to endorse an exodus from Boston, but agreed only that the colonies should recompense its residents if so radical a step became necessary. But to prevent that possibility from arising, Congress also wrote directly to Gage, warning him to halt the fortification and isolation of Boston, lest he risk "the horror of a civil war."

Samuel Adams was too seasoned a politician to let such little reverses unsettle him. As a former clerk of the Massachusetts assembly, he had learned a few lessons about the inefficiency of collective deliberation. Perhaps because he already knew, in the keep of his own counsel, how the dispute with Britain must end, he could accept decisions that fell short of his own assessment. Patience came less easily to John. With the temperament of a litigator, not a judge, he found it a struggle to tolerate his colleagues' oratorical excesses. Every delegate thought himself "a great Man," he complained to Abigail, "an orator, a Critick, a statesman" whose voice had to be heard "upon every Question," and if not in Congress, then during "the perpetual Round of feasting" which had finally grown "tedious" to endure but impossible to shirk. Decision making by consensus, however, is tedious, and that was the course to which Congress was committed.

The delegates had to labor over the precise wording of their various acts, not only because differences in language betokened real differences in policy, but also because they were acutely sensitive to how their statements would be read. They were appealing to multiple audiences: colonists anxious for their guidance, potential supporters in Britain, and, most important, the king and his ministers, who had to be convinced that they faced real opposition. Congress had to persuade beleaguered Massachusetts that it would be supported whatever the peril, while discouraging the other colonies from thinking that it was already bracing for war. Looking across the Atlantic, Congress had to demonstrate that Americans were united in defying the claims of Parliament, yet also sincere in their desire to remain within the empire.

The sharpest test of this balancing act came when Congress completed its Declaration of Rights in mid-October. A single issue provided the one true sticking point. Should the colonies allow Parliament to continue to regulate imperial trade? The delegates knew that American trade was essential to British prosperity. That was why the more optimistic believed that an effective commercial boycott would force the empire to retreat. But they also understood that the trade of a vast empire could not be effectively regulated by a squadron of individual legislatures, each with petty interests to protect. Even if Crown and Parliament agreed that the colonies were competent to manage their "internal police," some central institution would have to oversee trade among His Majesty's dominions. What institution other than Parliament could do that? But in a political culture that valued the authority of precedent, such a concession had its risks. If Congress agreed that Parliament had a *right* to regulate trade, it could undermine its claim that Americans could be governed only by laws to which their representatives had freely consented. How could the colonies prevent parliamentary duties levied to regulate trade from being treated as revenue from taxation? How could Congress seek a restoration of the *status quo ante* of 1763 without acknowledging the authority of all the earlier Navigation Acts that Americans had often evaded but never challenged?

After several days of debate, Congress answered these questions in a way that illustrated just how militant it was prepared to be. As a basis for conciliation, the fourth resolution of the Declaration of Rights proclaimed that Americans would "cheerfully consent" to obey such parliamentary acts as were "restrained" in good faith to securing the "commercial advantages" and "benefits" of a common trading system. "Cheerfully consent" was the key phrase. While the adverb implied that Americans were happy to remain within the empire, the verb declared that they were conceding this point of their own free will. This was the lone conciliatory gesture the delegates were prepared to advance—and the only one adopted by a split vote. The other articles of the Declaration of Rights were approved unanimously. Taken together, they constituted more of an ultimatum than a plan of negotiation.

The increasingly restless delegates had other work to complete before they could adjourn. Drafting an address to the king was the most demanding. Consistent with their underlying theory of empire, the delegates balked at submitting their grievances to Parliament or asking that body to repeal its offensive acts. It was the king who would have to take the initiative in urging his ministers and his Parliament to reverse course. Congress was asking George III to play the part that a young Virginia attorney had sketched in drafting instructions for the Virginia delegation. With astounding presumption, Thomas Jefferson wanted Congress to remind the king that "the whole art of government consists in the art of being honest. Only aim to do your duty, and mankind will give you credit where you fail." Jefferson would have the king remember that "fortune" had placed him "holding the balance of a great, if a well poised empire." It was his duty to "deal out to all equal and impartial right" and resist favoring the interests of one part of his empire over another.

Congress knew it could not address the throne quite that freely. As if to apologize for Jefferson's bold language, Richard Henry Lee drafted an address that interjected a suitably contrite "may it please Your Majesty" wherever possible. Congress eventually approved an address that was deeply respectful but hardly fawning. It contained the expected declarations of the colonists' "affectionate attachment to your majesty's person, family and government." But the head that wore the crown would be more likely to note the insult delivered to his ministers. "Those designing and dangerous men, who daringly interposing themselves between your royal person and your faithful subjects," Congress complained, "have at length compelled us, by the force of accumulated injuries too severe to be any longer tolerable, to disturb your majesty's repose by our complaints."

Nothing could be better contrived to perturb the royal repose. From boyhood George III had been trained to play the part of constitutional monarch, committed to the principle of parliamentary supremacy. He was hardly likely to abandon his duty as a king-in-Parliament to accept the constitutional whimsies of an upstart Congress in distant provinces he would never see. The king was just as committed to a policy of repression as the evil ministers whom Congress urged him to replace. Many delegates understood this. It was the logic of their theory of empire, rather than knowledge of their sovereign, that led them to place this burden on George. Whether he would graciously answer their petition, as a faithful monarch should, was beside the point. Congress had delivered an ultimatum consistent with its own theory of empire, which insisted that the colonial assemblies were virtually equivalent to Parliament in legislative authority. For a reformed British Empire to survive this crisis, the entire government—king, ministers, and Parliament—would have to accept this heresy.

The First Continental Congress adjourned on October 26, 1774, after seven weeks of debate and more banquets than the delegates cared to count. The last took place, like the first, at the City Tavern. John Adams spent the next day showing William Tudor, his law clerk, around town. Then, on October 28, the Massachusetts delegates left for home in a downpour, leaving behind "the happy, the peacefull, the elegant, the hospitable City of Phyladelphia"—adjectives they could not apply to desolate Boston. Adams doubted that "I shall ever see this Part of the World again, but I shall ever retain a most greatfull, pleasing Sense, of the many Civilities I have received, in it."

The notion that he might never return is curious. Before adjourning, the delegates had resolved that a second congress would convene at Philadelphia in May. Adams should have recognized that he would be reappointed. It was nearly as naive to assume that Parliament would simply repeal its offensive legislation and restore the *status quo ante* of 1763. Perhaps Adams felt guilty over the time he had spent away from his family and his affairs, a mood that his impending thirty-ninth birthday on October 30 could have reinforced. Abigail's most recent letter had opened with an unusual salutation to "My Much Loved Friend" and a heartfelt expression of the "fears and apprehensions" that "agitate my bosom." John would have been loath to contemplate taking another prolonged leave if the situation at home remained so perilous.

Back in Boston, Governor Gage marked Adams's birthday by writing two dispatches to Lord Dartmouth. The first opened with the strange saga of Samuel Dyer, a colonial seaman arrested for urging British soldiers to desert, something they seemed all too willing to do. After being sent to England, then released, Dyer made his way home, bent on revenge. Encountering two officers in the street, he snatched one's "hanger" (sword), slashing its owner, then pulled a pistol, which "missed Fire" twice. After fleeing to Cambridge, Dyer was returned to British custody by colonial officials fearful that his vengeful act exceeded the bounds of legitimate resistance. "He appears to be a vagabond and enthouiastically Mad," Gage concluded. But then again, it was hard to distinguish Dyer from his countrymen. After five months in Boston, Gage was still astonished "that the Country People could have been raised to such a pitch of Phrenzy as to be ready for any mad attempt they may be put upon." It was still possible that "The People would cool" in their support for resistance, Gage mused, "was not Means taken to keep up their Enthousiasm." But for now they seemed "so besotted to one Side" that he had little hope of being able "to convince them of their Errors."

Beyond popular frenzy, Gage faced three greater problems for which he saw no quick solution. One was the sheer inadequacy of the force he commanded. Three thousand men hemmed up in the "prison of Boston" (as Samuel Adams called it) were too few to restore royal rule to Massachusetts. Nor were significant reinforcements expected for months. Second, Gage understood that the results of the Continental Congress limited his options even further. "Affairs are at such a Pitch thro' a general union of the whole," he observed, "that I am obliged to use more caution than could otherwise be necessary, least all the Continent should unite in hostile Proceedings against us." Third, the patriot leaders in Massachusetts had already set up surrogate institutions to circumvent the legal government that Gage nominally headed. A provincial congress had just adjourned, after discussing, it was rumored, the formation of a military force of fifteen thousand, with additional support from "the Neighbouring Provinces."

Of these developments, Gage had the least to say about the last, which was perhaps the most significant. Congress had balked at allowing Massachusetts to resume *legal* government. But it did not object to the formation of *extralegal* institutions—bodies that would not be tolerated when government functioned normally, but which became permissible when it did not. Congress had itself summoned similar bodies into existence across the continent by instructing every local community to elect committees of inspection to enforce its boycott of British imports. The delegates did not regard the creation of this Association (as the boycott was called) as a revolutionary act—at least not yet. But so it potentially was, because it hastened the flow of power away from legal government to an emerging network of committees and conventions. Every county and township in America would now have its own revolutionary nucleus in place, drawing authority from Congress and local citizens alike. It was a formidable combination, and neither the beleaguered Gage nor any of the other royal governors could do anything to prevent it.

The Massachusetts delegates took nearly two weeks to make their way home. En route they again received "the most pressing Invitations" to dine but did their best to decline in order to return "as fast as possible" to their province and their families. Once home, both Adamses attended the provincial congress that had become the colony's effective, if extralegal, government. After it adjourned on December 10, John probably spent most of the winter at Braintree, where Abigail lived with their family since his departure for Philadelphia. He kept no diary for this period but instead busied himself on a different literary project, drafting a lengthy series of letters, under the pen name "Novanglus," meant to refute an effective loyalist writer writing as "Massachusettensis." (Only late in life did he learn that his opponent was the same Daniel Leonard who had been lured away from the General Court just before it elected its delegation to the First Continental Congress.) From his home in Boston, Samuel Adams remained more closely involved with the daily coordination of resistance. Under the instructions of Congress, the British presence had to be tolerated—barely—and measures taken to ensure that the angry townspeople and their neighbors in nearby communities continued to act "on the defensive" only.

Through the winter, Boston remained an occupied city but not yet a besieged one. On March 6 Samuel Adams sat in the moderator's chair at Old South Meetinghouse as Dr. Joseph Warren gave the annual address commemorating the Boston Massacre. British officers were present, and Adams, expecting them to "take that Occasion to beat up a Breeze," asked them to take "convenient Seats" so that "they might have no pretence to behave ill, for it is a good Maxim in Politicks as well as War to put & keep the Enemy in the wrong." Adams did not misjudge. The officers sat quietly while Warren spoke, but after he finished "they began to hiss," irritating the inhabitants on this solemn occasion, "and Confusion ensued." The disorder at Old South was only one mark of the tension in the town beyond. Once again the presence of British troops—parading, drilling, getting drunk on cheap rum, bearing endless abuse from the townspeople—fed the perception that an armed clash was inevitable. The soldiers' morale plummeted. A few drank themselves to death, others deserted, and a handful were executed for attempting to do so.

On the last Sabbath of February, Gage sent a party by sea to Salem, hoping to confiscate munitions and cannon he knew to be stored there. Once landed, the troops were quickly detected, and the alarm sounded as worshipers poured out of church and frantically dragged the cannon into any available hiding place. As defiant colonists obstructed the soldiers in their searches and as militia hurried in from nearby towns, a quick-thinking local minister arranged a sham exercise, allowing the British to pretend to look for the cannon before confessing failure and returning to their ship. Occasionally Gage sortied other troops on similar missions into the nearby countryside, where they met insults but no armed opposition.

In mid-April 1775 Gage finally received fresh instructions from London. His reports had not been read sympathetically. Impatient with their commander's caution and legal scruples, the king and his ministers wanted bolder action: the arrest of resistance leaders and the seizure of colonial munitions. This was not the first time that the British army had been used to impose civil order on unruly populations. It had done so repeatedly in Scotland, Ireland, even in England itself. As a young officer, Gage had fought at Culloden, where the king's army butchered the Scottish clans. Though Gage understood that the American protests were not a replay of the Jacobite rising thirty years earlier, he retained a commander's confidence that disciplined troops could always vanquish a provincial rabble.

Gage set out to implement his orders. Recognizing that the capture of resistance leaders was unlikely, but believing that lessons from previous forays might enable his troops to seize colonial arms, Gage methodically planned a raid on Concord, where large stocks of munitions were known to be held. His preparations were impossible to mask from the colonists, who carefully monitored the activity of British forces. On the fateful night of April 18, 1775, a force of seven hundred soldiers set out for Concord. At Lexington they met a hastily gathered force of American militia loosely deployed on the town green. Shots were fired, and eight Americans died as the others scattered. The British proceeded to Concord. Here the Americans proved better organized, and their fire more accurate. Now it was the British who broke, beginning a long, harried retreat to Boston. When it was over and the exhausted regulars regrouped in Boston to call their roll, they counted nearly 250 casualties, almost thrice those of the colonists with whom they were now assuredly at war.

Four weeks earlier, while the ministry's new instructions were still crossing the Atlantic, the House of Commons heard a far more acute analysis of the deteriorating situation in America than either Gage or his distant superiors ever offered. Its author was Edmund Burke, member of Parliament from Bristol, agent of New York, author of the aesthetic theory of the sublime, a known "friend of America," and political lieutenant to the Marquess of Rockingham, the king's chief minister when the Stamp Act was repealed in 1766. As London waited to learn how the ministry would respond to Congress, Burke prepared his own proposals for "conciliating" the American crisis, which he presented to the Commons in a speech that ranks high in the annals of political oratory—and futility. In November, before the results of the Congress could reach Britain, Lord North had called for fresh elections, and his government now stood secure in its command of both houses of Parliament. Once the news from Philadelphia became known, the lack of any conciliatory gesture from Congress steeled the government to hold to its prior decisions.

It was this fixed adherence to coercion that Burke challenged. The critical passages of his speech offer a brilliant exercise in political sociology, a model of analysis that modern statesmen could well study to their benefit. Burke recognized some provincial differences among Americans. Yet he also grasped the key fact that they were beginning to form a collective people—that is, a nation. There were a number of reasons why coercion must fail, Burke argued. Some of these reasons Britons should find flattering. The Americans were their descendants and shared their common ancestors' fierce devotion to their rights and liberties. Religion reinforced this attachment. The Americans were all Protestants, Burke reminded the Commons, and "All Protestantism, even the most cold and passive, is a sort of dissent." But American Protestantism hardly erred on the side of passivity, for "the religion most prevalent in our northern colonies is a refinement on the principle of resistance; it is the dissidence of dissent, and the Protestantism of the Protestant religion." True, south of Pennsylvania the Church of England was predominant. Its principles should favor loyalty to empire. But in the southern plantation colonies, a different social fact intervened: the existence of "a vast multitude of slaves." In such societies, perversely or otherwise, freedom became "a kind of rank or privilege." There, those who enjoyed it could not understand how elsewhere—as in Britain itself—an attachment to liberty "may be united with much abject toil, with great misery, with all the exterior of servitude" that enabled the laboring folk of rural and urban England to glory in their birthright of liberty even while a restricted suffrage denied them the formal political rights Americans took for granted. Owning slaves made planters perversely more devoted to preserving the liberty they carelessly stripped from others.

Burke added two other circumstances to the reasons why coercion must fail. One was the deep legalism of a society where William Blackstone's *Commentaries on the Laws of England* was selling as well as it was in England. Gage himself had noted that "all the people in his government are lawyers, or smatterers in law." This devotion to the law, Burke noted, "renders men acute, inquisitive, dexterous, prompt in attack, ready in defence, full of resources." Reasoning like lawyers, Americans anticipated an evil in principle before its real mischief was felt. "They augur misgovernment at a distance; and snuff the approach of tyranny in every tainted breeze."

Finally there was the critical factor of geography. Government could not expect distant provinces to be as compliant with its decisions as were the home counties. Government at a distance was necessarily weaker, less efficient, more vulnerable to defiance. "Seas roll, and months pass, between the order and the execution; and the want of a speedy explanation of a single point is enough to defeat a whole system."

To "these six capital sources" of the colonists' fierce attachment to liberty, rooted in the structure of colonial society, Burke added a last acute remark about the revolutionary situation he rightly grasped was emerging throughout America. We had once assumed that "the utmost which the discontented colonists could do, was to disturb authority; we never dreamt they could of themselves supply it; knowing in general what an operose business it is to establish a government absolutely new." That assumption now lay disproved. A "great province" had gone without legal government for nearly a year. Far from drifting into the disorder that would make solid citizens long for the restoration of law and order, "A new, strange, unexpected face of things appeared. Anarchy is found tolerable." Once the colonists learned the lesson the empire insisted on teaching them, that they could enjoy "the advantages of order in the midst of a struggle for liberty, such struggles will not henceforward seem so terrible to the settled and sober part of mankind as they had appeared before the trial." Nor was this insight limited to the flashpoint of Massachusetts. Other "provinces have tried their experiment, as we have ours; and theirs have succeeded."

A month before Gage faithfully executed his fateful orders, then, Burke already knew why the choice of coercion was a formula for failure. Like most members of the British ruling elite, Burke never thought to pay America an actual visit. Yet he grasped its politics much better than did Gage, who had served there for twenty years. Gage had been with Braddock on the disastrous expedition of 1755, commanded a regiment of light infantry, accepted appointment as His Majesty's commander in chief for North America, and married Margaret Kemble, a wealthy New Jersey heiress related to the powerful DeLancey family of New York. (Her stunning portrait by the American artist John Singleton Copley is also a famous example of the Turkish style of dress indulged by the fashionable.) Her husband did know America well enough to grasp that it could not be ruled in the mode of conquered Ireland or suppressed Scotland. Yet next to Burke's incisive account of the *political* obstacles to a policy of repression, Gage's belief that he faced a people gripped by an irrational "Enthousiasm" and "Phrenzy" suggests that he saw himself dealing with the zealous converts at a religious revival rather than an alienated population readying itself for revolutionary upheaval.

Burke's speech on conciliation is hardly his best-known political writing. He is better remembered for his later *Reflections on the Revolution in France*, the work that secured his reputation as a founder of modern political conservatism. Those reflections included an extended discussion of the deep differences between the temperaments of the French and English peoples, and can thus be read as a pioneering if primitive effort to fashion a theory of national character in terms that could not be reduced to simple visual caricatures of the kind that William Hogarth or his artistic heir, James Gillray, would purvey. Burke's earlier reflections on the looming revolution in America confirm that this knack for describing a people's political habits was an integral element of his genius.

There was, however, one significant difference between these two assessments of revolutions in the making. When Burke published his famous *Reflections* in 1790, he wrote in the idiom of reaction. But as spring came to the capital of empire in 1775, he spoke as a voice of moderation whose wisdom was lost on the king, his ministers, and most members of the Commons, who could not absorb the lessons of this brilliant political analysis.

2

The Revolt of the Moderates

When it came to debating the merits of conciliation or confrontation after 1774, there were probably many more voices for moderation in America than there were in Britain. But being a moderate proved a harder task and greater responsibility in the colonies than it was in London. Edmund Burke knew that his appeal for conciliation would fall on ears that would admire his eloquence yet ignore his insights. The American protests challenged too many convictions, threatened too many interests, to sway a government convinced that only arms could maintain British rule in America. Not every member of Lord North's ministry was equally confident this coercive strategy would work. Yet the ministry and most of Parliament saw no other alternative. Their dominant response when the news of war arrived in late May was not to question their strategy but to fault the hapless general charged with carrying it out. Yet like Gage, they tended to think that Americans were in the grip of a delusion, under the sway of demagogues bent on seizing power by misrepresenting the British government's reasonable plans and motives.

In Britain, then, moderate voices like Burke's were politically impotent—and knew it. But in America political moderates such as John Dickinson, Robert Morris, James Duane, and John Jay played a critical role in maintaining the consensus on which the resistance movement depended. The work of the First Congress in hammering out basic positions marked only the first step in this process. Soon after the Congress reconvened in May 1775, it plunged into a spirited debate over whether to cling to its earlier position or give the government an opening for negotiations. Dickinson was adamant on this point, and his prestige helped guarantee that Congress would send a second "Olive Branch" petition to a king who had never acknowledged the first. Yet to gain even this modest concession, moderates paid a steep price: active support for all the military preparations that could only confirm London's worst fear, that American resistance had moved from truculent protest to armed rebellion. Within weeks George Washington was en route to Boston to take command of the new Continental army, which was besieging Gage's forces. The moderates who fully supported his appointment were left to wrestle with the dilemma that vexed them for the next year: how to combine mobilization for war with their own desperate hope that Britain would come to its senses and open a pathway to serious negotiations.

That dilemma predated the news from Lexington and Concord. Moderates had left the First Congress in a mood of foreboding, hopeful that the uneasy truce around Boston would last long enough for the ministry to take the measure of American unity and retreat from confrontation, but recognizing that war could erupt first. John Jay captured this mood in a New Year's letter to his close friend Robert R. Livingston Jr., fellow alumnus of King's College in New York, former law partner, and future colleague in the New York delegation to Congress. Jay lamented their separation at a time when he felt "the want of some Person in whom I could repose absolute confidence." He recalled the exact day (March 29, 1765) when they had "particularly professed" their friendship and the initial differences in their "disposition" and personality: "you formed for a citizen of the world I for a college or a Village." These differences had since narrowed. "You have become less flexible and I rather less pertinaceous." The "counsel" Jay wanted doubtless extended to public affairs, but he balked at saying so directly. "I ought to say something to you about politicks," Jay almost sighed, "but am sick of the subject." His colleague James Duane entered the new year in a similarly morose frame of mind, "plunge[d] into the midst of a Tempest which I find myself unable to direct." Like Jay, Duane had played an active role at the First Congress, but he was not the happiest of political warriors. "This Trade of Patriotism but ill agrees with the profession of a practising Lawyer," he grumbled after six weeks of debate. "I have lost my Clients the Benefit of a Circuit and now despair of doing any thing the ensuing term."

The unease the two men felt that winter owed something to the special situation of New York. Marriage had made both men part of the extended Livingston clan, which strongly favored the patriot cause. But the New York legislature was dominated by the rival DeLancey "interest," and that family remained loyal to the empire. If the Crown asked the assembly to submit its own petition, rather than endorse the proceedings of Congress, it might well comply and thus fray the bonds of colonial union. Jay and Duane had to reassure correspondents that whatever the assembly did, the people of New York supported the cause as fervently as colonists elsewhere. Nor was New York the only target for this approach. In Pennsylvania a "decisive & determined" John Dickinson had to beat back an effort by the assembly speaker Joseph Galloway and Quaker deputies "to petition the King, on Principles different from those of the Congress." Moderates were just as fearful that hotheads across the colonies might commit provocative acts that would only confirm the worst British suspicions. Writing to Thomas Cushing in Boston, Dickinson noted his "inexpressible Pain of Mind" over reports that Massachusetts might resume legal government under its own authority. Such a step would "strike dumb our Friends in England," he warned, and produce "such a Provocation" that it would likely end "in the Calamities of War."

In April came the news that turned the fear of calamity into grim reality. Though leaders in Massachusetts were quick to gather depositions indicting the British as aggressors, the question of who fired first at Lexington Green and Concord's North Bridge did not matter. The British raid itself was a provocation, and the militia, as instructed, had properly acted on the defensive. But once war began, the noble bluster of 1774 had to give way to a sober assessment of the real costs Americans now risked incurring.

For the resistance and protest of 1774 to develop into the revolutionary movement of 1775–1776, moderates such as Dickinson, Duane, and Jay had to agree that these risks were worth taking. For the revolutionary movement of 1776 to end in the national liberation of 1783, they also had to agree that the mounting costs of a protracted war were worth bearing. That price in turn required a personal commitment that no one in 1774 had anticipated. Duane served in Congress every year until 1783. Jay went on a diplomatic mission to a forbidding Spanish court, then negotiated the treaty of peace that secured American independence. Dickinson, by contrast, had to struggle to redeem his reputation after opposing the decision for independence in 1776.

Historians have not found it easy to explain the commitments of these moderates, who were prudent men of affairs and property, merchants and landlords, rather than creatures of politics and ideology. They are often portrayed as conservatives who could easily have become loyalists, who were dragged into revolution less from fervor than fear, to protect their elite interests against populist aspirations that the turmoil of revolution was unleashing. They were, that is, just the kind of men whom the British government expected to come to their senses and reject the demagogic folly of radicals like the Adamses and Richard Henry Lee. There is some truth in this portrait. The young fop Gouverneur Morris snidely compared the New York City crowds protesting the Boston Port Act to reptiles learning to "bask in the sunshine, and ere noon they will bite"—let "the gentry" be warned. A year later Duane thought it vital that "Men of property and Rank" should accept military commissions so that they could "preserve the same Authority over the Minds of the people which they enjoyed in the Hour of Tranquility." Such sentiments hardly celebrate the democratic impulses that we associate with the revolutionary legacy. One might as well say that the true heroes among "the greatest generation" who waged World War II were the "dollar a year" executives who came to Washington to manage the war effort.

Yet the Revolution did depend on the moderates' commitment. Independence could not have been declared in 1776 without their active support. Their contributions became all the more important after the vote of July 2, 1776, was followed by a string of military defeats that brought the cause to the brink of collapse and many moderates to the point of despair, privately worrying that the estates they had labored to amass might prove forfeit to a vindictive Crown. Yet they emerged from these reverses with their commitment not only intact but reinforced, and with an anger against Britain that ran just as deep as the old revolutionary leaven of Puritan New England. For the enemy brought the bitter fruits of war into the moderates' heartland, the middle states, with a vengeance that New England was largely spared. Having plunged into politics in the hope of averting war, they were all the more distressed by its wanton brutality, and far more disillusioned by the enemy's conduct than the ideologues of New England could ever have been. This disenchantment was compounded by Britain's inability to recognize their own good motives, a failure that led the moderates to swallow their misgivings and accept independence, even when they privately believed another outcome had been possible all along.

To be a political moderate during this final crisis meant something more than holding a middle ground between radicals like the Adamses and future turncoats like Joseph Galloway. Moderation did not mean keeping an open mind about rival claims or seeking to balance clashing points of view. Nor were moderates the eighteenth-century version of undecided voters, unsure of their own views or trimming their sails to any passing political breeze. Moderation is better considered as a political position in its own right. Moderates were not ideologues, as historians might apply that term to Samuel Adams or Richard Henry Lee. They were not inclined to think that the king's ministers were waging a systematic campaign to turn the colonists into docile slaves, the better to subdue their liberty-loving countrymen at home in Britain. Rather, moderates believed that the ministry had badly miscalculated. If it could only gauge the depth of the American commitment, they hoped, it would abandon a mad policy that could be enforced only at great peril and enormous expense, and which was likely to destroy the mutually beneficial relations that the mother country and colonies had enjoyed for decades. But if the moderate political leaders were not ideologues in the most loaded sense of the term, they did possess a distinct set of attitudes.

Men of moderate views could be found throughout America. But in terms of the politics of resistance, the heartland of moderation lay in the middle colonies of New York, New Jersey, Pennsylvania, and Maryland. The roster of moderate leaders begins with John Dickinson, "the Farmer" from Pennsylvania whose writings against the Townshend duties of 1767 gave him unique political prestige. The larger group includes the Philadelphia merchant partners Robert Morris and Thomas Willing; James Wilson, the lowland Scots immigrant whom they sometimes used as their attorney; Jay, Duane, Robert Livingston, and Gouverneur Morris of New York; William Livingston (Jay's father-in-law) of New Jersey; and Maryland's Thomas Johnson and Charles Carroll of Carrollton, sole heir of Charles Carroll of Annapolis. Unlike Dickinson, who had been politically active for years, most of these men entered public life in response to the crisis of 1774—not to pursue the burning ambition that gnawed at John Adams but because events compelled them to do so.

Three factors lent coherence to this bloc of moderate leaders. First, they represented America's most diverse society, the one most likely to experience "intestine broils" amid the turmoil of revolution. Second, they shared the attitudes of men of property and affairs, entrepreneurs of regional development and improvement who dreaded the destruction of lives and property that war must ineluctably bring. Third, and perhaps most important, it was precisely because they entered the crisis eager to promote accommodation that they found Britain's obstinate commitment to repression and force so distressing.

In terms of religion, ethnicity, and language, the middle colonies constituted the most pluralistic societies in mainland North America. New York, originally a Dutch colony, was drawing wandering Puritans to Long Island even before the English conquest of 1664 opened it to British immigration. Swedes and Finns had settled along the Delaware River before the enlightened policies of William Penn made his sylvan colony a magnet for religious dissenters, beginning with his fellow Quakers, and opportunity-seeking German, Scots, and Scots-Irish immigrants. This was the region the former French army officer Hector St. John de Crèvecoeur had in mind when he asked his famous question "Who then is this new man, the American?" and imagined how the battered poor of Europe would become the productive citizens of a new world.

These differences in language and religion were accompanied by demarcations in class more complex than those found elsewhere in America. The bustling ports of Philadelphia, New York, Annapolis, and Baltimore had a class of aspiring artisans eager to participate in public affairs. Tenant farmers in New Jersey and New York were not docile yokels ever willing to tip their hats to the patroons of the great Hudson Valley estates owned by families like the Livingstons or to the developers of new townships like the one west of Albany that its founder, James Duane, charmingly named Duanesburg. These farmers had their own ambitions: to become independent landowners and to ensure that their children would never live as peasants.

These factors made this region potentially the most volatile in North America. They created a measure of social diversity far exceeding that of New England, where a single ethnic group of English descent predominated and where religious differences were still confined within a Calvinist tradition. To the south lay the great division between white and black in the slave societies of the plantation colonies. But slaves had no political voice, and the threat of rebellion implicit in their presence operated as a unifying force within the free community. By contrast, in the middle colonies social differences created potential for political disorder. If protest against the Coercive Acts developed into something more radical, this was the region that might become most gravely unsettled.

This was also the region where a modern vision of economic development was already taking hold. Hardworking as Crèvecoeur's aspiring immigrants might be, they still had to be settled in townships and provided with the mills and roads needed to bring their produce to market. Entrepreneurs (or "projectors") such as Duane or the Carrolls of Maryland—men who could look at a wilderness and envision a developed landscape—thrived here. By contrast, in New England, economic development was primarily a matter of subdividing existing allotments of land and providing the fourth- and fifth-generation sons of the original Puritan migration the resources with which to marry and start their own families. By the eve of the Revolution, New England was preparing to export population, as investors in Connecticut, for example, looked to the Wyoming Valley of northern Pennsylvania as a site for settlement. In the south the path to wealth still required the ownership of slaves, the chief capital factor in the means of production. Individual planters, such as George Washington, could be as entrepreneurial as their counterparts to the north, but the region remained wedded to the export of agricultural commodities, generally marketed by commercial agents (or "factors") representing firms in Glasgow and London. By contrast, the economy of the middle colonies depended on attracting free immigrants and harnessing their productive energy.

This constructive vision of economic development shaped the political views of the moderate leaders. Let Samuel Adams indulge his quaint notion of turning the town of Boston into "the *Christian* Sparta" and the planter-barons of the south seethe in resentment at the debts they permanently owed their British creditors and the Scots factors who kept such neat accounts of the exact sums they owed. A merchant-entrepreneur like Robert Morris knew that the small joys and comforts of consumption fit the American temperament better than Spartan self-denial and that British capital could help fund many a venture from which well-placed Americans could derive a healthy profit. Building that world was the real work that Americans and Britons should jointly undertake—not this mad project of imperial misrule into which the government of Lord North had rashly stumbled.

This strong commitment to the productive development and improvement of property helped distinguish the moderate political leaders of the middle colonies from their counterparts from other regions. Yet there was a deeper sense in which their attachment to the rights of property identified a value that all Americans shared. For *property* was one of the strongest words in the Anglo-American political vocabulary. Its security from unlawful taxation had been a dominant value of their common constitutional culture since the previous century. John Locke had grounded an entire theory of government—and the right to resist tyranny—on the concept of property in his *Second Treatise of Government.* But Locke only gave philosophical rigor to a belief that already permeated Anglo-American law and politics.

For Locke, as for his American readers, the concept of property encompassed not only the objects a person owned but also the ability, indeed the right, to acquire them. Just as men had a right to their property, so they held a property in their rights. Men did not merely *claim* their rights, but also *owned* them, and their title to their liberty was as sound as their title to the land or to the tools with which they earned their livelihood. Furthermore, Americans believed that they truly owned these rights because their ancestors had fairly purchased them through the arduous work of colonization. Just as Locke had grounded his theory of property on the labor through which men expropriate the fruits of nature for their personal use, making the earth more productive and thus fulfilling the divine injunction to preserve mankind, so the colonists looked back to their ancestors' pioneering and saw that it was good—and legal too. Property was a birthright, a legal entitlement and material legacy that one industrious generation transmitted to another. That was as true for the small farmers of New England, working their fifty or hundred acres and still fencing their fields with glacier-strewn rocks, as it was for the planters of the south, with their scores and hundreds of bondsmen, and for the merchants and land speculators of Pennsylvania and New York as well. Property, defined in this way, was the vital right that Parliament would infringe upon, even destroy, if it made good its claim to legislate for Americans "in all cases whatsoever."

It was not, then, sharp differences of principle that distinguished moderates from radicals. Instead, other calculations and attitudes shaped their position. Because their lives as merchant-entrepreneurs revolved around the assessment of risk, they were peculiarly disposed to view the costs and benefits of resistance in calculating terms. If the restrictions that the Navigation Acts placed upon the colonies occasionally irked the merchants among them, they nonetheless recognized that the British and American economies complemented each other nicely. They believed that Anglo-American harmony was essential to the continued development of their own economy. They did not regard America as an exploited hinterland that imperial overlords were systematically plundering, the way Spain had extracted the mineral wealth of Mexico and Peru. They saw Britain as a source of the goods they wished to consume and of capital investment for their enterprises. They understood that British arms promised an effective counterweight against Native Americans on the frontier and Spanish influence in the Mississippi Valley.

But these views presupposed that an enlightened British government would perceive its own true interest and act accordingly, allowing the colonies to remain largely autonomous while trusting the familiar ties of commerce, culture, affection, and patriotism to maintain the sinews of empire. In the early 1770s, that perception remained plausible. Moderates could still believe that the ministry of Lord North was consciously avoiding the miscues of the late 1760s. So long as it did, they could treat their own political involvement as a matter of convenience and calculation, but not a commitment or duty, much less a burning ambition merely awaiting its fulfillment.

These assumptions and hopes evaporated in May 1774. Whatever their personal desires, the moderates found themselves sucked into the political vortex by the sheer magnitude of the crisis. They shared the general outrage over the Boston Port Act that swept the colonies. Whatever offenses Bostonians had committed against property did not justify this blanket assertion of a punitive parliamentary sovereignty. But because the moderates believed that British and American interests were essentially harmonious, they were eager to prevent resistance in New England from escalating so rapidly as to rule out any chance of reconciliation. Recalling the problems the colonies had met in organizing opposition to the Townshend duties, they wanted to ensure that this new round of protests was effectively coordinated and broadly supported. Their goal was to disprove the reigning British assumption that Americans would not rally to the support of a single turbulent province once the real costs of defying the empire were made apparent.

That there could be real costs was, however, a fact that moderates never ignored. They did not underestimate the danger of opposing the premier power of the Atlantic, with its dominant navy and veteran regiments. They knew enough about the harsh justice that Britain had imposed on subjected Ireland and rebellious Scotland to foresee the fate of their own fortunes, should protest lead to violence and violence end in defeat. They were not itching to escape the restraints of the imperial navigation system in order to reach foreign markets legally barred to Americans. If events created such opportunities, they would capitalize on them. But whether such opportunities would offer greater rewards than the ones they already enjoyed was not a risk they could readily calculate. No one familiar with America's seaborne trade could ignore the threat that the Royal Navy posed to commerce. War was not the risk they preferred, but it became the risk they came to accept, and in doing so, they became full-fledged revolutionaries in their own right, however difficult it may appear to cast them in that light.

From this bloc of moderate leaders three names deserve special attention, not only because they reveal the dilemma of those most eager for accommodation, but also because they illustrate how the Revolution created political opportunities and commitments that would have been impossible to anticipate even in 1773. The most obvious name belongs to the "celebrated Farmer," John Dickinson, whose prestige made him the moderate bloc's natural leader. But after July 1776 Dickinson fell into a political eclipse from which it took him years to recover. In his place two other men who originally resisted the decision for independence rose in prominence. One was John Jay, who eventually served as president of Congress, minister to Spain, leading negotiator of the treaty of peace securing independence, secretary of foreign affairs, and first chief justice under the Constitution. The other was Robert Morris, the ambitious and avaricious Philadelphia merchant whose service as superintendent of finance (1781–1784) brought the nation's first sustained efforts to place public credit on a sound and permanent footing, influencing and anticipating the even more ambitious program that his young supporter, Alexander Hamilton, would successfully promote a decade later.

Dickinson was one of a handful of Americans who enjoyed a truly "continental" reputation before the crisis of independence began creating new political celebrities. His *Letters from a Farmer in Pennsylvania* played a crucial role in energizing colonial opposition to the Townshend duties of 1767, closing a gap that Benjamin Franklin had opened the year before. While testifying before the House of Commons in favor of repealing the Stamp Act, Franklin hinted that Americans might not object to all taxes levied by Parliament. "Internal" taxes, such as the stamp duty, were improper, but "external" taxes—such as duties on trade—might be acceptable. That was the loophole Charles Townshend, chancellor of the exchequer, proposed to exploit with his duties on paint, paper, lead, and tea. In his *Letters*, Dickinson argued that it was the *purpose* of the tax, not its *form*, that mattered. If the Townshend duties were designed to raise a revenue—and not, say, operate as a protectionist tariff—then they were clearly as much a tax as the detested Stamp Act, and equally impermissible. Nor did Dickinson confine himself to the learned constitutional pleadings for which his training at the Inns of Court and active participation in the contentious politics of Pennsylvania ably prepared him. He also set his political sentiments to music, composing new lyrics for the immensely popular "Hearts of Oak," a chantey celebrating the Royal Navy, and publishing the results as the stirring "Liberty Song," which soon became "America's first hit song," as his keenest biographer quips. "Not as slaves, but as freemen our money we'll give," each refrain concluded, and Dickinson invoked the same theme of birthright liberty, secured by the honest toil of ancestors, that resonated in formal political argument.

By reputation and celebrity, Dickinson was thus a force to reckon with. He was also fiercely independent in judgment and action. Though not a member of the proprietary party (that is, those who supported the policies of the Penn family), Dickinson led the successful opposition to Franklin's campaign to strip the proprietors of their rights of government and turn their colony into a royal province. Unlike his parents, wife, and two daughters, Dickinson was never formally a member of the

Society of Friends. But except for his years in London, he always lived within a devoutly Quaker environment. Its theology and its distinctive approach to politics were deeply imprinted on his personality, and in provincial politics he was allied with the Quaker party, which opposed the legislative alliance of Benjamin Franklin and Joseph Galloway. His marriage in 1770 to Margaret Norris, daughter and heiress of a Quaker merchant, made the wealthy Dickinson wealthier still, and left observers to wonder whether a concern for preserving his estate might temper his political views. From that unique and complex Quaker tradition, Dickinson derived a commitment not to pacifism in the strict sense of refusing to take up arms, but to the moral duty to seek peaceful resolution of all civil conflicts. That commitment was in turn reinforced by two other elements in Dickinson's political thinking. One was the affection he still openly expressed for the mother country, an attachment that increasingly set him apart from those who were wholly exasperated with British conduct. The other stemmed from his plausible doubts as to whether a harmonious American nation could ever be built on the sole foundation of opposition to British misrule. Remove the superintending authority of empire, Dickinson worried, and Americans would quickly fall into internecine quarrels of their own.

Opposition from Speaker Galloway kept Dickinson out of the Pennsylvania delegation to the First Continental Congress—at least until his election to the provincial legislature led to his belated appointment in mid-October. But he was effectively a party to its deliberations nonetheless. He first visited the Massachusetts delegates on August 31, arriving "in his Coach and four beautifull Horses." Recovering from "Hectic Complaints," he was not the picture of health. "He is a Shadow—tall, but slender as a Reed—pale as ashes," John Adams noted. "One would think at first sight that he could not live a month." Other meetings followed. On September 12 John Adams recorded dining at his "fine Seat" just north of the city at Fair Hill, with its "beauty full Prospect, of the City, the River and the Country—fine Gardens, and a very grand Library" that impressed Adams, himself a great bibliomaniac. He saw Dickinson on the twentieth, and they dined again on the very next day and on the twenty-fourth. "Sweet Communion indeed we had," Adams noted in his diary. "Mr. Dickinson gave us his Thoughts and his Correspondence very freely."

While Dickinson missed the opening weeks of the First Congress,

John Jay came to Philadelphia fully prepared to enter its debates from the start. Like his colleagues and friends James Duane and Robert Livingston, Jay was a New York lawyer. The American founder of his family was his grandfather Auguste, part of the great Huguenot diaspora that fled France during the persecutions that accompanied the revocation in 1685 of the Edict of Nantes. The Jays had been merchants in the Huguenot stronghold of La Rochelle, and so they remained in their New York refuge. Auguste married well, into the prominent Dutch-Huguenot Bayard family, and so did his son, Peter, when he took Mary Van Cortlandt as his bride in 1728. John, born in 1745, was their sixth son and eighth of ten children. Three died in infancy, two were blinded by smallpox in 1739, and two others suffered developmental disabilities. Jay spent much of his childhood at the family estate at Rye, at the western end of Long Island Sound. After completing his studies at King's College, he quickly became a rising member of the New York City bar, sometimes serving as co-counsel with Duane and Livingston. Nothing in his extant papers before 1774 suggests a strong interest in politics. His few biographers have been more impressed by the stuffy pride with which he kept the membership rolls of one of New York's dancing assemblies.

Like his forebears, Jay married well. When news of the Port Act reached America in May 1774, Jay and his beautiful new bride, eighteen-year-old Sarah Van Brugh Livingston, were visiting relations along the Hudson while the groom conducted legal business in "the northern Counties." Sarah blamed the "scrawls" in a hasty note to her mother on having to write in a "room where there is such a collection of country-men talking politics." Only upon returning to New York City did Jay learn of his election to the Committee of Fifty-one, formed to prevent a smaller committee controlled by radicals from directing the city's response to the Port Act. But Jay initially thought the crisis might soon pass. Writing to a college friend who had just sailed to London, Jay briefly summarized events but spent more time discussing the judicial appointments that he might obtain from the Crown. Even so, he was elected to the First Congress, where he served alongside his father-in-law, William Livingston, who soon dominated New Jersey politics as its revolutionary governor.

Jay was not the youngest member of Congress—John Rutledge was four years junior—nor did he allow inexperience to consign him to respectful silence. He was the second speaker up when Congress first discussed the sources of colonial rights on September 8, 1774, taking the floor three times in a heated debate. It was Jay who later moved that Congress offer to pay for the tea destroyed at Boston. With his New York colleague James Duane, Jay was a consistent advocate for conciliatory measures. Patrick Henry "has a horrid Opinion of Galloway, Jay, and the Rutledges," John Adams noted after an evening spent discussing the royal petition. "He is very impatient to see such Fellows, and not be at Liberty to describe them in their true Colours." But delegates not only had to see each other daily; they also had to collaborate actively. Thus when Congress appointed a committee to draft addresses to the peoples of America and Britain, it elected Jay and his father-in-law to serve with Lee. Jay wound up writing the British address, Lee the American. No British reader would suspect that its author privately hankered for reconciliation. Nor could anyone accuse Jay of trying to appease British feelings. The address vividly described "the ministerial plan for inslaving us" while warning Britons that they too might soon "fall into the pit that is preparing for us."

Robert Morris, the third of the major moderate leaders, did not join Congress until September 1775, but once he did, he threw himself into the part of its business he knew best, which was business, or rather, the commercial connections needed to sell American exports to procure the munitions the country now needed. Morris was only thirteen when his father, a factor (agent) for a Liverpool tobacco firm, brought him to Maryland in 1747. While his father started a new family with a common-law wife, Robert was conveniently sent to school in Philadelphia. Orphaned by his father's accidental death, Morris was apprenticed to Charles Willing, twice mayor of Philadelphia and owner of three ships at a time when most traders bought fractional shares in individual voyages. Morris soon began serving as a supercargo on voyages to the West Indies; on one trip, at the start of the Seven Years' War, he was captured by a French privateer. Upon his release, he went into partnership with Thomas Willing, who had inherited the family firm. At first, Morris had little capital to invest in their joint ventures. His entrance into the partnership was a reflection of both his enterprise and his friendship with the younger Willing. Their commercial correspondents circled the Atlantic, from London, Lisbon, and Madrid to Jamaica, Barbados, and Antigua. With additional ships at their disposal—ten by

1773, an enormous fleet by colonial standards—the firm's trade continued to diversify. And as it did, Morris became the more active partner. Thomas Willing had nine years of formal English schooling, culminating in legal studies at the Inns of Court. That much education presumed a life not confined to a merchant's dock and countinghouse, and after his return to Philadelphia, he began collecting public offices. Having Morris for a partner made it easier for Willing to pursue public service; but as the rising generation of an established family, he also had a greater incentive and inclination to enter public life. In both city and province, Willing played an active part in the opposition to the Stamp Act and the Townshend duties. Morris signed the requisite protests but took no leadership role. He had more important and immediate objectives to pursue.

Like other men of their class, Willing and Morris invested significant portions of their profits in land. Just before the crisis broke in 1774, for example, they were working on an ambitious scheme to start a plantation (Orange Grove Estate) on the Mississippi River, to be manned by slaves under the oversight of a Willing kinsman. In their wealth and ambitions, men like Morris and Willing were hardly typical of the broader colonial population. The artisans they employed in their ships, shops, and mills would never enter the front door of their handsome townhouses or country "seats." Although members of the elite had to know how to "condescend"—to treat their social inferiors with candor and sympathy—they never forgot the disparity in manners, culture, and aspirations that distinguished one class from another. Their quest for gentility, measured in the elegance of their handwriting and domestic furnishings, was meant to establish visible badges of superiority that the "lower orders" would naturally respect. In the early 1770s, their desire for wealth and refinement beat as strongly as ever. Yet once the crisis broke in 1774, they cast their lot with the American cause, and they did so out of the same zeal for property that fired their private ambitions. The moderates saw themselves, or their forebears, as self-made men. Their wealth was the fruit of their own and their families' diligence and enterprise. The independence they sought was personal, not collective; economic, not political. If attained, they could leave the bustling worlds of litigation and commerce, with their attendant tedium and risk, and live secure on their investments in the real property of land and buildings. But their concept of security also had a profoundly political dimension. No less than Samuel Adams, who had somehow pulled off the neat trick of ruining the family brewery, they genuinely feared that the rights they treasured most would be rendered insecure in an empire that subjected their property to the legislative control of others.

Still, the idea that Britain could be so mad as to overlook the good intentions that made men like Dickinson, Jay, and Morris yearn for accommodation remained difficult to accept. That commitment was sorely tested by the news from Lexington and Concord. "What human Policy can divine the Prudence of precipitating Us into these shocking Scenes," Dickinson wrote to Arthur Lee in London on April 29, 1775. "Why have We been so rashly declared Rebels?" Why had Gage not waited "till the sense of another Congress could be collected?" Some of its members were already resolved "to have strain'd every nerve at that Meeting, to attempt bringing the unhappy Dispute to Terms of Accommodation," he observed. "But what Topicks of Reconciliation" could they now propose to their countrymen, what "Reason to hope, that those Ministers & Representatives will not be supported throughout the Tragedy as They have been thro the first Act?"

Dickinson's despair was one mark of the reaction that swept the country as the news of war spread. Another was the tumultuous reception that the Massachusetts delegates to the Second Continental Congress received en route to Philadelphia, which exceeded in fervor the more earnest and anxious support they had enjoyed in September. The welcome they enjoyed in New York amazed John Hancock, the delegation's newest member, to the point of embarrassment. "Persons appearing with proper Harnesses insisted upon taking out my Horses and Dragging me into and through the City" under their own power. The New Yorkers had a point to make. Back in February their provincial assembly had indeed confirmed patriot fears by submitting its own petition to the Crown, without endorsing the decisions of Congress. But now that hostilities had begun, the people of New York were eager to prove that they supported the cause even if their legislative representatives did not.

From whatever direction they came, the arriving delegations were hailed by well-turned-out contingents of militia. Some were newly formed "independent" companies, such as the one that greeted the North

Carolina and Virginia delegates at Port Tobacco, Maryland. They made "a Most Glorious Appearance," Richard Caswell wrote to his son, "68 Men besides officers all Genteelly drest in Scarlet & well equiped with Arms & Warlike Implements." Fresh companies appeared in each county, "with their Colours Flying, drums Beating & Fifes playing." At Baltimore they joined Colonel George Washington in reviewing the troops, no doubt meekly trailing behind his vigorous stride. Arrival in Philadelphia was even more exhilarating. "Here a Greater Martial Spirit prevails if possible," Caswell marveled. Twenty-eight companies were drilling twice a day, and "Scarce any thing But Warlike Music is to be heard in the Streets." Caswell ended his letter by instructing his son to "become a Soldier & risk your life." But he hedged this martial outburst by adding that the boy's two uncles should not yet "engage" so deeply in militia activities as to risk their Crown offices as county clerk and sheriff.

In May 1775, as a year earlier, the dominant fact of American politics was the trauma Britain had inflicted, then by legislation, now by arms. The initial shock of 1774 had produced a broad consensus about the calling of a congress and the measures it should adopt. Now the rampant martial fervor of the spring of 1775 established even deeper agreement. The eruption in Massachusetts was a willful act of aggression, provoked by Britain, from which Americans could not flinch. The people of Massachusetts had upheld their promise to act only on the defensive.

Military preparations thus became the first task of the new session of Congress, but not its sole duty. Congress convened on May 10, but serious discussion did not begin until Tuesday, May 16. It started as Richard Henry Lee introduced "proposals for raising an army," seconded by two South Carolina delegates, Thomas Lynch and John Rutledge. But Rutledge immediately insisted "that previously some other points must be settled, such as do We aim at independency? or do We only ask for a Restoration of Rights & putting of Us on Our old footing." John Adams sought to deflect this basic question by restating the position of 1774. Americans desired independence only from Parliament, "but a dependance on the Crown is what We own." This no longer satisfied Dickinson. In a prepared speech lasting over an hour, he insisted that a "Vigorous preparation for Warr" and a "Vigorous prosecution of it" had to be buttressed by a fresh plan of reconciliation requiring Americans to yield "intirely the Regulating of Trade" to Parliament.

Before Congress could pursue this last question, it took up more urgent concerns. The first involved New York. By early May, Americans knew that the royal garrison in New York City would soon be reinforced. Responding to a request from the city's Committee of One Hundred, Congress issued instructions similar to those it had given Boston in September. New Yorkers should "act on the defensive so long as may be consistent with their safety and security" but not allow the British to fortify or isolate the city. Should the British "commit hostilities or invade private property," the townspeople should "repel force by force" and in the meantime take other measures essential "for protecting the inhabitants from insult and injury."

The Congress that adopted this resolution no longer regarded New York as the weak link in the American union. "New York has been converted almost as instantaneously as St Paul was of old," one North Carolina delegate reported, while Richard Henry Lee exulted that "there never was a more total revolution at any place than at New York." James Duane offered the best testimony to this conversion. Military inexperience did not prevent Duane from describing how exposed his colony was: its capital easily blockaded by sea, the Hudson open to navigation all the way to Albany, its scattered frontier settlements, perhaps even Duanesburg, "exposed to the ravages" of Indians. But rather than conclude that these dangers warranted rethinking the costs of defiance, Duane endorsed the need for urgent preparation. Delegates from every colony had to take these concerns to heart. With their naval supremacy, the British could carry war wherever they chose.

But while there was no denying the urgency of military preparations, "the Question shall we treat" (that is, offer to negotiate) proved far more controversial. A week passed before Congress found time to discuss "a further plan of reconciliation." Dickinson used the time to draft a heartfelt speech that assumed that the North ministry remained fully committed to war and that a united Parliament and people supported its policy. The news of war would not shake this resolve. The British were used to losing early battles. The "Dishonor to their Arms returned from despised America" would only make them more eager "to restore their Lustre by revenging their Disgrace." The colonists thus had no alternative but to "prepare vigorously for War."

But their efforts could not stop there, Dickinson added. Congress should petition anew, less because a fresh appeal would have any great effect, but simply because the horrors awaiting both sides demanded it. The real thrust of his proposal was to urge Congress to send a peace mission to London. "We have not yet tasted deeply of that bitter Cup called the Fortunes of War," Dickinson observed. Countless misfortunes yet awaited the Americans:

> A bloody Battle lost—The peculiar Distress of the brave Defenders of American Liberties in the Massachusetts Bay—The Multitude thrown out of Employ by the Loss of their Fisheries & the stoppage of their other Trade—Disease breaking out among their Troops unaccustomed to the Confinement of an Encampment—Divisions in any one province which might interrupt provisions or Relief going to their Aid by Land—the Enemies superiority forbidding it by Sea—The Difficulties from Distance if no other Objections—The Danger of Insurrection by Negroes in Southern Colonies—Incursions of Canadians & Indians upon the Northern Colonies—Incidental Proposals to disunite Us—false Hopes—selfish Designs—Giving up the Point of Taxation or some other partial Concession that might be greedily catched at by a peaceable People jaded out with the tedium of Civil Discords—may all operate hereafter to our Disadvantage.

Americans were lucky that Britain had so badly gauged the depth of their resistance. A bolder ruler "would have sent over 20,000 Men & a formidable naval Armament to Boston." Should the fortunes of war turn, "We may pretty well guess the Consequences"—a warning of the loss of liberty and property each delegate faced if the cause crumpled.

Nothing in this sober assessment placed Dickinson at odds with his colleagues. What did set him apart was the affection he still harbored for Britain. He had "not yet found Cause to change a single Iota" of the principles that had guided him through "this unhappy controversy." He simply could not imagine "any Idea of Happiness for these Colonies for several ages to come, but in a State of Dependence upon & subordination to our Parent State." This last phrase, like "mother country," was a shopworn cliché used to describe Britain's relation to its colonies, but for Dickinson, with his warm memories of England, it retained a personal meaning. The filial duty that Americans owed the parent country obliged Congress to send a diplomatic mission to London. The emissaries must be freely empowered to negotiate, and carry instructions "suited as nearly as may be to every possible State of Affairs that may exist on their arrival in England."

Dickinson's proposal "to treat" was roundly supported by Jay and Duane, and just as sharply criticized by other delegates. Patrick Henry argued that "the Bill of Rights must never be receded from," meaning that Congress should not concede that Parliament could regulate trade as a matter of right. When "The old affair of the right of regulating Trade" was debated on May 24, tempers frayed badly. Dickinson's plan was attacked "with spirit" by his own colleague, Thomas Mifflin, with "sever[ity]" by Lee, and "the utmost contempt" by John Rutledge, who declared that "Lord North has given Us his Ultimatum with which We cannot agree." At some point the discussion proved "so disagreeable that one half of the Congress withdraw," as Silas Deane noted in his running summary of the debate.

This lone remark is the only recorded evidence for this apparent walkout. However high emotions ran, no delegate was about to reveal anything about these internal divisions. In the end, Dickinson's plea for a bolder gesture of reconciliation failed. On May 26, Congress adopted a set of resolutions stating its general policy. The only one not to pass unanimously proposed "that measures be entered into for opening a Negotiation" with the government, "and that this be made a part of the petition" to be submitted to the king. The petition was drafted by a committee dominated by three moderates—Dickinson, Jay, and Thomas Johnson. They were joined by Benjamin Franklin, who had sat mysteriously silent in Congress since returning from London but who must have offered a firsthand account of the abuse heaped on anyone in London who dared criticize the ministry's policy of repression. That had been the fate of America's great "friend," William Pitt, after he and Franklin had discussed just this subject during the winter. In its report, the committee did not suggest how a negotiation might be opened.

Over the next two months the Second Continental Congress took a series of decisions that effectively committed the colonies to war. On June 9 it authorized Massachusetts to resume government under the vacated royal charter of 1691, with an elected council replacing the royal governor. The next day it adopted resolutions encouraging the manufacture of munitions. On June 14 it began the process of transforming the provisional forces outside Boston into the Continental army, to be led, it decided the next day, by George Washington. In obedience to republican principles, which prohibited the open pursuit of any office, Washington described his appointment as "an honour I neither sought after, nor desired." But the fact that he had taken to attending Congress in uniform sheds some doubt on that disclaimer.

Many of the delegates rode along for a few miles when Washington and his entourage left for Boston on June 23. The day before they had learned of the carnage at Bunker Hill on June 17, when entrenched New England troops inflicted heavy casualties on a British landing in Charlestown before exhausting their ammunition and fleeing to Cambridge. The report reminded John Adams of the Powder Alarm of September, but the sight of the newly commissioned and handsomely uniformed officers heading off amid "the Pride and Pomp of War" inspired a moment of self-pity. He "must leave others to wear the Lawrells which I have sown," he wrote Abigail; "others to eat the Bread which I have earned.—A Common Case."

Few other delegates were thinking about laurels—except, perhaps, John Hancock, who coveted the appointment Washington won. The very fact that Adams could feel these pangs at this early moment in the conflict, or somehow equate the cause he was advocating so well with the plaudits he believed were his due, is stunning evidence of the extent to which the vanity that tormented him was already nagging at his tenuous equanimity. There was satisfaction but no glory to be earned from his daily tasks in Congress: persuading his colleagues to do everything necessary for war while assuaging their misgivings about the role of Massachusetts in creating the current calamity. When the Second Congress "first came together, I found a strong Jealousy of Us" still persisted. "Suspicions were entertained of Designs of Independency—an American Republic—Presbyterian Principles—and twenty other Things." With each passing day, Adams felt he could speak more plainly. But he still chafed over the diversionary measures that moderate members insisted on pursuing. His frustration came to a boil in late July. "A certain great Fortune and piddling Genius whose Fame has been trumpeted so loudly, has given a silly Cast to our whole Doings," he grumbled in a letter to James Warren. He obviously meant Dickinson, and he then went on to complain that "the Farmer's" insistence on a second petition to the king was retarding other measures Congress should be taking.

But a British patrol vessel intercepted the letter at a Rhode Island ferry and sent it on to Boston, where General Gage was all too happy to publish it and enjoy the embarrassment it caused.

Adams received his comeuppance when Congress reconvened in mid-September after a five-week recess. Walking to the State House one morning, he encountered Dickinson "in Chesnut Street. We met, and passed near enough to touch Elbows. He passed without moving his Hat, or Head or Hand. I bowed and pulled off my Hat. He passed hautily by. The Cause of his Offence, is the Letter no doubt which Gage has printed in Drapers Paper." That letter was as unfair in its judgment as it was ill-advised in its shipment (because mail often went astray). Dickinson sincerely thought a second petition was necessary, not only to give the government a last chance to relent but also to persuade Americans that their Congress was acting prudently. In terms of diplomacy the petition might prove superfluous; in terms of politics it had its uses. Having pushed so hard to give peace a chance, Dickinson felt equally obliged to honor his other commitment to "prepare vigorously for War." That was why he joined Thomas Jefferson, a newly arrived Virginia delegate, in drafting the Declaration of the Causes and Necessity for Taking Arms, which Washington was instructed to publish upon his arrival in Boston.

For someone with Dickinson's Quaker ties and spiritual leanings, that was not an easy assignment to accept—which is exactly why it symbolizes the dilemma of the entire bloc of moderates. There were no pacifists in Congress, and no one opposed the military preparations it was compelled to take. The further those preparations went, the more likely it became that Britain would view American motives and aims in the worst light possible. Yet the closer Americans moved toward independence, the more anxiously moderates looked to Britain to break the diplomatic impasse. Against their private inclinations, they found themselves caught up in the business of revolution, lamenting its demands while their own involvement in the cause only deepened, yet still believing their hearts were in the right place in favoring accommodation.

After a recess of six weeks, Congress reconvened in the second week of September 1775. Save for two brief adjournments occasioned by the approach of British troops, it thereafter met continuously until the war ended. All thirteen colonies were present; a delegation from distant

Georgia appeared at last. Unlike its previous session in May, the new session of Congress felt no need to review the ends of resistance. For one thing, it had little fresh news to consider, the commercial boycott of the Association having brought a sharp decline in transatlantic sailings. The intelligence that mattered most was reports of the troops and vessels the ministry had ordered to America. As yet there was no official response to the second petition that Dickinson had prevailed on Congress to dispatch—not that a favorable answer was expected. Then on November 2 two ships brought news that in late August the king had declared all his American colonies to be in a state of rebellion. "This has a most happy effect here," Samuel Ward noted, "for those who hoped for Redress from our Petitions now give them up & heartily join with us in carrying on the War vigorously."

Such a conflict had to be fought with America's limited war-making capacity. In committee meetings and general debates, the delegates wrestled with the task of preparing for war. Some of their concerns were broadly strategic. Should expeditions be sent to capture the British fort at Detroit or to occupy Nova Scotia? Could Britain's traditional Indian allies be persuaded to remain neutral? Other issues, though more prosaic, were equally urgent. How to obtain or make saltpeter, essential for manufacturing gunpowder, was a recurring topic of discussion. Like Cato the Elder, the ancient Roman senator who ended every speech with the words *"Delenda est Carthago!"* ("Carthage must be destroyed!"), John Adams pledged "never to have Salt Petre out of my Mind but to insert some stroke or other about it in every Letter." If it and other munitions had to be purchased elsewhere, as the firm of Thomas Willing and Robert Morris (the first partner already a member of Congress, the second soon to join it) contracted to do, should the merchants be allowed a hefty profit of twelve thousand pounds?

Though the fall of 1775 was hardly a moment of political calm, neither was it a time of deep agitation. Congress had crucial issues to discuss: the formation of a navy; the advantages and potential drawbacks of the non-exportation part of the Association, which took effect on November 1; the benefit of permitting individual provinces to restore legal government. Moderate delegates were as fully engaged as their radical colleagues in these urgent tasks. Robert Livingston was one of a three-member committee sent to upstate New York to survey defenses along the northern frontier. Robert Morris entered Congress in early November and soon replaced his partner, Thomas Willing, on the Secret Committee, which was charged with obtaining the military supplies the colonies so desperately needed. That committee is not to be confused with the five-member Committee of Secret Correspondence appointed in November 1775 "for the sole purpose of corresponding with our friends in Britain, Ireland, and other parts of the world"—such as France. In addition to Franklin, the country's sole experienced diplomat, the latter committee included Dickinson, Jay, and Thomas Johnson. When the New Jersey assembly was reported to be drafting a separate petition to the Crown, Dickinson and Jay led the deputation that Congress hastily dispatched to head off this latest threat to colonial unity.

If there was no lull in the business of mobilizing for war, Americans still seemed loath to contemplate the aims and consequences of resistance. That reluctance soon faded. The Royal Navy rang in the New Year by bombarding Norfolk, Virginia, thus demonstrating that the war would carry beyond New England. A week later members of Congress read the king's October speech opening Parliament. Americans might still be unsure of their ultimate objectives, but an all-wise sovereign evidently knew his subjects' hearts better. "The rebellious war now levied is become more general," the king reported, "and is manifestly carried on for the purpose of establishing an independent empire." The king did not condescend to explain how this could be "manifest" when Americans still dreaded uttering the fateful word *independence.* But within two days of the arrival of the text of the king's speech in America, the word lost its power to shock. For on January 9, 1776, the first edition of Thomas Paine's *Common Sense* appeared, and the vocabulary of politics was permanently altered.

Paine was an unlikely catalyst for this transformation—but so an iconoclast must be. He had debarked in Philadelphia in November 1774, leaving behind in England a failed marriage and two dismissals from His Majesty's excise service—the second for publishing a pamphlet demanding higher wages for excise collectors. Sporting a recommendation from Franklin, who quickly spotted the talents of a self-taught writer like himself, Paine soon became the editor of the monthly *Pennsylvania Magazine.* As the year closed, he began drafting *Common Sense.* Like other pamphlets, it was published anonymously—but its author's identity was soon known, as was the new spelling of his name, which added a final *e* to the original *Pain.*

One source of its remarkable success was Paine's vivid imagery. Though his train of thought sometimes meandered, Paine always knew how to regain readers' attention with arresting turns of phrase. But the deeper impact of *Common Sense* came from the candor with which Paine derided the whole idea of monarchy, dismissed prospects for reconciliation, and portrayed independence as the only logical outcome of the struggle. Paine's flair for the well-turned phrase is exemplified in his wry rejoinder to the claim that America still needed British protection: "Small islands not capable of protecting themselves, are the proper objects for kingdoms to take under their care, but there is something very absurd, in supposing a continent to be perpetually governed by an island." The charge that American produce would not find a market if excluded from imperial ports drew an equally pithy response. Americans "will always have a market while eating is the custom of Europe," Paine quipped.

In urging independence, Paine also repeatedly disparaged the prospects for reconciliation. Its advocates, he charged, were either "interested men, who are not to be trusted; weak men, who *cannot* see; prejudiced men, who *will not* see; and a certain set of moderate men, who think better of the European world than it deserves; and this last class, by an ill-judged deliberation, will be the cause of more calamities to this continent, than all the other three." Real moderates certainly rejected this characterization. Their anger at British policies was no less pronounced than the radicals' ire. But it was not the upstart Paine whose criticism most distressed them. The graver challenge came from the "royal brute" himself, George III. For his claim that the Americans were "manifestly" bent on independence impugned every motive that moderates acknowledged to justify their own involvement in politics. The charge seemed personal as well as political. Almost as a matter of honor, many moderates felt that this insult to their conduct had to be refuted.

Their immediate response came from James Wilson, who urged Congress to draft a fresh address to the colonists, disclaiming any desire for independence. A fortnight passed before Congress appointed a committee of five moderates—including Wilson, Dickinson, and Duane—to prepare a suitable message. Nor was this the only response that moderates explored. Jay also took umbrage at the "ungenerous & groundless Charge" that the Americans had been "aiming at Independence" all along. Amid his other duties, Jay drafted an essay detailing all the statements of Congress "which abundantly prove the Malice and falsity of such a Charge." Dickinson characteristically believed that yet another petition should be sent to the king, this one carried by a congressional delegation bearing new conciliatory proposals and negotiating instructions.

Moderates were eager to refute the king's allegation because they entered the new year anxiously hoping that the government might yet come to its senses. "If Parliament should offer [reasonable?] Terms of Conciliation," Duane wrote to Robert Livingston, explaining why he was remaining in Congress, "I shoud never forgive myself for being absent on so great and interesting an Occasion." In his October speech, the king had also declared that he would appoint "certain persons upon the spot to grant general or particular pardons and indemnities" to individuals who returned to their allegiance. Although this hardly sounded like a formula for negotiations, the arrival in Philadelphia of Thomas Lundin, Lord Drummond, offered hope that diplomacy could still have a chance. A young Scots nobleman who had lived in New Jersey since 1768, managing family lands, Drummond had returned to England in November 1774, where he spoke extensively with Lord North and Lord Dartmouth about ideas for reconciliation.

Drummond carried no commission from the ministry and acted only as a private person well informed of the views of North and other high officials. The delegates who met with him in Philadelphia—including Jay, William Livingston, Duane, and Wilson—likewise acted unofficially. Unencumbered by formal instructions, their conversations ranged widely but ended indecisively. Jay hoped Drummond would return to London and lobby the government to send a true emissary to America; Drummond replied that Congress should send its own mission to London. That was an idea that Dickinson might have endorsed, but which a Congress that the king's government had never formally recognized was unlikely to approve.

In affairs of state, occasions often arise when private diplomacy and back-channel contacts seem preferable to formal or public negotiations. Prior to his return to America, Franklin held just such discussions with Admiral Richard Howe, who would soon come to America to act (with his brother, General William Howe) in the dual role of commanders in chief and peace commissioners. But Congress needed something more than informal hints from a self-appointed emissary to believe that negotiations were possible, much less imminent. Most delegates thought they had given Britain ample time to answer their petitions. A government that was serious about reconciliation would act forthrightly. The lack of a diplomatic channel was not the problem, moderates thought, but rather the government's refusal to address American grievances directly.

In mid-February 1776, however, word arrived that Lord North had told Parliament that the government did plan to appoint an American peace commission. Its powers, composition, and instructions remained uncertain. Moreover, this proposal was part of a Prohibitory Bill extending the partial bans on colonial commerce into a sweeping measure subjecting all American ships and cargo to confiscation. Amazingly, North believed these measures were conciliatory because the commissioners would be empowered to exempt from this severe decree any colony that returned to obedience. But in both Britain and America, most observers were more impressed by the blanket threat the Royal Navy could now execute. As John Hancock wryly observed, "making all our Vessells lawful Prize don't look like a Reconciliation." The commissioners' errand might be worthwhile if they came armed with "proper Powers," Robert Morris wrote to Charles Lee, the eccentric dog-loving former British officer commanding American forces in New York City. "But to come limited to terms inconsistent with Freedom will be doing nothing." Three weeks later, Morris heard a report that the commissioners "can only Treat with the Colonies separately & will have nothing to do with the Congress. If this is the case," he snorted, "they may as well stay where they are."

From the moderates' perspective, a fully empowered peace commission was something the government should already have dispatched, not left hanging. The lack of urgency in London undercut the concern that Congress needed to defend itself against the charge of being intent on independence. Since late January, James Wilson had been laboring over his refutation of this canard. But his address met a cool reception when it was read in Congress on February 13. It was, the New Jersey delegate Richard Smith noted, "very long, badly written, & full against Independency." It was also, he might have added, pointless. Not only had Wilson tediously rehearsed arguments that no longer merited repetition. He also felt bound to rebut charges that only outright loyalists were by now likely to make. Wilson's apologetic excess was a tribute to the resentment that moderates felt over having their own motives besmirched. In their case, the king had the story exactly wrong. It was the desire to preserve the possibility for accommodation, not the grail of independence, that had brought them into politics.

As spring came, moderates clung to the hope that emissaries would miraculously appear. "Where the plague are these commissioners," a restless Robert Morris exploded in early April; "if they are to come, what is it that detains them?" But even for Morris the puzzling delay was irritating not because he expected genuine negotiations but because the delay could have divisive effects. The time had come to "know positively whether the Libertys of America can be established & Secured by reconciliation, or whether we must totally renounce Connection with Great Britain & fight our way to a total Independence." United, Americans could attain either goal. But unless commissioners "appear soon" to open the "path" to negotiation, Americans could still find themselves divided over their ultimate aims.

This fear of "intestine division" ran strongest in the middle colonies, with their pluralist, polyglot, multidenominational populations. Even if public opinion favored independence, as it promised to do by April and May, moderate leaders from this region still worried that their constituents would not remain unified for long. But the shift in sentiment initiated by *Common Sense* in January made a decision for independence only politically possible. Two other conditions had to be met before independence could be declared.

First, it had to be perceived as morally legitimate. Americans had to believe that, as a matter of right, they were entitled to repudiate the bonds of allegiance to the king who remained their last acknowledged link to the empire. That the Crown had already forfeited that allegiance was not in dispute. In the familiar definition of the contract of government, a people rendered allegiance to their sovereign in exchange for his protection. The civil war Americans were now experiencing was the opposite of protection. As winter gave way to spring, fresh intelligence confirmed the ministry's commitment to war. First came reports that Britain had signed treaties with several small German states to supply mercenaries who would be used as combat troops and not (as once thought) to replace British troops sent to America. Items in the British press suggested that forty thousand soldiers would sail for America by summer. Only a government bent on restoring its rule by force would resort to so great an expense and effort. Then in early June word came of another bellicose pronouncement from the king, brusquely rejecting an address from the City of London urging the Crown to negotiate with the Americans. "I confess I never lost hopes of reconciliation untill I saw this Answer which in my opinion breaths nothing but Death & Destruction," Morris wrote to Silas Deane, the former Connecticut delegate who was now the congressional agent in Paris.

The question whether Americans still owed allegiance to George III was thus easily answered: obviously they did not. But did they have to say so publicly now? If the colonies were independent in fact, as John Adams liked to observe, did they also have to declare it in name? The second condition that had to be met before independence could be declared was that it had to be perceived as strategically necessary. If Britain pursued a military solution in America, the colonists would need to secure supplies and succor from foreign sources—above all, France. A formal decision for independence, with a suitable public declaration, would be essential to that objective. And that is why, when Congress began its final preparations for independence in June, it simultaneously appointed three committees: one to draft a declaration of independence, one to prepare a plan of foreign treaties, and a third to write articles of confederation binding the thirteen commonwealths that would cease to be colonies once independence was proclaimed.

No one grasped the inexorable logic of independence better than Samuel Adams. As an ideologue, he already knew how events were destined to turn out. But he also believed that the exact timing of this decision was not critical. He was disappointed that the colonies had not yet established new governments, "but I bear it tollerably well," Adams wrote on the last day of April. "If I do not find my self chargeable with Neglect I am not greatly chagrind when things do not go on exactly according to my mind. Indeed I have the Happiness of believing that what I most earnestly wish for will in due time be effected. We cannot make Events. Our Business is wisely to improve them." Reviewing how events had unfolded since 1774, Adams noted that British actions had repeatedly stiffened colonial opinion. One more battle to the southward "would do more towards a Declaration of Independency," he predicted, "than a long chain of conclusive Arguments in a provincial Convention or the Continental Congress." Another veteran politician had reached the same conclusion. "The Novelty of the Thing deters some" from supporting independence, "the Doubt of Success others, the vain Hope of Reconciliation many," Benjamin Franklin wrote in mid-April, while en route to Quebec on a mission with Charles Carroll and Samuel Chase of Maryland. "But our Enemies take continually every proper Measure to remove these Obstacles," daily providing "new Causes of increasing Enmity, and new reasons for wishing an eternal Separation."

If Morris, Dickinson, or Jay had any notion of "improving events," it did not involve bringing America to the verge of "independency" as quickly as possible. It would instead require the British government to admit the futility of clinging to policies that ignored the new realities of American politics. The first of these was the consensus on American rights that had emerged by the fall of 1774—a consensus moderates fully shared. A second was the legitimacy of Congress itself—a body that was now a national government, yet which British leaders refused to recognize, much less approach as an authority, with whom they could (or indeed must) "treat." Every conciliatory scheme the North ministry conjured was premised on persuading individual colonies to defect from the general union that Congress embodied. From an American vantage point, this refusal to deal directly with Congress only confirmed the government's lack of interest in conciliation.

These were strategic realities that a more perceptive government could have grasped. To reckon with either, however, meant defying the orthodox creed of parliamentary supremacy to which virtually every member of the British political elite subscribed. That was why the government's dismissal of the constitutional claims of the First Continental Congress and the refusal of the Second Congress to reconsider the same claims made independence nearly inevitable. Yet there was one other political reality that the ministry could have exploited: the existence within Congress of a moderate bloc that was not predisposed to favor independence. Admittedly, unlike other realities the ministry chose to ignore, this was one it might never have perceived. It required better information and greater discrimination than it possessed to recognize that the American consensus on principles and tactics masked narrow but significant differences about the ultimate ends of resistance. Lacking useful political "assets" of its own in the colonies—other than impotent royal governors and a small platoon of Anglican ministers—the government could only adopt a strategy of repression to teach Americans the real costs of defiance.

In fact, Britain's optimal strategy would have been to appeal to those moderates who best understood the great risk America was running because their own pursuit of property made the calculation of risk a daily concern. Had the government ever presented substantive proposals for genuine negotiations, the moderates in Congress—and much of public opinion "out-of-doors"—would have leapt at the offer. Being prudent, the moderates would have continued pursuing preparations for war. But neither would they have allowed military exigency to cut off a real prospect of negotiations. Lord North never issued the one clear conciliatory signal they awaited. "If any equitable Terms of Accommodation had been offered us, and we had rejected them," James Wilson wrote in his abortive address of February, "there would have been some Foundation for the Charge that we endeavoured to establish an independent Empire." But no such terms were tendered.

In London, such claims were easily dismissed as self-serving. But Wilson was not protesting too much. With other moderates, he believed that the real dynamic of the movement toward independence came from the British government's aversion to diplomacy. On this critical point, no words carried greater weight than the king's. His rebuff of the London petition urging conciliation made independence "inevitable," Robert Morris observed in early June.

> Great Britain may thank herself for this Event, for whatever might have been the original designs of some Men in promoting the present Contest I am sure that America in general never set out with any View or desire of establishing an Independant Empire. They have been drove into it step by step with a reluctance on their part that has been manifested in all their proceedings, & yet I dare say our Enemies will assert that it was planned from the first movements.

By "some Men," Morris meant Samuel Adams and his allies, who he and other moderates suspected had harbored ulterior designs of independence. But that suspicion matters less than the fact that figures as different as Robert Morris and Samuel Adams ultimately understood the revolution in similar if not quite identical terms. Where the ideologue Adams believed that a raw lust for power was driving Britain's leaders to seek dominion over America, Morris preferred to blame obtuse stupidity and miscalculation. But both agreed that British missteps, rather than American desires, had brought the colonies to the point of independence.

"Every Post and every Day rolls in upon Us Independance like a Torrent," John Adams wrote to James Warren on May 20. Township and county meetings across the colonies were endorsing independence, providing solid evidence that the radical conclusions Tom Paine had voiced in January were indeed becoming the common sense of public opinion. But Adams's exultation had a poignant undertone. "What do you think must be my sensations," he asked, "when I see the Congress now daily passing Resolutions, which I most earnestly pressed for against Wind and Tide, Twelve Months ago?" And then, on an even more personal note: "What do you think must be my Reflections when I see the Farmer himself, now confessing the Falshood of all his Prophecies and the Truth of mine, and confessing himself, now for instituting Governments, forming a Continental Constitution, making alliances with foreigners, opening Ports and all that—and confessing that the Defence of the Colonies, and Preparations for defence have been neglected, in Consequence of fond delusive hopes and deceitfull Expectations?"

Dickinson's confession took him to the threshold of independence, but even in June 1776 he was not quite prepared to cross it. These preparations should still precede a final vote for independence, he thought, positioning America for sustained conflict without shutting the door to reconciliation. For two years he had wrestled with the conflict between his legacy as a leading advocate for colonial rights and the fears, concerns, and Quaker principles that drove his own moderation. The advanced stand "the Farmer" had taken against the Townshend duties was not the sole or truest mark of his political attitudes. He had always believed that Americans needed the empire to mediate the disputes that would arise if the separate provinces were left to their own devices. In 1774 he had rallied with every patriot to the support of Massachusetts. To maintain his bona fides, he also felt obliged to endorse every military preparation. In February 1776 he even volunteered to command a contingent of Pennsylvania militia that Congress thought of sending to New York when it received reports (false, as it turned out) of a British landing. And as chair of the committee to frame articles of confederation, he took the lead in drafting a plan of union that would reduce the potential for "intestine division."

Yet Dickinson also did everything in his power to defer a vote on independence. Long after other delegates had given up on petitioning, he wanted to make one more diplomatic gesture, and then another. To bolster his position within Congress, he had the Pennsylvania assembly issue instructions he himself had written, ordering its delegates to "dissent from and utterly reject any propositions . . . lead[ing] to a Separation from our Mother Country." For Dickinson that last phrase still retained an affectionate meaning. Others now relished Paine's rejoinder: "Even brutes do not devour their young." The instructions to the Pennsylvania delegation were lifted only in late May, after Congress authorized every province to establish a new government, drawing authority directly from the people. Here at last John Adams got the better of Dickinson, for it was Adams who wrote the fiery preamble that empowered Dickinson's radical critics within the province to argue that Penn's charter of government was no longer binding. Dickinson's views remained unchanged, however. In the final congressional debate of July 1, he was the main speaker against independence, arguing that Americans should not sever their last link with Britain until they better knew the intentions of France. But by then the time for deliberation was over.

Congress approved independence the next day, July 2—the day that John Adams wrongly predicted Americans would celebrate ever after. Dickinson abstained on the final vote, as did Robert Morris. Dickinson used this decision as a pretext to leave Congress, even though that meant he would not personally present the Articles of Confederation, which he had spent the past weeks drafting. His long campaign as a voice of moderation had exhausted Dickinson; so had the "burthen assigned me" of living up to his reputation. His opposition to independence, he told Congress, "will give the finishing Blow to my once too great, and my Integrity considered, now too diminish'd Popularity." A month later he told his friend Charles Thomson that "no youthful Lover ever stript off his Cloathes to step into Bed to his blooming beautiful bride with more delight than I have cast off my Popularity."

There survives in Dickinson's papers the draft of a letter that pursues this subject in greater detail. With its deletions and addenda, it is one of those documents that a historian can spend hours transcribing yet not wholly decipher. Dated August 25, it answers an unknown recipient who had candidly explained to Dickinson the sources of the animosity he had engendered among some of Pennsylvania's rising leaders, notably Joseph Reed and Benjamin Rush. Both were linked to the Presbyterian "interest" that solidly backed the patriot cause and reportedly viewed Dickinson's family ties to the Quakers with mistrust. Even so, Dickinson still hoped that "the Purity of my Intentions would so far appear from the Integrity of my Actions in general that I should retain the Character of an honest Man, the only Name I think worth my Ambition." He was infuriated to think that anyone could ascribe his conduct to any motive other than "Conscience" and concern for "the true Interest of my Country." "Any man not more than half an Idiot must perceive" that he had far more to gain by following the tide of opinion, Dickinson sputtered, than clinging to his reservations. If popularity was his god, what benefit could he gain by courting Quakers, of all people, when "All Things were verging to a Revolution in which they would have little Power"?

Popularity was no burden for Bob Morris. Debate, public speaking, pamphleteering never appealed to him; nor did the adulation of voters. For a man of business, attending the daily sessions of Congress, with their tedious wrangling over petty details, was time ill budgeted. As a member of the Secret Committee, Morris already had his hands full, figuring out how Americans would obtain and pay for the military supplies they badly needed. With his extensive commercial ambitions, he knew that the war would offer great if necessarily risky opportunities—and as a private citizen with established patriotic credentials and access to the best information, he and his Willing partner would be well positioned to pursue them. There would be fortunes to be made and expanded in this war, Morris knew, and he had every intention of seizing the main chance and treating his political connections as one more advantage to exploit.

Recognizing that independence was both inevitable and justifiable, Morris still believed the decision was premature. His chief concern, he wrote in late July, was that the decision "has caused division when we wanted Union." Morris thought "my Conduct in this great Question would have procured my dismission" from Congress "but find myself disappointed" that the Pennsylvania provincial convention, though dominated by radical upstarts, had just reelected him. Even then Morris could easily have followed Dickinson into retirement simply by resigning his seat. But he did not, and his professed reasons for staying in Congress illustrate why his behavior is a better guide to the shifting position of the moderates than is Dickinson's adherence to the claims of conscience. Although "interest & inclination prompt me to decline the Service," Morris wrote, he could not "depart from one point that first induced me to enter in the Public Line." That was "an oppinion that it is the duty of every Individual to Act his part in whatever Station his Country may call him to, in times of difficulty, danger & distress." And this remained true even when "the Councils of America have taken a different course from my Judgements & wishes."

Evidence for his sincerity comes from an unlikely source: John Adams. Despite their different views, Adams rendered Morris the respect Dickinson had forfeited. Morris had "a masterly Understanding, an open Temper and an honest Heart," Adams observed, and if he also had "vast designs in the mercantile way," he was still "an excellent Member of our Body." So he remained after his new term in Congress began. If anything, his commitment grew deeper even as events resolved one lingering doubt while confirming other fears.

The resolved doubt was a gift from the peace commissioners whose arrival Morris had awaited so anxiously. General William Howe had replaced Gage as the British army's commander in chief in October 1775 and led the British evacuation of Boston in March 1776. In mid-July his older brother Admiral Richard Howe anchored near New York City with a vast armada ferrying over thirty thousand British and German troops. Americans who had just absorbed the Declaration of Independence were soon reading Lord Howe's proclamation as peace commissioner, which said nothing about negotiations but only indicated that he could grant a pardon to any American who returned to allegiance. This disclosure of the feeble powers of the peace commission so pleased Congress that it reprinted Howe's proclamation for the general public. By early September, however, British victories in combat around New York convinced Congress that it could not ignore the Howes' request for a meeting, even if the delegates it sent on this futile mission would be received only as private individuals, not the representatives of a new nation. John Adams, Franklin, and Edward Rutledge accordingly went to meet the Howes at Staten Island. (This was the occasion when Adams and Franklin had to share a bed, and Adams fell asleep to

Franklin's lullaby lecture on the virtue of sleeping with windows wide open.) Once again the commissioners offered far less than what moderates had pined for in the spring and despaired of in the summer.

As summer gave way to fall, however, events on the battlefield threatened to make the question of the commissioners' powers irrelevant. In a series of engagements fought around New York City, imperial troops repeatedly pummeled the Continental army. Individual American units fought valiantly, but the campaign was confirming that the leadership and discipline of veteran British regiments would overmatch Americans' patriotic enthusiasm and the limited military skills George Washington had acquired two decades earlier. Though subordinates faulted General Howe for failing to pursue his victories, an American collapse seemed only a matter of time. That impression gained force in December, when the approach of British troops across New Jersey led Congress to flee to Baltimore, leaving behind a three-member executive committee in Philadelphia.

Robert Morris was its leading member. For him, as for other moderates, the autumn of 1776 was a moment of desperate soul-searching. In late December, he unburdened himself to Silas Deane in Paris. "Our people knew not the hardships & Calamities of War when they so boldly dared Brittain to arms," Morris wrote. "Every man was then a bold Patriot, felt himself equal to the Contest and seemed to wish for an opportunity of evincing his prowess, but now when we are fairly engaged, when Death & Ruin stare us in the face and when nothing but the most intrepid Courage can rescue us from Contempt & disgrace, sorry I am to Say it, many of those who were foremost in Noise, Shrink coward like from the Danger and are begging pardon without striking a Blow." Unless Deane and his fellow commissioners—Arthur Lee and Franklin, just arrived in Paris—persuaded France to enter the war, the Americans would have to sue for peace.

Five days later, Washington crossed the Delaware River to surprise and destroy a Hessian garrison at Trenton, then quickly withdrew. A week and another crossing later came a second victory at Princeton. British lines contracted, Philadelphia escaped occupation, and the New Jersey militia began regrouping and harassing enemy foraging parties with some success. The Revolution had survived its darkest hour yet—"the times that try men's souls," in Paine's never-forgotten phrase opening *The Crisis*.

Morris was not one of those "sunshine patriots" whom Paine disparaged in that pamphlet. After seeing to the safe removal of his family and books from Philadelphia, he stayed on, discharging so many tasks that he became, he told Jay, "the veriest Slave you ever saw." He did not fault Congress for its flight. It would have been "criminal & rash in them to the last degree" to risk capture. Equally revealing, however, were his views of two colleagues in the Pennsylvania delegation. One was Andrew Allen, who had also opposed independence and then taken refuge with Howe. The other was Dickinson, who had been discovered warning his brother Philemon not to accept continental paper money—a misstep that implied, at the very least, a loss of faith in the cause, and at the worst smacked of political apostasy. Morris did not so much condemn the two as "pity" them. The nerves of both, he wrote Jay, "gave way" as Howe marched across New Jersey. But Morris also resented the taint their actions had placed on men of their class. The "mortal stabs" they had given "to their own characters" also threatened to "pierce . . . into the vitals of those who have similar pretensions to Fortune & good Character," he complained. "The defection of these men is supposed to originate in a desire to preserve their Estates & consequently glances a suspicion on all that have Estates to loose." Morris could easily reckon the penalty he would pay if the cause collapsed. But he did not allow that reckoning to guide his political conduct.

Whatever misgivings Morris felt that gloomy fall did not shake his commitment. If anything, his responsibility for managing the affairs of an absent Congress produced the opposite result. He had been "very usefull on many occasions, both to this State & to the Continent," Morris believed, "and in every instance I have exerted myself to the utmost." He took pride in efficiently handling the routine tasks over which Congress often haggled. "I believe we dispatch about 7/8ths of that damn'd trash that used to take up 3/4ths of the debates in Congress; and give them no trouble about the matter," he wrote to Jay, just before Congress returned to Philadelphia. Morris drew a broader conclusion. "If the Congress mean to succeed in this Contest they must pay good Executive Men to do their business as it ought to be & not lavish milions away by their own mismanagement," he observed. "I say mismanagement because no Men living can attend the daily deliberations of Congress & do executive parts of business at the same time."

This division of labor could never be as neat as Morris desired—and perhaps that was for the better. Had the responsibilities of the moderates been limited to debating policies, they all could have followed Dickinson into retirement after Congress voted for independence. But of course their duties were far more extensive. Having entered politics to preserve the possibility of reconciliation, they could remain plausible advocates for diplomacy only by collaborating in the conduct of the war. And having assumed their fair share and more of this burden, they could not abandon the cause without compromising their own sense of honor. They felt this obligation both individually and collectively, egging one another on at those moments when private concerns might have tempted them to shirk the business of making the revolution work.

None of the moderates better illustrates this aspect of the revolutionary experience than John Jay. He remained an active and valued member of Congress until the spring of 1776. Early in the year he managed to pry two months away from Philadelphia to attend Sally for the birth of their first child. "Nothing but actual imprisonment will be able to keep me from you," he promised. Jay managed to return to Congress in March. But when rheumatism and other ills kept Sally ailing for months after delivery, his anxiety over her health and that of his parents brought Jay home again in early May. He stayed away even after friends like Edward Rutledge urged him to return and assist "The Sensible part" of Congress who opposed declaring independence prematurely. Robert Morris similarly asked, "Why are we so long deprived of your abilitys in Congress?" But he already knew the answer: Jay was needed closer to home, not only to care for his family, but because there was equally pressing work to do at the state level of government. In Morris's view, "such Men as You, in times like these, should be every where."

Jay would have said the same of Morris. Though they had met only in the fall of 1775 and were a good decade apart in age, the stout merchant and the lean, hollow-cheeked lawyer formed a natural bond as they jointly resisted a decision for independence. It was to Jay that Morris felt comfortable expressing the disturbing thoughts that arose after reading one of Silas Deane's early dispatches on the progress of his mission to Paris (written in the invisible ink invented by Jay's vexing brother Sir James Jay): "Its a horrid consideration that our own Safety should call on us to involve other Nations in the Calamities of Warr," Morris observed. "Can this be morally right or have Morality and Policy nothing to do with each other?"

Morris deflected his own question with a merchant's sense of priorities: "Perhaps it may not be good Policy to investigate the Question at this time." If Jay answered this letter, his reply is lost. But whatever doubts he felt about the morality of seeking allies or the timing of independence did not alter his view of the moral rightness of the American cause. As a member of New York's Committee for Detecting Conspiracies, Jay gathered intelligence about persons of doubtful loyalty and subjected suspects to interrogation and punishment. His legal training shaped his conduct but did not outweigh his political commitments. "We have passed the Rubicon," he reminded one prominent suspect, Beverly Robinson, whose wife, Susanna, was Sally's cousin; and those who could not renounce their loyalty to the king deserved close scrutiny. After Robinson confirmed these suspicions by fleeing to British lines and raising a loyalist regiment, Jay privately urged Susanna not to follow his example. If America prevailed, Jay asked her in March 1777, where would you "spend the remainder of your Days and how provide for your Children?" But "suppose Heaven unjust, Britain victorious, and the Americans bound in all cases whatsoever," Jay added. "Will you ever Madam be able to reconcile yourself to the mortifying reflection of being the Mother of Slaves?" Jay had spoken in similar terms a few months earlier when the provincial convention instructed him to write an *Address* to the people of New York, urging them to keep up their spirits even while the cause was reeling from one defeat to another. Jay struck a decidedly religious note, with repeated biblical references and a pointed reminder that "the King of Heaven is not like the King of Britain, implacable. If his assistance be sincerely implored, it will be obtained."

Jay did some imploring of his own when he coaxed Robert Livingston to take a more active part in the struggle. When his oldest friend reported the deaths, first of his father, then of his brother-in-law, General Richard Montgomery, shot through the face at the failed American siege of Quebec, Jay added a patriotic appeal to his condolences. "As soon as the roads will permit you, go to the Camp, to Philadelphia, in short anywhere," Jay wrote. Domestic moping and private mourning were greater threats to health and equanimity, he warned, than activity. With Livingston the problem was not honor but indolence: "that lazy he is too many know and all his friends regret," Jay reminded Gouverneur Morris. Livingston came in for a good upbraiding after he gave Jay a self-pitying account of how badly his immersion in the "little party politicks" of the state assembly was interfering with "the tranquil pleasures of a rural life" that he might otherwise be enjoying. This was too much for Jay. "In such rugged Times as these other Sensations are to be cherished," he rejoined. "Rural scenes, domestic Bliss, and the charming Group of Pleasures found in the Train of Peace, fly at the approach of War, and are seldom to be found in Fields stained with Blood, or Habitations polluted by outrage and Desolation."

By the time he wrote these lines, these two sets of images—one pastoral, the other martial—were equally vivid in Jay's thinking. But one operated as a fading memory, to be renewed in happier times. The other now represented the most urgent reality, a set of pressing demands to be acted on daily. There had been a time when martial images could only be imagined, when bucolic scenes lay far closer to the day-to-day life of Jay, Livingston, and other members of their favored class. Now that state of affairs was inverted. The British and Hessian troops and loyalist auxiliaries who reveled in ransacking their property felt no qualms about the desolation they were wreaking. The "patriots" had renounced their allegiance, after all, and such destruction was the familiar price of rebellion.

Perhaps none of this truly surprised the American moderates, who had spent the past two years reminding everyone how dangerous a war against Britain would be. Yet the experience of war could still shock, even if it did not wholly surprise. The carnage of the first battles of 1775 could be forgiven, or at least discounted, as early tests of strength and applications of the mistaken assumptions that had brought the empire to this sorry state of civil war. The new and destructive turn the war took after July 1776 revealed intention and strategy, the military extension of the British government's judgment that Americans were fit only to be pardoned or punished. Knowing that their desire for reconciliation was sincere, the moderates took this as an affront to their own sense of political morality and a damning indictment of the empire they reluctantly renounced. They may have thought that stupidity and misinformation offered a better explanation of British errors than the ideology-laden belief that evil ministers were mounting a sinister campaign against human liberty. But the results of such errors, as evidenced by the escalating campaign of 1776, were what mattered. For men who had made the improvement and accumulation of property the great goal of life, the war's wanton and needless destructiveness best explained why Americans were right to fight.

Thus moderates like John Jay and Robert Morris, though skeptical about the timing and even the wisdom of the decision for independence, came out of the crisis with their commitment to the cause not only intact but reinforced. That they did so, however, also owed much to the actions and example of George Washington, the commander who faced the daunting task of waging war with an army he could barely keep together.

3

The Character of a General

AS CHAIR OF THE EXECUTIVE committee in Philadelphia, Robert Morris wrote often to George Washington while the hard-pressed commander tried to rally his dwindling forces in late 1776. The letter Morris sent on Christmas Eve was particularly direct. It opened by observing that the chronic woes of the Continental army—short-term enlistments of soldiers and the "appointment of Officers" by the separate state legislatures—had "long been known to such as wou'd open their Eyes." The problem, he noted, was that an ingrained New England prejudice against standing armies had prevented Congress from forming the kind of regular force Americans now needed.

Washington received Morris's letter on Christmas Day, at his Bucks County headquarters ten miles above Trenton, and dictated a lengthy reply. He agreed "that it is vain to ruminate upon, or even reflect upon the Authors or Causes of our present Misfortunes." He worried that "the late Treachery and defection of those, who stood foremost in the Opposition, while Fortune smiled upon us" could infect others, so that "whole Towns, Counties, nay Provinces" might "follow their example." Yet it was still better, he hinted, that "we should rather exert ourselves, and look forward with Hopes, that some lucky Chance may yet turn up in our Favour."

Washington did not reveal that one such "chance" was at hand. The British army's harsh treatment of American prisoners and civilians was finally goading the New Jersey militia to rally against the invaders. General Howe's "ravages in the Jersies exceeds all description," Nathanael Greene wrote to the governor of Rhode Island. "Men slaughterd, Women ravisht, and Houses plunderd, little Girls not ten Years old ravisht, Mothers and Daughters ravisht in presence of the Husbands and Sons." Exposed to repeated ambush, the British and their Hessian hirelings were hunkering down in scattered positions across the state. Reliable intelligence led Washington's staff to sense an opportunity. "We are all of Opinion my dear General that something must be attempted to revive our expiring Credit give our Cause some Degree of Reputation & prevent a total Depreciation of the Continental Money," his adjutant general, Joseph Reed, wrote on December 22. "Our affairs are hasting fast to Ruin if we do not reprieve them by some happy Event." Reed was another Philadelphia moderate, a lawyer and a confidant of Morris who had supported independence only reluctantly. Like Morris, Reed felt free to speak candidly to the American commander, for "the Love of my Country, a Wife and 4 Children in the Enemys Hand, the Respect & Attachment I have to you—the Ruin & Poverty that must attend me & thousands of others will plead my Excuse for so much Freedom."

The council of war that Washington quickly convened endorsed Reed's proposal to attack the enemy garrison at Trenton. Washington let Reed know that the crossing of the ice-strewn Delaware River was set for Christmas night, with the assault to begin an hour before daybreak. "For heaven's sake keep this to yourself," he excitedly added, a caution his adjutant surely did not need. Three days of planning for the operation went for naught on Christmas night as a raging storm drenched the troops waiting to cross the river, where wind-whipped waves added to the peril of floating ice. Two of the detachments in this three-pronged plan failed to cross in numbers adequate to pursue the attack. Only Washington's own northernmost contingent got over, drenched but still spirited—and hours behind schedule. It was past dawn when they deployed around Trenton, but surprise had been preserved. By midmorning the battle was won, and most of three Hessian regiments captured. Had the other crossings succeeded, Washington believed, the entire Hessian force would have been taken, and an even greater victory attained.

The Americans quickly withdrew to Pennsylvania. Then Washington learned that one of the southern contingents had belatedly made it to the Jersey side: Philadelphia militia led by John Cadwalader, a merchant as enterprising as Morris and just as wealthy. Their presence persuaded a skeptical council of war that further opportunities in New Jersey awaited. On December 29, Continental troops began to move again. Trenton was reoccupied, and there a second battle was fought when an enemy force under Lord Charles Cornwallis stormed the town on January 2. In a skillful defense, the Americans repulsed a series of attacks before the enemy finally withdrew.

Again Washington and his staff pondered the aftermath of victory. Staying in Trenton invited fresh assault, and one better prepared for the resistance the Americans had shown they could mount. Withdrawal to Pennsylvania was one option; the other was to seize the initiative and attack again. The decision was taken to slip the American force southeast and then strike north toward enemy depots at Princeton and Brunswick. Approaching Princeton on January 3, the Americans glimpsed a British column marching away toward Trenton. But they had also been spotted. In the ensuing battle, the more numerous Americans prevailed over the first column and a second force that sortied from the usually somnolent college town at the sound of combat. Both sides fought well, but British casualties were far greater.

These winter victories fully merit their storied place in our historical memory. Their military impact was immediate. In mid-December New Jersey was nearly an occupied province. A month later British forces were huddling within a small patch of territory between Brunswick and Perth Amboy. An emboldened militia was again harassing enemy foraging parties, inflicting serious casualties. The political consequences were just as profound. The fear of impending "ruin" that peppered the letters, conversations, and meditations of the autumn suddenly evaporated. The year of independence closed with the cause battered but intact. A new campaign, offering new opportunities, was waiting to open.

While Washington was seizing the initiative, Congress was acting to expand his authority. In a lengthy dispatch of December 20, Washington had expressed his anxiety about the continuing difficulty of maintaining an adequate force in the field. The removal of Congress to Baltimore would make it even more difficult to coordinate its decisions with his. In this crisis, the general wanted authority to do whatever seemed necessary to maintain "the existence of our Army." Some might object that he was seeking "powers, that are too dangerous to be intrusted," he observed, but the obvious answer was "that desperate diseases require desperate remedies." Then he added a personal note he hoped Congress already grasped: "that I have no lust after power but wish with as much fervency as any man upon this wide extended Continent for an Opportunity of turning the Sword into a ploughshare."

Congress read this letter just as the Americans were corralling their Hessian captives at Trenton. It responded with rare alacrity. The next day it adopted a raft of measures to expedite the conduct of the war, culminating in a general resolution declaring its "perfect reliance on the wisdom, vigour, and uprightness of General Washington" while granting him "full, ample, and complete powers" for six months to do whatever he deemed necessary. Loyalists treated this resolution as tantamount to making Washington a "dictator." This term then had a meaning very different from its modern association with murderous tyrants. The prototypical dictator was Cincinnatus, the fifth-century B.C.E. soldier twice called from his farm to repel threats from Rome's enemies. By insisting that he sought only to return to his own plow, Washington thus invoked both the Old Testament vision of Isaiah and the ideals of classical republicanism. John Adams thought the comparison overdrawn. "Congress never thought of making him Dictator, or giving him a Sovereignty," he sputtered. But others recognized it for the tribute it was. "Happy it is for this Country," Morris's executive committee wrote to Washington, that its commanding general "can safely be entrusted with the most unlimited Power & neither personal security, liberty or Property be in the least degree endangered thereby."

The committee was just as flattering in saluting the commander after Trenton. "We rejoice," they wrote, that the victory "will do justice in some degree to a Character we admire & which we have long wished to appear in the World with that Brilliancy that success always obtains & which the Members of Congress know you deserve." But where did this confidence come from? Before Trenton, Washington's sole success was the British evacuation of Boston. But that was more a matter of topography, numbers, and patience than tactical ingenuity. His record in the late-1776 battles in lower New York hardly proved that the former Virginia colonel had the makings of a victorious American general. One defeat followed another, and the survival of the American army was primarily due to General Howe's failure to exploit his victories. Even after the military situation stabilized in 1777, doubts about Washington's capacity for strategic command festered within Congress and among some of his rivals. Now Washington was compared to Fabius (known as *Cunctator*, or Delayer), another Roman dictator who gave his name to the strategy of retreat he employed against Hannibal (and his elephants) in the Second Punic War. Sometimes the Fabian comparison was used to applaud a commander who understood the limits of his resources and the danger of risking all. Yet it could also disparage a general who failed to bring the enemy to a decisive battle. That was the doubt that insinuated itself into John Adams's diary on September 21, 1777, after learning that Philadelphia lay open to British occupation. "Oh, Heaven! grant Us one great Soul! One leading Mind would extricate the best Cause, from that Ruin which seems to await it, for the Want of it."

Yet it is the Cincinnatus in Washington that we continue to venerate, and the Fabius who persists as a footnote for academic debate. Between his great battlefield successes at Trenton and Princeton and then at the 1781 siege at Yorktown, Washington endured nearly five years when political and administrative concerns far outweighed occasions for combat. In the American military tradition he founded, he bears closer comparison to George Marshall, the "organizer of victory" in World War II, than to dashing combat commanders like George Patton. More remarkable, the respect and confidence Washington generated apparently took hold before he had acquired a record of accomplishment. They were conditions, not consequences, of his capacity to command. When admirers such as Reed and Morris wrote to him, as they did before the great stroke at Trenton, they almost seemed to understand the role they expected him to play better than he did himself. Almost—except that the knack of fostering that expectation in others was one of the traits that enabled Washington to pursue the destiny he sought to attain in himself.

One last allusion to antiquity, involving a very different form of dictatorship than that of Cincinnatus or Fabius, also helps explain how Washington gained the thorough confidence of those around him. Washington's managerial skills were those of the proprietors of Roman *latifundia*, the vast agricultural estates, worked by hundreds and thousands of slaves, which developed into an early yet highly sophisticated form of industrial-scale agriculture. Plantations like the storied Mount Vernon were not only the American colonies' closest approximation to that ancient form of production. They also represented the most complex form of economic organization and planning that colonial society yet knew, requiring a supervision of work and workers for which there was as yet no counterpart north of the Chesapeake. As the architect and administrator of the Continental army, Washington demonstrated an attention to detail that only his most talented subordinates, like Nathanael Greene, could begin to rival. No aspect of military administration escaped his scrutiny. But this was less a skill that he had to develop afresh than an extension to army life of habits he had perfected at Mount Vernon. That conscientiousness was an example to everyone he commanded as well as a source of their belief that no one cared more deeply for the army's welfare than its commander. That conviction was repeatedly confirmed throughout the war, but it was never more sorely tested than in the months following the wintry successes at Trenton and Princeton.

The commander whom moderates such as Bob Morris so admired was no moderate himself. Not that Washington was ever part of the network of colonial radicals who manned the political watchtowers of the early 1770s while others slept. He had far more pressing concerns in the management of his estate. As a planter and land developer, he was as much an entrepreneur as Morris was. The two had indirect dealings early in 1774, when one of Washington's agents sought advice about his scheme to charter a ship to obtain German tenants for his lands. Better to obtain them in Philadelphia, Morris suggested, adding "that he would gladly give his Advice and Assistance in procuring them upon the easiest Terms." (Late in 1776, Morris offered fresh advice about a new type of German immigrant. Don't exchange your Hessian prisoners yet, he suggested, before they had seen "the situation & Circumstances of many of their Country Men who came here without a farthing of property" and now lived in prosperity.) Washington's entrepreneurship extended to the management of his lands and the exploitation of the broad Potomac, which flowed just beyond Mount Vernon's veranda. He was one of the first Chesapeake planters to cultivate wheat in place of soil-depleting tobacco. He also built a profitable fishing business that aided his quest to avoid the chronic indebtedness to British merchants that ensnared many planters. The bulk of the labor was done by his own slaves and those he controlled through his advantageous marriage. As a master, he valued efficiency and productivity above humanity. That meant inflicting punishments as well as offering a master's notion of incentives, and he did not shrink from the whip when other inducements failed.

Throughout this period Washington also sat in the Virginia House of Burgesses. But for him public service was more a mark of social status than a vocation. He was a model backbencher who served on the occasional committee but left no paper trail of legislation. This modest record had one notable exception, however. In 1769 Washington actively urged the assembly to endorse the non-importation agreement against the Townshend duties. Writing to his neighbor George Mason, he denounced "our lordly Masters in Great Britain" and asserted that "no man should scruple, or hesitate a moment to use a-ms in defence" of "the liberty which we have derived from our Ancestors." (The missing *r* in *arms* was a convention for conveying a seditious thought without fully spelling it out.) Once the Townshend duties were lifted, his interest in the imperial controversy dissipated. His papers for the early 1770s record his economic ambitions and family concerns, including the desponding death of his epileptic stepdaughter Martha. They are devoid, however, of any reference to politics.

Yet when Parliament decreed its punishment of Massachusetts in 1774, Washington again adopted a militant voice. Accompanied by Martha, he was attending the spring session of the assembly when word of the Port Act reached Williamsburg. He was one of the members who stayed behind, after Governor Dunmore dissolved the assembly, to join in endorsing Boston's call for a general congress. "Not that we approve their cond[uct] in destroy[in]g the Tea," he informed George William Fairfax, the patron of his youth. But "the despotick Measures" against Boston were "the cause of America," and "we shall not suffer ourselves to be sacrificed by piecemeal." He was even more explicit with Fairfax's half-brother Bryan, who wanted Americans to petition first and defer overt acts of resistance. "Is there anything to be expected of petitions after this?" Washington asked. "Does it not appear, as clear as the sun in its meridian brightness, that there is a regular, systematic plan formed to fix the right and practice of taxation upon us?" In July he presided over the Fairfax County meeting that adopted twenty-three militant resolutions drafted by Mason. In August he attended the first provincial convention, which named him to its distinguished delegation to the Continental Congress. (He placed third in the balloting, behind Peyton Randolph and Richard Henry Lee.)

Washington did not figure prominently in the work of the First Congress. Surprisingly, John Adams did not describe him in his diary. His "countenance" seemed "hard," Silas Deane thought, "yet with a very young Look, & an easy Soldierlike Air, & gesture." Both men cited the apocryphal story that Washington had offered to raise a thousand men at his personal expense to march under his command to relieve Boston. But he served on no committees, nor was he recorded as speaking on any point. He did find time to win seven pounds playing cards. The shopping was good too. The "sundries" he bought included a bell, Irish linen, snuff, shoes, ribbed hose, a sword chain, four nutcrackers, six knives for gutting mackerel, and "a chaize for my Mother" costing forty pounds in Pennsylvania currency.

Back in Virginia, Washington resumed his usual concerns. Any foreboding that the sword would soon replace the plow went unexpressed, though hardly unfelt. Nearly two decades had passed since he last took the field. His youthful resentment over his failure to obtain a British commission had long since passed. Nor had he spent the intervening years pining for military command. He knew that his experience leading a homegrown regiment of Virginians was poor training for the command of a national army. No fewer than five independent militia companies that formed in Virginia during the winter of 1775 asked Washington to be their captain. But should their services be needed, he knew that his own would be summoned at a higher level.

That summons came five weeks after Washington returned to Philadelphia for the Second Continental Congress. Years later John Adams recalled that some delegates worried about asking a Virginian to command a New England army. If they did, their doubts soon faded, because another calculation argued as strongly in his favor. "This Appointment will have a great Effect, in cementing and securing the Union of these Colonies," Adams told Abigail, releasing his usual flood of adjectives to praise "the modest and virtuous, the amiable, generous, and brave George Washington." In point of experience, Eliphalet Dyer of Connecticut thought that the Virginia veteran of the 1750s might no longer have the edge over New England officers. But Dyer agreed that his appointment "more firmly Cements the Southern to the Northern," removing the fear that a New England general might use "his Victorious Army [to] give law to the Southern & Western gentry." Dyer also liked Washington's modesty. He was "no harum Starum ranting Swearing fellow but Sober, steady, & Calm."

That was true outwardly, at least. To his "dear Patcy," he worried that this was "a trust too great for my Capacity." But "a kind of Destiny" seemed at work. Whether destined or not, "it was utterly out of my power to refuse this Appointment without exposing my Character to such censures as would have reflected dishonour upon myself, and given pains to my friends." In other letters Washington imagined himself "Imbarkd on a tempestuous Ocean" with no "friendly Harbour" in view. He owed his appointment "to the partiallity of the Congress, joined to a political motive"—the same "cement" factor cited by Adams and Dyer. That political motive was not ambition, as we use the term. Propriety forbade announcing or celebrating one's intentions and aims so directly. True, he did appear in Congress in uniform (though how often is not known). He had kept it well tailored, and if it exposed a slight paunch, it still accentuated his stature and bearing. But he did not need to advertise his availability so blatantly, for he was already the most likely candidate for command. Even the Massachusetts delegates deemed Artemas Ward, the current commander of the provisional army, unsuitable. Two former British officers, Charles Lee and Horatio Gates, were potential rivals, but their English birth weighed against them. Washington attended Congress in uniform not to court its "partiality" but to express the same military enthusiasm that surged across America that spring.

His private qualms were a measure of his belief that a prolonged military struggle was likely. At first he hoped that a successful attack on Boston might bring the war to a quick end. But his army could never launch such an attack, and in March 1776 Washington permitted General Howe to evacuate Boston peacefully. A deeper calculation depended on his assessment of British aims. He had no stomach "for the dainty food of reconciliation" that tempted his moderate admirers. Nor did he think Congress should stoop to "artful declarations, or specious pretences" to state its grievances. It should rather address the royal "Tyrant & his diabolical Ministry" with "Words as clear as the Sun in its Meridian brightness"—the exact expression he applied to the Coercive Acts. He dismissed reports of a peace commission as a chimera. The ministry was "practising every strategem which Human Invention can devise, to divide us, & unite their own People." That opinion was only confirmed when the Howes issued their official proclamation as peace commissioners. Their mission, Washington quipped, was "to dispense pardons to Repenting Sinners."

This commitment to the cause also dictated his personal sense of duty. Now and again, Washington mentioned other courses of action he might have taken. "I have often thought, how much happier I should have been," he wrote to Reed in January 1776, had he simply "taken my Musket upon my Shoulder & entered the Ranks." More fancifully still, he mused that he could have "retir'd to the back Country & livd in a Wig-wam." But there was a catch. Sitting out the crisis in some imagined tent would be possible only "if I could have justified the Measure to Posterity & my own Conscience." Merely stating this condition was to eliminate the choice it was supposed to justify. "It is not sufficient for a Man to be a passive friend & well wisher to the Cause," he wrote to his brother in March. Everyone "should be active in some department or other, without paying too much attention to private Interest." Washington had already acted on that principle by declining any salary as commander in chief. But he did not expect his example to inspire others. The great task of building an army, he quickly realized, depended on appealing to the personal interests of the men—officers and soldiers alike—whom he intended to lead to victory.

Washington reached the army at Cambridge on Sunday, July 2, 1775, his trip "retarded" by the welcome he received in one town after another. He first lodged in the house of the Harvard president but soon moved to the mansion of John Vassal, a loyalist who had fled to British protection in Boston. (In 1843 the house was bestowed on Henry Wadsworth Longfellow and his bride as a wedding present, and there he wrote "Paul Revere's Ride.") Washington had visited Boston only once, in 1756, in futile pursuit of a royal commission. Now he led a makeshift force of volunteers against veteran regiments he described as "the flower of the British troops."

The army he inherited formed almost spontaneously after the engagements at Lexington and Concord. Militia companies throughout New England descended on Boston, as they had during the Powder Alarm of 1774. Officers beat the countryside for recruits, knowing that the number they harvested would affect their rank. Under the direction of the Massachusetts Provincial Congress, these units were hastily formed into regiments, and lines were established to contain the enemy in Boston. Their major test came on June 17, when the British awoke to discover that the colonists had fortified Breed's Hill in Charlestown overnight. Sleepless and unfed, the New Englanders still repulsed two charges and inflicted heavy casualties before lack of powder and the terror of British bayonets forced them to retreat across the isthmus connecting Charlestown, now aflame, to Cambridge. From nearby hills, thousands of civilians watched the horror.

The prevailing images of New England fortitude did not prepare Washington for the unsettling reality he now surveyed. His first impression of New England soldiers was harsh: "an exceeding dirty and nasty people," marked by "an unaccountable kind of stupidity in the lower classes" and a "levelling spirit" that offended his Virginia sensibilities. Convinced, as any veteran officer would be, that obedience to orders was the first rule of discipline, he was put off by the democratic camaraderie between officers and men. He was equally disturbed by "the dearth of Publick Spirit, & want of Virtue" and the "stock jobbing, and fertility in all the low Arts to obtain advantage" that he met in Massachusetts. "There is no nation under the Sun (that I ever came across) pay greater adoration to money than they do." Had he foreseen such behavior, "no consideration upon earth should have induced me to accept this Command." The soaring cost of provisions was also vexing. As Nathanael Greene noted, Washington was unprepared for the commercial spirit of New England, where every farmer fancied himself a merchant. In Virginia he was more familiar with the task of getting enslaved and free laborers alike to do a solid day's work. He may also have been an overly credulous consumer of the plaudits heaped upon Massachusetts since 1774. "His Excellency has been taught to believe the People here a superior race of Mortals," Greene added. Now he had to overcome his "uneasyness that they should take this Opportunity to extort from the necessity of the Army such Enormous prices."

In those first months in Cambridge, however, training and discipline were more urgent concerns than were the cost of supplies. "The abuses in this army, I fear, are considerable," Washington wrote to Richard Henry Lee, "and the new modelling of it, in the Face of an Enemy, from whom we every hour expect an attack exceedingly difficult, & dangerous." Each passing week made him more aware of these difficulties.

Twice within his first fortnight of command, he inserted the phrase "the General hears with astonishment" into his daily orders. The first noted his surprise that "not only Soldiers, but Officers" were "continually conversing" with the enemy at a time when the armies were only a mile apart. On the second occasion he remarked on that same proximity to ask, in amazement, how soldiers and officers could keep requesting furloughs "while the Enemy is in sight, and anxious to take every Advantage, any Indiscretion on our side may give them." The answer, he learned, was that soldiers who had enlisted out of enthusiasm for the cause still honored the New England tradition whereby military service was viewed as decidedly contractual in nature. Having enrolled for short terms, they believed they had a legal right to leave camp when these expired—regardless of how close the enemy might be.

Washington's disparaging comments of 1775 testify to the palpable cultural differences that distinguished the Chesapeake from New England. Colonial society was still provincial in fact and as yet American only in name. Beneath his complaints, however, lay a striking ambition. From the start of his command in 1775 until its conclusion eight years later, he idealized the army as a national institution and a nationalizing project. His second set of general orders, issued on July 4, informed his soldiers that "They are now the Troops of the United Provinces of North America: and it is hoped that all Distinctions of Colonies will be laid aside." His first mission, as he saw it, was to disband the provincial army he had inherited and create a truly national force through new recruitment and intensive training. That formal transformation took place on January 1, 1776, when the old regiments were dissolved. "This day giving commencement to the new-army, which, in every point of View is entirely continental," Washington declared in his general orders, "The General flatters himself, that a laudable Spirit of emulation, will now take place, and pervade the whole of it." If provincial differences persisted, as they must, they should prove neither a source of jealousy nor recrimination but rather an inspiration for a higher, truly American standard of conduct.

Taking leave of the army in 1783, Washington reminded his soldiers what a bold project this was. History had never seen anything like these events, "nor can they probably ever happen again. For who has before seen a disciplined Army form'd at once from such raw materials?" the general asked. "Who, that was not a witness, could imagine that the most violent local prejudices would cease so soon, and that Men who came from the different parts of the Continent, strongly disposed, by the habits of education, to despise and quarrel with each other, would instantly become but one patriotic band of Brothers?"

That had been the hope and ideal, and in victory it assumed the trappings of memory and myth. The reality had been otherwise, and with every passing month Washington's plans became more calculated and pragmatic. By the spring of 1776 he knew that the problems he faced ran well beyond dealing with New Englanders' avarice. The difficulties were systemic, inherent in the situation: communities, provinces, and the Continental army were competing to recruit and retain soldiers, and Americans were grappling with the unparalleled demands and burdens that the war imposed.

Three main concerns dominated Washington's understanding of those demands. The first, most urgent, and most persistent revolved around the recruitment, training, retention, and supply of the army. The second, perhaps most principled, governed his relations with Congress and other civil authorities. His third set of concerns lay with his own character and reputation, and their place in the cause of national independence.

Day in and day out, year in and year out, his most onerous tasks were to enlist, train, and maintain an effective fighting force. From the outset he urged Congress to recruit soldiers for the duration. They would form a "standing army"—a term fraught with dangerous connotations in the American vocabulary. The short-term enlistments of 1775 and 1776 retarded every effort at instilling the requisite discipline, much less the experience, that enabled veterans to stand and fight when "Raw, and undisciplined Recruits" would break and run. To get men to reenlist, it was often necessary to "relax in your discipline, in order as it were to curry favour with them," thereby "undoing" everything you had labored to accomplish. There were other costs. Men whose terms were expiring grew careless with their equipment. The cost of provisioning the militia who were summoned to fill gaps in manpower outweighed the scanty services they rendered before they restlessly broke for home. The only solution was to offer adequate bounties to induce soldiers to enlist for the duration. But Washington's first serious effort in February 1776 to persuade Congress of that point led only to a spirited but inconclusive debate. Most of the soldiers whom he commanded in 1776 were eligible to leave the army by year's end—and many of them did.

Washington frequently appealed to his men as patriots who were defending "Life, Liberty, Property and our Country" against mercenaries who "are fighting for *two pence* or *three pence* a day." Yet he knew better than to think this was enough. "A Soldier reasoned with upon the goodness of the cause he is engaged in and the inestimable rights he is contending for, hears you out with patience, & acknowledges the truth of your observations," he informed Congress in September 1776. But the same soldier would respond "that it is of no more importance to him than others." An officer would reply similarly, adding only "that he cannot ruin himself and Family to serve his Country, when every member of the community is equally Interested and benefitted by his Labours." Washington never assumed that the patriotic fervor of 1775 would survive a protracted struggle. To expect anything other than "interest" to motivate ordinary soldiers "is to look for what never did, and I fear never will happen." Men would accept the discipline and leadership needed to make them effective soldiers only if they were given material incentives—freehold land in the amount of 100–150 acres being the most attractive.

Similar concerns applied to officers. They could not be asked to follow Washington's example and serve without salary. Yet Washington thought about officers and their men in notably different ways. With ordinary soldiers he was primarily concerned about the *conditions* of their service. With officers the critical issue—indeed the key word—was *character*, which he in turn associated with the status of *gentlemen*. He used these terms interchangeably and synonymously. He wanted "Gentlemen, and Men of Character to engage" in the cause, and to give them "such allowances as will enable them to live like, and support the Characters of Gentlemen." Or again, the goal was to "induce Gentlemen of Character, and liberal sentiments to engage" and then to acquire "the Character of an Officer." He did not assume that only men of property could meet his standard, for there were not enough to go around. But poorly paid officers would persist in "the Sloth, negligence, and even disobedience of Orders" that caused their commander "inexpressible trouble & vexation."

When Washington conveyed these thoughts to Congress in the early fall of 1776, they reinforced his warning that the army was on the verge of dissolution. Congress in fact had just conceded his basic point, resolving to recruit eighty-eight battalions to serve for the duration. Washington was not in the happiest mood when, from Harlem Heights on October 4, he answered President Hancock's letter enclosing this resolution. His orders for that day lambasted his troops yet again for their "shameful inattention" to removing "the Offal and Filth of the Camp." Again warning that the army was near "political dissolution," Washington patiently explained why the new plan was both inadequate in the terms it offered officers and soldiers alike and overly ambitious. "No time is to be lost in making fruitless experiments," he wrote. "If we have an army at all," he concluded, "it will be composed of materials not only entirely raw, but if uncommon pains is not taken, entirely unfit."

Washington wrote this letter in his own hand, in as firm a tone as he ever used with Congress. But however much he chafed at the restrictions under which he labored, he never scorned the authority of Congress. From the beginning of his command he accepted the principle of civilian supremacy over the military that has always been an axiom of American political thinking. His long experience in the Virginia assembly exerted greater influence over his political values than did the military service of his youth. Most of his correspondence with Congress went through the proper channel to its president, not to individual delegates. He carried on similar exchanges with state governors, prompting, urging, and wheedling them into providing the men and supplies the army needed.

At the same time, Washington accepted that members of Congress had their own sources within the army. "A free correspondence between the Members of Congress and the Officers of the Army," John Adams wrote to Colonel Daniel Hitchcock, would provide "Advantages to the Public by improving both the Councils and Arms of America." Many officers used such channels to complain about being slighted in matters of rank. But others reinforced Washington's concerns with a pungency he could not deploy in his official letters. Adams, for example, corresponded regularly with a squad of well-placed New England officers, all of whom vigorously supported their commander's ideas about discipline, recruitment, and pay. The best defense Adams could then plead was that although he agreed with the officers, a majority in Congress as yet did not. Nor did it seem likely that its members would soon change their minds, Adams wrote to Henry Knox, the stout former Boston bookseller and now the chief of artillery, on August 25, 1776—unless Washington and other officers made the point more strongly "or unless two or three horrid Defeats, Should bring a more melancholly Conviction."

The next night General Howe did his best to supply the evidence Adams lacked, initiating the Battle of Long Island, which brought the Americans their first calamitous defeat. The ensuing four months were the severest test of character that Washington ever experienced. The challenge began with the rout of American troops on Long Island on the night of August 26, 1776. A long flanking movement around the left of the overextended American lines produced complete surprise. Though a few units fought valiantly, the American forces shattered into a desperate retreat. Its only redeeming feature was that the Americans outran their pursuers, who gave not quarter but rather the bayonet to the wounded or those who sought to surrender.

The Americans regrouped on Brooklyn Heights. The Howe brothers began a cautious siege, gradually advancing their lines toward the well-fortified American positions. With British ships capable of blocking retreat to Manhattan, the American force risked isolation and capture. On the twenty-eighth and the twenty-ninth, a heavy storm drenched both armies and turned the landscape to mud. But the miserable weather also created an opportunity to slip the army back over the East River to Manhattan. Most of the troops crossed that evening. Those left behind, vulnerable to attack at daylight, were saved when a thick fog arose to cover their crossing.

This first major defeat was a grim portent for the next three months. Though individual units often fought ably and valiantly, weak spots in the American positions typically gave the edge to the enemy, with its superior firepower, discipline, and leadership. The advantage in leadership perhaps did not extend to the highest level of command. Beginning at Long Island, British officers repeatedly faulted General William Howe for failing to pursue the advantages they were gaining on the battlefield. But little in Washington's defense of New York would have persuaded an impartial observer that the Howe brothers had met their match. In engagement after engagement, the principal question was not who would prevail but whether the bulk of the Continental forces would manage to escape.

After a fortnight's respite the British renewed operations on the night of September 14–15. An intense naval bombardment and amphibious landing at Kip's Bay (where Thirty-third Street now meets the East River) shattered the nerves of the defending Connecticut troops. The British occupied New York City at the southern tip of Manhattan while the Americans regrouped at Harlem Heights to the north. "His Excellency: the General was enraged," Philip Vickers Fithian, a New Jersey chaplain, wrote in his journal. "It is said for a short Time he was exceeding angry." The pious Fithian read the event as a providential judgment that raised a more troubling question. "We are a sinful Nation, O Lord," he confessed. "But is it written in thy Book concerning us that we must always fly before our Enemies?" A month later Fithian, just twenty-nine, was dead of the dysentery that ravaged American camps, immune to Washington's strictures on sanitation. But his lament lingered in the mind of his former commander. At the same time, Washington sought to implant an equally powerful question in the conscience of his troops. "How much better will it be to die honorable, fighting in the field," he asked (or rather urged) the army in his orders of October 13, "than to return home, covered with shame and disgrace; even if the cruelty of the Enemy should allow you to return?"

On the battlefield, though, the choice was rarely that simple. When the enemy renewed operations in mid-October, Washington and his council of war agreed to abandon Manhattan, lest the British cut the northern connection to the mainland at King's Bridge. Washington conducted a skillful withdrawal north to White Plains, and his troops displayed skill and fortitude in repulsing renewed British assaults. Over his own misgivings, however, he left a garrison at Fort Washington (just north of the modern George Washington Bridge, on the Manhattan side of the Hudson). This became Howe's next objective. On November 15 word came that the enemy were massing for an assault. Washington first hurried to the American position at Fort Lee, directly across on the New Jersey side, then started east across the river to assess the threat for himself. Mid-river he met another boat carrying two of his generals, Nathanael Greene and Israel Putnam. Both assured him that Fort Washington could hold. Washington returned with them to Fort Lee. The next morning they recrossed the Hudson and were still within the overextended American lines as the British assault began. His subordinates persuaded Washington to return to Fort Lee. From there he watched first in dismay and then in horror as the British overwhelmed Fort Washington. American losses neared three thousand, including surrendering soldiers to whom "the cruelty of the enemy" again gave no quarter.

The sight, one later account suggested, left Washington in tears, a display of feeling very different from the occasional outbursts of temper for which he was also known. It is perhaps noteworthy that Washington made no mention of having been present at any point at Fort Washington when he reported its loss to President Hancock or even in a personal letter to his brother. To risk capture by the enemy would have been foolish. Yet to leave a post to which he had committed substantial troops against his better judgment, and then to see his captured men slaughtered, was a tormenting ordeal. It was his own indecision, rather than a bold stroke by his opponent, that led to this needless sacrifice.

The loss of Fort Washington made the retention of Fort Lee "of no importance," and Washington ordered Greene to remove the supplies and artillery stored there. Greene dithered in his task and failed to identify all the potential approaches to his fort. On the night of November 20 a British force scrambled undetected up a steep, rain-drenched path along the Palisades and then occupied the fort, meeting little resistance. This time American casualties were few. But the loss of sorely needed cannon and thousands of cattle only aggravated Washington further.

A fortnight later, the American commander was deploying his ever-dwindling army along the western bank of the Delaware River separating New Jersey and Pennsylvania. Had the Howes not sent Sir Henry Clinton to occupy Rhode Island as a source of provisions, they could have pursued the campaign to take Philadelphia. To prevent that, New Jersey militia and ships of the Pennsylvania navy scoured the eastern side of the Delaware, transferring every boat they found to the Pennsylvania shore. Their thoroughness had a dual benefit. It not only fed Howe's reluctance to press the campaign with winter near but also laid the groundwork for the counterstroke Washington longed to launch.

Washington worried, though, that this was only a stopgap measure. "The enemy wait for two events only" to take Philadelphia, he warned Morris: "Ice for a Passage, and the dissolution of the poor remains of our debilitating Army." The political situation seemed equally bleak. New York, New Jersey, and Pennsylvania were all rife with "disaffection"—the catchall term that covered not only outright disloyalty but also the indifference or disillusionment with which too many Americans now viewed the war. Writing to family members on December 15, Washington worried that "the game is pretty near up"—or soon would be "if every nerve is not straind to recruit the new Army."

Yet even this brief flash of despair was more prospective than immediate, more concerned with the next campaign of 1777 than the expiring one of 1776. Five days later, Washington wrote the famous letter that led Congress to grant him dictatorial powers. Urgent as his situation was, with only untrustworthy militia to rely upon "Ten days hence," his focus was already set on the next campaign. Trenton and Princeton were the brilliant tactical strokes that converted that campaign from hope to certainty. New Jersey, though still the seat of active fighting, was largely recovered, while Philadelphia was spared. Whether its occupation would have mattered militarily was irrelevant. Washington rightly feared that loss of the capital, coupled with British control of New York City and much of New Jersey, would have been politically disastrous.

What is equally striking about his leadership during this period was his capacity to look beyond his daily struggle against impending catastrophe to the restructuring of the army for future campaigns. This was the repeated motif of his dispatches to Congress and to other leaders. It also resounded in the rare private letters that best reveal his state of mind. In early fall he complained to his cousin Lund Washington about the obstacles he faced: "I see the impossibility of serving with reputation, or doing any essential service to the cause by continuing in command, and yet I am told that if I quit the command inevitable ruin will follow." That was the unhappy price he paid for the plaudits that Morris and others showered upon him: to submit to the bonds of duty while forfeiting the opportunity for "reputation"—read "glory"—and all this because of the "unaccountable measures" Congress had adopted against his advice. "A pecuniary rew[ar]d of 20,000 £s a year would not induce me to undergo what I do," he grumbled in early November, "and after all perhaps to loose my Character."

Yet this rare venting of inner turmoil seemed to suffice. "But I will be done with the subject," he wrote to Lund, "with the precaution to you that it is not a fit one to be publicly known or discussed." Washington had neither the time to sulk nor the temperament to brood. We might say he had the ability to compartmentalize—as when, writing to Lund, he abruptly turned from an account of the loss of Fort Washington to discuss the trees he wished to have planted at Mount Vernon. But the neologism *compartmentalize* conveys neither the discipline nor the complexity of his mind. However mixed his record as a battlefield tactician, he was already a skilled planner who grasped the multiple levels on which the war would be fought. Here, again, we cannot overlook the latent connection between the ambitious peacetime entrepreneur and the wartime commander. His capacity to plan the recruitment of an army for 1777 when the force he commanded in 1776 was evaporating before his eyes arguably provides the single most telling clue to the confidence invested in him by other leaders, such as Morris, who also grasped the enormous challenge that lay ahead.

Washington ended the year of independence by publicly urging the New Jersey militia to prove their love of country "by boldly stepping forth and defending the Cause of Freedom." In early 1777 they did so far more effectively than he could have foreseen. The British had to keep scouring the countryside, seeking forage for their livestock. In one encounter after another, the New Jersey militia and the Continental units whom Washington sent to support them inflicted serious casualties that regularly exceeded their own losses. These engagements gave regulars and militia alike valuable experience, while teaching field officers much about tactics and leadership.

From the victory at Princeton, Washington took his dwindling force into winter quarters at Morristown, secure from assault in the hilly terrain of northern New Jersey. While monitoring the fighting, he focused his attention on recruiting and training an army for the campaign ahead. Though he applauded the New Jersey militia for their effectiveness, his underlying doubts about their value remained. The problem was not that militia scattered at the first whiff of powder. Rather, they came and went at will, consuming badly needed supplies without fully earning their keep. Providing them with weapons, he complained to Hancock, "scatters our Armoury all over the World" because officers allowed "their men to carry home every thing that is put into their hands." To rely on them over an entire campaign was equally foolish. The only safe assumption about the militia, *"whose ways, like the ways of Providence, are, almost, inscrutable,"* was that they "are here today, & gone tomorrow."

The greater concern was the recruitment of regular soldiers. The war had created a market for soldiers, sparking competition for manpower between the Continental army and the states, which were recruiting their own units, and even with the British. Howe's offer of a three-pound bounty (four pounds with a weapon) to deserters attracted a worrisome number of American soldiers. In the early months of 1777, recruitment progressed slowly, an early indication that the plan for an eighty-eight-battalion army was wildly optimistic. A mid-April report to Congress turned into a disheartening summary of the recruiting results, state by state, and a lament over the "total depression of that military Ardor" among his officers, even with the pay increase Congress had given them. Then the tide suddenly turned. By late May Washington had roughly ten thousand men in camp, mostly new recruits in need of training.

The Americans would thus be able to field some kind of army, newly trained but substantially reinforced, for the 1777 campaign. But how and where they would fight was for the enemy to determine. Two British armies would conduct the 1777 campaign: Howe's, and a new Canadian army under General John Burgoyne. As seen from the London offices of Lord George Germain, secretary of state for America, the grand strategy was simple: to gain control of the Hudson River axis and thereby isolate New England, the supposed hotbed of rebellion, from the other states. The view from New Jersey greatly complicated this strategy. Since December Howe had been committed to an invasion of Pennsylvania. Once he learned that he would receive only minor reinforcements, he abandoned a land approach across New Jersey and instead opted to rely on his brother Richard's naval transports. This decision had fateful implications for the second prong of British strategy. Burgoyne was to move south from Quebec, making Ticonderoga and Albany his objectives. Germain and Burgoyne expected Howe to lend support by pushing up the Hudson and by harrying southern New England. Germain thought that Howe could pursue his Pennsylvania plans and still have time to aid Burgoyne. But that belief rested on two conditions: that Howe begin the campaign quickly and prosecute it vigorously, and that he have forces ready to aid Burgoyne whenever opportunity arose or necessity required. Howe failed to do either.

Through the spring Washington believed that Howe's objective remained Philadelphia, the capital to which Congress returned in March.

But that was still a matter of conjecture. "What Mr. Howes present Plan is, no Conjurer can discover," John Adams marveled, "no Astrologer can divine." On June 13 Howe made a foray toward Philadelphia in hope of drawing the Americans into open combat. Undeceived, Washington rested content with harassing operations against the British flank. After five fruitless days Howe withdrew to Brunswick. To Washington's surprise his stay there was brief. On June 22 the enemy headed back toward the Jersey shore, "Robbing, Plundering, & burning Houses as they went." On the twenty-eighth, after one last feint inland, they abandoned New Jersey to regroup on Staten Island.

Summer had come, and with it Howe's thoughts finally turned to war. (Not that he ignored other seasonal feelings, enjoying an amorous liaison with Elizabeth Loring, the beautiful young wife of a loyalist officer whom he conveniently sent away as commissary of prisoners.) Within days his men began the tedious process of loading supplies and livestock aboard the transports that would carry them to Pennsylvania. They had no idea of their destination, nor would anyone else, other than senior British officers, for weeks to come. On July 17 Howe learned that Burgoyne had taken Fort Ticonderoga, the key American post guarding southern Lake Champlain. But Ti, as Americans called it, did not really fall. It was simply abandoned when the British surprised the garrison by hauling heavy cannon up the slopes of Mount Defiance, which overlooked the fort. To Howe this success indicated that Burgoyne should easily take Albany. In that case the bulk of his own forces could proceed with the invasion of Pennsylvania.

Once loaded, the transports dropped down to Sandy Hook, the beachy extension of the Jersey shore. Contrary winds detained them another week. On July 23, this impressive armada of 260 vessels finally set out eastward and dropped from American view. A sighting on August 1 off the Delaware Capes suggested it was bound for Philadelphia; then it disappeared again. On the tenth the fleet was seen off the Delmarva Peninsula, perhaps heading for South Carolina. Another week and a half passed before Washington learned that Howe was sailing up Chesapeake Bay. By now Howe's men had kept cramped shipboard quarters for more than a month—"long enough to have gone to London," John Adams observed. The same thought must have struck the enemy troops sailing past England's first colonial plantations. Their officers were even more critical. They simply could not understand why

Howe was pursuing his Philadelphia whim while ignoring the imperative advantage of linking with Burgoyne.

Washington shared that assessment of his adversary's strange strategy, but he had to keep his own army on the move to parry Howe, whenever and wherever he landed. Assuming Howe would sail up the Hudson to seek a juncture with Burgoyne, he first marched his army northward. The sighting of August 1 sent them back toward the Delaware River. Howe's next disappearance suggested Albany was again the object, so it was back to the Hudson. The next sighting, on August 10, sent the soldiers once again to the Delaware. There they remained, recovering from blisters and resting from the summer heat, while their perplexed commander waited to learn Howe's intentions.

While Howe was wasting weeks at sea, Burgoyne was meeting and creating delays of his own. His six-thousand-man army—half British regulars, half Brunswick mercenaries—came heavily accoutered, with more artillery than they could use, a few dozen wagons for Burgoyne's wardrobe and larder, and a German marching band, complete with trumpets. Burgoyne had a mistress along for diversion, while the German commander, Baron Riedesel, brought the fetching Baroness Frederika and their three toddlers. Having reached Ti by early July, Burgoyne was only seventy miles from Albany. Rather than continue down Lake George, he sent his army slogging toward the Hudson, twenty-three densely forested miles to the west. With Americans felling heavy trees in their path and battalions of patriotic mosquitoes feasting on the road builders, it took twenty-four days to reach the east bank of the river. Further dithering worsened the shortages of food and forage, which threatened the army's progress. In mid-August a German detachment sent to seek supplies at Bennington, Vermont, took nine hundred casualties from the American militia.

Washington closely followed affairs in the Northern Department. The disposition of Continental units depended on both Burgoyne's progress and the mystery of Howe's destination. On August 20, when Charleston appeared to be the British objective, Washington informed Congress that he would not hazard taking his own army southward on an exhausting march during the unhealthy conditions of summer. A council of war agreed that the army should instead return to the Hudson to act against Burgoyne. That raised a delicate issue, however. Congress had divided command of the northern troops between Philip

Schuyler, the scion of a prominent Albany family, and Horatio Gates, the former British officer. The ensuing rivalry was exacerbated by the deep suspicion with which New Englanders viewed Schuyler, a wariness he reciprocated. Rather than intervene personally in this quarrel, Washington sent his talented young aide, Alexander Hamilton, to Congress with the request that it first endorse his efforts to mediate the dispute.

Howe's appearance in the Chesapeake mooted this issue. Once the British landed at Head of Elk, the far northeastern inlet of Chesapeake Bay, Washington knew he had time to array his troops to defend Philadelphia. Seven weeks shipboard had left Howe's men anxious for action but less than fit. Their draft animals were in worse shape. Washington instructed militia from Maryland and Delaware to "distress" the enemy by removing all the "articles of subsistence" they could, right down to stripping gristmills of "the Runners" on which the grindstones revolved. If possible, the militia should "hang constantly" on Howe's flank and rear if he marched toward Philadelphia, rendering "all the annoyance in your power." Washington meanwhile put his army in motion to take up a position between Howe's forces and Philadelphia.

The army that paraded through Philadelphia on Sunday, August 24, hard on the heels of a crackling thunderstorm, bore no resemblance to the small force Washington barely held together after Princeton. With secret aid now coming from France, it was, John Adams thought, "extreamly well armed, pretty well cloathed, and tolerably disciplined." By "tolerably," Adams meant this: "They dont step exactly in Time. They dont hold up their Heads, quite erect, nor turn out their Toes, so exactly as they ought." They had the characteristic slouch that would always distinguish an American army from its Old World counterparts—the "Willie and Joe look" that the great World War II cartoonist Bill Mauldin later captured. But it took them two hours to pass through town, in ranks twelve deep, and that counted for more than the fact that "They dont all of them cock their Hats." Adams thought Philadelphia *could* be successfully defended; whether it *should* be was another matter. If Howe did occupy the capital, Adams wrote to Abigail, it would only make it easier for the Americans to prevail elsewhere. And prevail they would, Adams and most members of Congress believed. By late August, they were confident "that fop Burgoyne" would be lucky to retreat to Canada with even "half his troops." If Howe could also be checked, Henry Laurens added, "I hope in a few Weeks you will learn that these blustering Heroes have ended the Campaign by escaping with fragments of their pretended omnipotency."

Washington arrayed his army behind Brandywine Creek, west of Philadelphia, cautiously waiting for Howe to "unfold his true Designs, which I trust we shall be able to baffle." In this he was disappointed. On September 11 Howe launched his attack, sending Hessians under General Knyphausen to feint against the American left flank at Chadds Ford while he and Cornwallis swung well north in a bid to turn the American right, using fords over the Brandywine of which Washington was unaware. Washington was slow to ascertain Howe's intentions, but Howe also took hours to ready his assault. As a result the main engagement began only in midafternoon. The combat was intense, with attack and counterattack occurring amid dense and rapid volleys from both sides and fierce bayonet charges. On both lines of attack the enemy ultimately prevailed. By early evening the Americans were retreating toward Chester, southwest of their capital. British exhaustion and Howe's caution saved the Americans from greater defeat, though their casualties were roughly twice the enemy's. As in his New York campaign a year earlier, Howe proved a superior battlefield tactician. Yet the critical fact remained that this newly trained, untested American force had fought for hours, retreating not in panic but only after an experienced enemy gradually took control of the fighting.

Far from broken, the American army and its commander looked forward to renewed combat. A major engagement nearly occurred on September 16, as the British detected the Americans near Goshen in western Chester County. Just as it began, however, a torrential storm broke, soaking munitions and bogging down men, animals, artillery, and wagons in the muck. Eighteen drenching hours later, the Americans trudged off along muddy roads to gain fresh ammunition. Washington now had to choose between protecting his supplies at Reading or blocking Howe from Philadelphia. There was no real issue to debate. The army could survive the loss of the capital but not its supplies.

In a rare Sunday meeting, on September 14, Congress resolved to adjourn to Lancaster if necessary. Late on the eighteenth President Hancock received an urgent note from Hamilton, whom Washington had sent to destroy flour mills along the Schuylkill River. Surprised by British troops, Hamilton and his men came under fire as they grabbed a flatboat and fled across the river. Hamilton's horse and a soldier were killed, and two boats, large enough to ferry fifty men at a crossing, were lost. Philadelphia is "no longer a place of safety for you," Hamilton wrote to Hancock. Duly warned, the delegates left town in the wee hours. John Adams was awakened at 3 A.M. and told that other delegates had already departed around midnight.

Regrouping with the other members of Congress at Bristol, New Jersey, Adams wondered whether Hamilton had not given a "false alarm." He was again questioning whether American commanders possessed the initiative required for victory. His doubts covered both Washington and Gates, whom Congress had placed in sole command on the northern front after holding Schuyler accountable for the loss of Ticonderoga.

Howe, as usual, lacked any sense of urgency in making his next move. Few British officers thought he possessed "the active masterly Capacity" for which Adams pined in an American commander. But the road to Philadelphia was now open, and on September 26 Cornwallis led three thousand troops into the capital. Within days that feat was offset by reports from the north. On September 13 Burgoyne's army had crossed to the west bank of the Hudson. Retreating to Canada would have been wiser, because Gates now had Continental reinforcements and a flood of New England militia, fresh from the harvest, at his disposal. At Freeman's Farm on September 19, Burgoyne's men came under intense fire from Daniel Morgan's North Carolina riflemen. With casualties double those of the Americans, Burgoyne's army was evidently near defeat. For the next four weeks its men camped out in the open—hungry, cold, and soaking wet, under continuous harassing fire, with carcasses of draft animals rotting around them and their retreat cut off by the dense militia blocking access to the Hudson.

Washington did not need this news to prod him into planning an attack on Howe. He was right to assume that Howe might be vulnerable and that an assault could be launched, as his council of war recommended, as soon as "a further Reinforcement" and "favourable Opportunity" permitted. By October 3 he felt that opportunity had arrived. Invoking Gates's success, he challenged "This Army—the main American Army" not to "suffer itself to be out done by their northern Brethren." That night the army set out on its first real attack since Princeton, nine months earlier. Its objective was the smaller half of the British force that Howe had placed at Germantown, eight miles above Philadelphia. Numbers favored the Americans. But sturdy stone houses and fences gave the defenders cover, while numerous fields and crisscrossing roads made it difficult to execute Washington's complicated plan of attack. The battle was fought in a heavy fog made denser by smoke from gun and cannon. Washington made a crucial error when he halted his advance to roust a British company holed up in the mansion of Benjamin Chew. Its thick walls proved impervious to Henry Knox's cannon, and the infantry who managed to reach its doors and even clamber inside left their corpses "stretched in the doorways, under the tables and chairs, and under the windows." The delay enabled the British, now reinforced by Cornwallis, to regroup and push the increasingly disorganized Americans to retreat.

American officers did not regard this setback as a defeat. The "almost midnight darkness" that obscured the battlefield loomed as large as any tactical mistakes in their analysis of what had gone wrong. Had they regrouped quickly, a second assault launched before the British set up effective defenses around Philadelphia might still have succeeded. But the opportunity passed and the occupation of the capital began in earnest. To maintain it, the British had to bring their supplies up the Delaware River, past several forts and the massive barriers known as chevaux-de-frise that the Americans had anchored in the riverbed, with iron-tipped timbers capable of slashing open a ship's bottom.

In late October Howe shifted his attention to opening the Delaware, and Washington began to respond accordingly. That response was framed in the wake of Burgoyne's surrender to Gates on October 17. That decisive event promised a new opportunity to press Howe. At month's end Washington dispatched Hamilton on a delicate mission to Albany, to explain "to Genl. Gates the absolute necessity that there is for detaching a very considerable part of the Army at present under his Command to the reinforcement of this." If he did, Howe could soon find himself in Burgoyne's situation. But do not press the request, Washington added, if Gates was contemplating a fresh offensive of his own. No such initiative was planned, Hamilton learned in a tense interview on November 5. Yet Gates would release only one of the three brigades Washington requested. To send more, he told Hamilton, would expose New England "to the depredations & ravages of the enemy." Hamilton knew better, but his precocious political acumen dissuaded him from pressing the case. The "intire confidence" that Gates enjoyed in New England, Hamilton wrote to Washington, along with his "influence and interest elsewhere"—meaning Congress—rendered it "dangerous to insist" on a measure that Gates "so warmly opposed."

The very fact that Hamilton was sent not to convey an order but to negotiate with a fellow commander illustrates a central ambiguity in the American war effort. Being commander in chief of the Continental army did not place Washington in a position to peremptorily order the forces about as he wished. Congress, not its leading general, was the ultimate source of command authority. The surprise and daring Washington had shown at Trenton and Princeton renewed the reservoir of support he enjoyed among the delegates. But detractors and skeptics were also present in Congress, and until or unless Washington could gain victory in a fully engaged combat, their doubts would remain.

Gates's attitude and celebrity rankled with Washington and his subordinates. "General Gates has immortaliz'd himself," Nathanael Greene wrote archly, while noting that it was the unpopular Schuyler who actually laid the groundwork for victory. "Great credit is due to all the northern army, but that army has been much stronger than ours and a far less force to contend with," he added. Had southern militia given Washington the same support Gates gained from New England, Greene thought, the British would never have been able to occupy Philadelphia. By the time significant northern reinforcements reached Washington, Howe's effort to open the Delaware had succeeded. It had been hard work, first to clear the chevaux-de-frise, then to mount a prolonged artillery bombardment of the key American position, Fort Mifflin, on the aptly named Mud Island, just below the point where the Schuylkill flowed into the Delaware. But by late November, the river was open, and Philadelphia safe for a continuing occupation.

A speedier arrival of reinforcements, Washington claimed, would have saved Fort Mifflin and thereby "rendered Phila. a very ineligable situation for them this Winter." Perhaps, or perhaps not. Whatever aggressive desires he still harbored were offset by the prudent advice of his subordinates. On November 24 Washington held a council of war to hear an ambitious plan, prepared by John Cadwalader, for a direct attack on Philadelphia. The council was decidedly cautious in assessing its prospects and equally subtle in weighing a broad array of political and strategic concerns. The most candid and balanced assessments came from the two generals Washington trusted most, the New Englanders Greene and Knox. Greene was keenly alert to Washington's concern with reputation, both in Congress and among "an ignorant & impatient populace," as he bluntly put it. But that could never warrant ignoring the sage judgment of "all military Gentlemen of Experience" as well as "your own mind." For his part, Greene concluded, "The cause is too important to be trifled with to shew our Courage, & your Character too deeply interested, to sport away upon unmilitary Principles." Knox took a different tack on the matter of reputation, reassuring his commander that "the people of America look up to you as their Father and into your hands they intrust their *all* fully." No one versed in the science of war—and Knox, the former bookseller, was a well-read soldier—could endorse a direct assault against the secure redoubts that now sheltered the enemy.

The critical campaign of 1777 thus ended with the British in possession of the American capital and the Americans in possession of an entire British army. There was no question which side got the better of the exchange. Philadelphia had no strategic value; nine months after its occupation, Congress would again meet in the State House there. By contrast, Burgoyne's surrender at Saratoga led directly to the signing of the Franco-American alliance in February 1778. Through its navy, Britain retained the strategic initiative: it could always decide when to fight, and where, and its soldiers could always be expected to fight effectively. But once France entered the war, Lord North's government could no longer afford to commit to distant North America anything like the force necessary to subdue the rebellion. North himself despaired of victory—though Lord Germain and their sovereign king did not.

As 1777 drew to a close, Washington and his staff had to decide how to dispose the army for the winter. One option was to pursue a winter campaign to dislodge Howe from Philadelphia—an alternative that Washington's generals had already dismissed but some in Congress still favored. A second was to take winter quarters in Lancaster and Reading, safely west of Philadelphia, where the army could prepare for the next campaign. This was the logical choice for an eighteenth-century army fighting a conventional war. But the Revolution was not such a war. Maintaining an army capable of fighting another year was only one element of attaining victory. It was equally essential to retain the loyalty of the civilian population, which was already showing signs of war weariness. This was especially the case in Pennsylvania, with its significant pockets of loyalist sentiment and its pacifist Quaker community. The radical constitution the state had adopted in 1776 had created further divisions, which left its government insecure and vulnerable. Withdrawing the army into the interior would expose the Revolution's supporters to British impositions and loyalist reprisals that could weaken their allegiance to the cause. That concern was confirmed on December 4–5, when enemy forces sallied north toward the entrenched American positions around Whitemarsh. After Washington declined to be drawn into open combat, the British burnt nearby villages and again ransacked civilian property as they withdrew to Philadelphia.

Encamping near Philadelphia, on terrain secure from assault but close enough to monitor Howe, was the third option and the one that Washington and his generals finally endorsed. Whether the presence of the army would strengthen or sap the loyalties of civilians was a conundrum. Against the urgent need to "cover the country" against the British, Washington had to weigh the burden that his soldiers would impose on the local population. During the fall Congress again gave him the authority to commandeer the supplies the army required—a revival of the so-called dictatorial powers he had received in 1776. This was an authority he was reluctant to wield. "I have felt myself greatly embarrassed with respect to a rigorous exercise of Military power," he wrote to Henry Laurens, the new president of Congress, in mid-November. "A reluctance to give distress may have restrained me too far." But he was also "well aware of the prevalent jealousy of Military power," an attitude deeply rooted in Anglo-American law and history. While prepared to impress vital supplies when necessary, he wanted the state legislatures to take the lead in filling his requisitions. But in Pennsylvania the needs of the army and the weakness of the government left little choice. Moreover, with the British ensconced in Philadelphia, better housed but no less eager to eat, it was imperative to prevent disaffected farmers from carrying their produce to town.

The decision to encamp at Valley Forge was taken in mid-December. On December 17, when it still feared that the army might take quarters to the west, the Pennsylvania council sent Congress a "remonstrance" in protest. Two days later Congress forwarded it to Washington, inquiring what he had decided but also implying that the exposed situation of New Jersey merited his special attention. Even as Congress was acting, the first units of Continental soldiers were arriving at Valley Forge. That was where Washington received this dispatch on the twenty-second, just as he was completing a lengthy report to Congress. "It would give me infinite pleasure," he wrote to Laurens, "to afford protection to every Individual and to every Spot of Ground in the whole of the United States." But that goal was clearly unattainable. Washington then reviewed his reasons for encamping around Valley Forge, while avoiding dispersing his forces against threats that might never materialize. He sent a more urgent dispatch the next day. This famous letter opened with the bleak warning that unless "some great and capital change suddenly takes place" in the commissary, his army must "starve—dissolve—or disperse, in order to obtain subsistence in the best manner they can." Only yesterday, Washington noted, the complete lack of provisions had prevented him from opposing a large British foraging party that had descended on Darby.

Whether out of political calculation or self-righteous anger, Washington justified his blunt account of the dire conditions of his men with a direct appeal to "my own reputation." He had sounded the same theme a year earlier in private letters citing frustration over soldiers' pay, enlistments, and discipline as a threat to the "character" he wanted to establish. Hitherto he had been "tender" about "lodging complaints" against the various administrative reforms that Congress had imposed against "my Judgement." But "finding that the inactivity of the Army" had now been "charged to my own account, not only by the common vulgar, but those in power, it is time to speak plain in exculpation of myself." There followed not only a detailed review of the needs of the army, from the basics of food and clothing down to soap and vinegar, but also pointed digs at the local officials who had dissuaded him from using the confiscatory powers Congress had vested in him, with assurances that Pennsylvania would do its duty. "I can assure those Gentlemen, that it is a much easier and less distressing thing, to draw Remonstrances in a comfortable room by a good fire side, than to occupy a cold, bleak hill, and sleep under frost & snow without Cloaths and Blankets." Yet even at this moment of urgency, his concern lay as much with the future campaign as the immediate crisis. "We have not more than three months to prepare a great deal of business in," he estimated; "if we let these slip or waste, we shall be labouring under the same difficulties all next Campaign, as we have done this."

Thus began the Valley Forge winter, that episode in privation that summons images of gaunt, half-clad, barefoot soldiers huddled in the snow around flickering fires, staring vacantly into gruel-filled cups, bitterly aware that the enemy was sleeping warm and well fed in nearby Philadelphia. The first crisis followed hard upon the army's arrival at Valley Forge, when spontaneous chants of "No meat! No meat!" rang across the encampment. A second, graver one occurred in mid-February 1778, when drenching rains washed out roads, swelled rivers, and badly disrupted the flow of supplies to camp. Over two thousand soldiers died that winter. Outright starvation was not the immediate killer, but rather the lowered resistance to disease that repeated food shortages and inadequate clothing fostered in the overcrowded huts the army had hastily erected.

Yet in both its origins and outcome, the Valley Forge saga is not a simple story of suffering soldiers, incompetent politicians, and indifferent civilians. The strategic and political considerations that led to the army's unusual winter encampment strained to the breaking point a supply system upon which Congress had recently imposed well-meant but misguided reforms. Back in June, Congress had attempted to deal with the rising cost of provisions by reforming the Commissary, the department responsible for keeping the army provisioned. Rather than allow the commissaries to be paid by commission, and thus allow their earnings to track the inflation that was already plaguing the economy, Congress placed its agents on fixed salaries. This led to the resignation of Commissary General Joseph Trumbull—son of the Connecticut governor Jonathan and brother of the painter John—and numerous subordinates. Trumbull's successor proved inept, and the department was clearly struggling even while the army was operating in the breadbasket of southeastern Pennsylvania. Similar reforms to the Quartermaster department discouraged teamsters from providing the wages and animals needed to transport provisions.

Compounding these difficulties was the attitude of significant segments of the local population. Area farmers were little help. Many of them preferred dealing with British agents, who paid hard coin while their American counterparts forced farmers to accept the scrip that passed for currency. Acting either from "disaffection" or simple greed, farmers squirreled away their produce and livestock in the woods and other hiding places until they could make their way to British-protected markets or Philadelphia itself.

As always, Washington was deeply attentive to every facet of camp life. His general orders for January 8, 1778, for example, noted that "gaming is again creeping into the Army," and prohibited the playing of "Cards and Dice under any pretence whatsoever." Learning "that many men are render'd unfit for duty by the Itch," the commander also ordered regimental surgeons "to look attentively into this matter" and have the sufferers suitably "anointed." But from early January onward, he was preoccupied with preparing for the visit of the special committee that Congress had appointed "for reforming the abuses which have too long prevailed in the different departments belonging to the army."

That committee had an unusual cast. It consisted of three members of Congress, including Washington's former aide, Joseph Reed, and three members of the Board of War, which Congress had established the previous fall to ease its own burden of military administration. The contingent from the Board of War included Horatio Gates, its president, and the former quartermaster general Thomas Mifflin, a known critic of Washington. The board also included Thomas Conway (though he was not a member of the committee), whom Washington detested. Conway was one of a steady stream of Europeans to whom Silas Deane, the American agent in Paris, had granted military commissions, much to the annoyance of Washington and the jealousy of other officers who were inordinately sensitive over matters of rank and quick to take offense whenever someone jumped the queue for promotion. With Washington's lukewarm endorsement, Congress made Conway a brigadier general, and Washington gave him command of a Pennsylvania brigade. Conway drilled them well, and at Brandywine they fought bravely in slowing the British assault. But he played a strange part in the councils of war on which Washington relied, declining to give advice while his sour looks and body language expressed his dissent. Worse still, Washington learned that Conway had written a snide letter to Gates, remarking that "Heaven has been determined to save your country; or a weak general and bad counsellors would have ruined it." In the epistolary skirmish that followed, Washington coldly pressed both men for apologies, but the breach was irreparable.

To his loyal subordinates, perhaps to Washington himself, the Board of War seemed to be an alternative command staff in waiting, and its participation in the committee sent to the camp at Valley Forge looked like a challenge to his preeminent stature within the army. That may have girded up the army's determination to present the committee with a coherent plan of reform. Extensive discussions among Washington and his subordinates led to the drafting of a comprehensive memorandum—a virtual state paper—which he gave to the committee on January 29. These consultations among his staff illustrated not only Washington's mode of command but also the deep sources of influence that attracted men to his leadership. The reliance he placed on their views conveyed a sense of trust that gave his authority a foundation broader than superiority of rank. He repaid their candor by placing their own concerns first on his list of the "important alterations" that "must be made." The current distress of the army did not shake his original conviction that securing adequate pay for officers remained "the basis of every other regulation and arrangement, necessary to be made."

This had always been his view. But the decision to rank it at the top of the reform agenda also reflected the vigor with which Nathanael Greene pressed the officers' case, and the surges of complaints and resignations that the army's concentration around Valley Forge encouraged. The son of a Rhode Island anchor maker, a Quaker by birth but no longer by conviction, Greene was a self-taught soldier whose original knowledge of war came from extensive boyhood reading. He was one of the greatest military talents the war produced and one of his commander's keenest supporters and disciples. Three years of combat had taught Greene lessons no book could offer. When his brother lamented the mild discomforts Rhode Island soldiers had met in an aborted plan to dislodge the enemy from Newport, Greene set him straight. "If you could see a defeat, follow a long and tedious nights march, hear the screams of the wounded that are going of[f] the field, see the labour and difficulty of geting them off, have to march forty miles without victuals or sleep, you would hardly think your sufferings worth nameing," he replied. Like all combat veterans, he now knew better than to bridge the gap between what he had observed and what he could possibly describe—or what he had once read. That same experience revealed how little reliance could still be placed on "the glow of patriotism and the honors of the field." To retain officers, Greene argued, they had to be given material rewards comparable to what their British counterparts received: "a regular establishment for life," meaning military pensions.

Washington agreed completely and urged the congressional committee at camp to support his proposal to grant half-pay pensions to officers willing to serve throughout the war. He told the committee, as he had previously instructed Congress, that "interest is the governing principle" of all mankind. The "motives of public virtue" that originally brought men to serve could not withstand the hard evidence that his officers were becoming "losers by their patriotism." Confirmation for this truth lay in the "frequent resignations, dayly happening, and the more frequent importunities for permission to resign, and from some officers of the greatest merit." Washington had to belabor these points because the idea of military pensions threatened the core republican belief that patriotism should be enough, and which regarded half-pay pensions for officers as a dangerous step toward creating a permanent military caste. Washington's concern for his officer corps extended to the delicate matters of rank and promotion. Both had provoked "numerous bickerings and resignations," which had caused him "infinite trouble and vexation." Only after discussing these fundamental problems with the structure of the army did he turn to the urgent logistical matters that were the immediate source of the soldiers' winter misery.

The discussions with the congressional committee consumed a good part of the winter, and they went on even during the crisis caused by the February deluge. On the critical question of reforming military logistics, the committee made a clear choice. The Board of War had proposed placing the supply departments under its own direction, but instead the committee recommended the appointment of two new department heads, both to serve directly under Washington. As commissary general, the committee nominated the Connecticut merchant Jeremiah Wadsworth, and as quartermaster general, none other than Nathanael Greene. The prejudice Washington had expressed against New Englanders in 1775 had long since dissipated. If it took a New England sense of business to bring order to these essential departments, so be it. Greene accepted this new duty grudgingly. The tedious task of quartermaster would take him "out of the Line of splendor," he reminded Joseph Reed. But if the welfare of the army and the stature of his commanding officer demanded that sacrifice, it was a price this lapsed-Quaker general was ready to pay.

In the ongoing consultations between Washington's staff and the congressional committee, the commander in chief was the decisive winner. By early spring the distracting competition between the main army and the Board of War was over. Gates resigned as president of the board, and when Conway seemed to tender his commission in late April, Congress accepted his offer posthaste. Even then Conway remained a pest, protesting that he had not intended to resign. It took the Irish-descended Charles Carroll to let Conway know his fatal defect. "He told me a few Days agoe almost Literally," Conway wrote to Gates in early June, "that any Body that Displea'd or did not admire the Commander in chief ought not to be Kept in the army." Not long after, Conway found himself lucky to survive a duel with John Cadwalader, the merchant and general from Philadelphia who had prompted Washington to make the second Trenton crossing in December 1776. Adding injury to insult, Cadwalader shot a ball through Conway's cheek.

Had Conway confined himself to his assigned duty of inspector general in charge of training soldiers, he might still have served the cause as he professed he wished to do. That essential task instead fell to Friedrich Steuben, another foreign volunteer sent on by Silas Deane. He arrived at Valley Forge just as the subsistence crisis of February was ending. Steuben had extensive experience in the Prussian army—though rather less than he led American admirers to believe. Unlike many of Deane's recruits, he did not stand on pride and was willing to serve where needed. The most crying need was to teach soldiers the drills on which their battlefield deployment and effectiveness depended. Today such drills are used chiefly to instill the rudiments of discipline in raw recruits. In the eighteenth century, however, the outcome of battle hinged on how quickly and cohesively formations could take and maintain positions and direct their fire against the enemy. American officers had been avidly reading European manuals of arms and even writing some of their own. But the turnover in the ranks of the army of 1776 and the tramping to and fro of 1777 had limited the opportunity for sustained training. Washington's decision to entrust this task to Steuben, and the zeal that the Prussian brought to his work, made a material difference in the fighting capacity of the army.

So, however, did the consolidation of Washington's authority, which was also a major result of the Valley Forge winter. Whether that authority was ever in jeopardy seems doubtful. Most scholars agree that the existence of a sinister "Conway Cabal" was more a projection of his injured pride than a real movement to displace him from command. Washington did have his critics in Congress. The most ardent was James Lovell, a sardonic former Boston schoolteacher who complained to Gates that under Washington "our Affairs are fabiused into a very disagreeable posture." But Congress itself was too beleaguered even to imagine, much less impose, a change in command, and Washington had his own channels there as well. One ran to President Laurens, whose son John was one of his closest aides. Another was the new Virginia delegate, John Banister, to whom he wrote a long, patiently reasoned letter in April clearly meant for wider readership at York. After urging Congress to master the "indecision" blocking some of its committee's key proposals, Washington attacked "the *jealousy* which Congress unhappily entertain of the army." Such jealousy might be proper when directed against standing armies "in time of *peace*," especially those composed of "mercenaries, hirelings." But regrettably "It is our policy to be prejudiced against them in time of *War*—and tho they are Citizens, having all the ties—& interests of Citizens." These distinctions were specious. "We should all be considered, Congress—Army &c., as one people, embarked in one cause—in one interest; acting on the same principle, and to the same end." The best proof of this lay in the army's conduct the past winter. For "no history, now extant, can furnish an instance of an army's suffering such uncommon hardships as ours have done, and bearing them with the same patience and Fortitude."

When he wrote like this, Washington was his own best advocate and his army's as well, even if his common sense and plain speaking could not wholly dissolve attitudes so deeply embedded in American culture. In the jockeying for influence between the board and the army, his position was fortified by the open support he received from the committee at camp. It was his former adjutant, Joseph Reed, who stated the case for Washington most directly. Back in February, in a lengthy letter tacitly meant to bring Congress to its senses, Reed reminded another Pennsylvania delegate how often "the fatal Consequences of a Division in Councils" and "the jarring Interests of ambitious Men have blasted the fairest Hopes of political Happiness." America now risked just such a "political Suicide." The rivalry among the commanders was well known in the army, among the general public, even "publickly talk'd" about in occupied New York and Philadelphia. Reed also worried that Congress had seemed "to throw a Weight into one Scale" by approving an abortive winter expedition against Canada, which the Board of War had planned. Like all the officers at Valley Forge, Reed thought that project wholly impractical. But if it was pursued, Reed warned, it would have two baneful consequences. Not only would the resources it consumed fatally weaken the main army in Pennsylvania, but it would also undermine the "great Character" on whom the cause depended.

Much like Washington, Reed used the word *character* to refer to the man and his qualities. His manifest partiality for Washington was not mere hero worship. It rested instead on a shrewd assessment of his ability and stature. "Whatever may be thought by some Gentlemen, the Attachments of this Army to its Commander are extremely strong, & very natural," Reed noted.

> A long Connection, winning Manners, unspotted Morals, & disinterested Views cast a Lustre round him which a Want of Success cannot obscure. They are even strengthned by that Circumstance of Character, which some deem a Blemish: I mean a Diffidence of his own Judgment & Reliance upon that of his Officers. In supporting him, they support their own Opinions & are interested in vindicating the Measures which they have advised.

This attitude sometimes produced "an unhappy Unanimity of Councils" where more open debate was needed. But it also indicated that Washington's cautious management of his theater of war was not due to any failure of nerve or strategic vision. To replace such a commander, Reed implied, would require an extraordinary and as yet wholly undeserved confidence in his rivals.

In Reed's judgment, Washington was a Fabius more by circumstance and calculation than from character or (worse) cowardice. The discipline that enabled him to avoid being drawn into combat on unfavorable ground never deterred him from looking for occasions when it seemed advantageous to strike—as he had at Trenton, Princeton, and Germantown. His own impulses were aggressive, but his view of the war was never impulsive. That was the genius, or rather the inner tension, in Washington's generalship. However much he would have welcomed a decisive clash with the enemy, a chance to win the battlefield laurels all commanders covet, he regularly subordinated his impulses to his broader understanding of the revolutionary struggle, the limitations of his troops, and the counsel of his advisers.

On April 20, 1778, he asked his council to consider the coming campaign. These discussions began amid reports, emanating from occupied Philadelphia, that Lord North had introduced two conciliatory bills in Parliament, one renouncing its authority to tax America, the other for sending a new peace commission to "the colonies." Nearly a year had passed since Congress last heard from its envoys in Paris—Deane, Franklin, and Arthur Lee. American leaders initially wondered whether the handbills Howe was distributing by the cartload might be a ruse "to poison the minds of the people & mislead the wavering." Only in early May did they discover a better explanation. The bills were authentic. They had been introduced in response to the dramatic news that Congress received on May 2. Three months earlier, France had concluded a treaty of alliance with the United States.

Before this development became known, Washington had presented his council of war with three options for the coming campaign. Two were aggressive: to attack either the substantial British garrison in Philadelphia or the lesser force in New York City. The third was to remain "quiet in a secure, fortified Camp, disciplining and arranging the army," until the enemy's operations clarified the situation. Characteristically, Washington did not express his own views. Instead, he reminded his generals that he trusted them to "make their opinions the result of mature deliberation." They responded, individually, with the candor he expected, not wholly of one mind but on balance in favor of the third option. On May 8, with the intelligence from France in hand, a second council reviewed the options. This one included his ostensible rival, Gates, and his known detractor, Mifflin, whom Congress had sent to camp to join the council of war it had gratuitously instructed Washington to convene. He again expected each of the ten generals present to furnish his own recommendations in writing. Instead the entire council (Gates and Mifflin included) signed a joint memorandum endorsing the strategy of "remaining on the defensive." According to their collective judgment, the costs of defeat in an assault on Philadelphia or New York outweighed the potential benefits of victory.

The American command thus seemed Fabian to the core, but their commander still hoped for an opportunity to strike. By mid-May Washington had rightly concluded that the British were preparing to abandon Philadelphia and consolidate their forces at New York. If they moved by land, as seemed likely, it might be possible to bring them to battle. Washington now had to reckon with a new British commander. Sir Henry Clinton had succeeded Howe, who had been recalled to London. On May 18 Howe's departure was honored by the most extravagant pageant that either army enjoyed during the entire war. It was "a strange kind of entertainment, wch the Projectors styled a Meschianza or Medley," an embarrassed British official wrote, "consisting of Tilts & Tournaments" along with dancing, fireworks, and dinner for over a thousand officers, dignitaries, and their wives and lady friends. The revelry lasted until dawn. But no enchanted evening of this kind could calm the Philadelphia loyalists who now wondered what other entertainments would await them when the patriots reoccupied their capital. Howe's parting advice—to "make [their] peace with the States, who, he supposed, would not treat them harshly"—left them in a state of "melancholy."

The order to abandon Philadelphia came from Lord Germain in London. Lord North's first inclination upon learning of the Franco-American alliance was to submit his resignation to his king. But George III, made of sterner stuff, insisted he stay at his post. With war against France imminent, Britain had to reallocate its American forces to protect its valuable possessions in the West Indies. That was the chief concern that made Philadelphia expendable, for New York City was the greater asset and a far more defensible headquarters. British officers could rightly wonder what future the American war now had. After three years their army still seemed capable of pummeling the Continentals on any given day, but that superiority had gained them nothing.

The British evacuation began with naval transports carrying baggage and three thousand loyalist refugees down the Delaware River. On June 18 the last troops left the city in two main columns. Washington set his army in motion, shadowing the British across New Jersey while advanced detachments and militia were deployed to impede their march. Though Washington worried that Clinton might wheel to launch his own attack, his inclinations grew more aggressive as the days passed, to the point where he was willing to risk a major battle. His generals, however, were less united on strategy than they had been in May. The leading skeptic was Charles Lee, who had just rejoined the army after spending the past year and a half as a British captive before securing release in an exchange. His capture was a tribute to his own carelessness—he had been snatched, unguarded, from a New Jersey tavern—and Lee numbered few admirers among American officers. But Washington gave him the command of a division. Then, on June 27, with the Americans catching up to the enemy near Monmouth, he granted Lee's request to replace the ardent but inexperienced Lafayette, Washington's young protégé, at the head of the substantial force he proposed to unleash on the British rear guard the next day.

The ensuing battle brought one of those indecisive outcomes that military historians love to dissect. Whether because Washington failed to instruct Lee adequately, or because Lee did not share his superior's views, or because other commanders left Lee in the lurch, the Americans crumpled before the counterattack that Clinton launched to support his rear guard, commanded by Lord Cornwallis. Washington arrived at the engagement to discover Lee's men in retreat. After a heated but brief exchange he sent Lee to the rear and undertook to halt the retreat and arrange the defense. Fortunately for the Americans, the immediate terrain favored them. Fighting continued throughout the sultry afternoon, with exhausted soldiers wilting from heatstroke on a day when the temperature was in the nineties. The American lines held and the British withdrew. "The commander in Chief was every where," Greene wrote to his brother a few days later. "His Presence gave Spirit and Confidence and his confidence and authority soon brought every thing into Order and Regularity." Heartened by his success in rallying the defense, Washington hoped to resume combat the next day. But rather than wait around, Clinton stole a march and put his troops on the move again at midnight. The exhaustion of battle and heavy rains on June 29 prevented the Americans from following.

For all practical purposes, the 1778 campaign began and ended at Monmouth. Fittingly, both sides could claim a victory of sorts—the British for warding off attack, the Americans for halting a retreat and holding their ground. Within days Clinton's troops were safely lodged at New York, which they occupied until war's end. Washington began erecting a new defensive arc around Manhattan island, with a strong anchor at West Point to prevent British ships from cruising up the Hudson to Albany. The one clear loser at Monmouth was Charles Lee. With faithful Washington supporters such as Greene and Hamilton lambasting his failure to press an attack, Lee demanded a hearing to clear his name. He might as well have pled his cause to a jury of British officers. It was folly to expect an American court-martial to vindicate his reputation at the expense of its hallowed commander.

The difficulty of classifying Monmouth as a victory for either side serves as an apt commentary on Washington's reputation and stature. As a battlefield tactician, he was not a giant of his age. Perhaps Lee let Washington down by not pressing an attack, but perhaps he had reason to be uncertain of Washington's intentions. To allow Lee to lead an assault whose wisdom he doubted was arguably another error of judgment. Yet in the chaos of retreat Washington had personally rallied the men streaming down the road from Monmouth, quickly assigned them to defensible positions, and turned a potential rout into an even fight. Probably no other American officer could have done that, because none commanded the same confidence and respect. This respect stemmed not from military genius, but from character, those qualities of personality and endurance that Washington sought to instill in others and that he embodied to so many. Whatever his limitations as a tactician, Washington grasped the strategic dimensions of the war as well as anyone. More important, he never lost sight of the priority of turning his army, with its jealous officers and its restless soldiers, into a genuine institution, with its own integrity and coherence. Amid the trials and tribulations that distinguished the desperate campaign of 1776 from the to-and-fro marching of 1777 and the sufferings of early 1778, he clung to that objective.

That army was one of the two genuinely national institutions created in the early years of the Revolution. The other was Congress, the political council to which Washington formally looked for guidance and to which he routinely deferred, though he and his nucleus of loyal officers judged it increasingly wanting in the performance of its duties. In 1777 and 1778, Congress was often composed of barely two dozen members, most of them coming and going for a few months at a time; often a state would be represented by one single duty-bound delegate. It was not a decision-making body that military men were disposed to admire. "Folly, caprice, a want of foresight, comprehension and dignity characterise their actions," Alexander Hamilton complained to Governor George Clinton of New York, doubtless reflecting a view common among Washington's aides and the commander himself. "Their conduct with respect to the army is feeble indecisive and improvident." Yet however much grumbling went on in the officers' tents and the farmhouses that served as headquarters, however many bitter jokes about congressional foibles have been lost to the historical record, Washington never allowed the army to disdain its civilian superiors. Fabius was the role circumstances forced him to play, Cincinnatus the character his own more closely resembled.

PART II

CHALLENGES

4

The First Constitution Makers

BEFORE GEORGE WASHINGTON resumed his military career in 1775, his closest political associate was George Mason IV, his neighbor in Virginia's Northern Neck. For a few months after Washington assumed command at Cambridge, they continued to write. Then Mason let the correspondence lapse. "Knowing how little Time you had to spare, I thought it wrong to intrude upon it," he apologized four years later. They had been friends since Washington first leased Mount Vernon from his brother Lawrence in 1754. Their families were already connected in the ambitious land speculations of the Ohio Company. Mason was its treasurer, and the Washington brothers were also members. It was on behalf of the Ohio Company that Lieutenant Governor Dinwiddie (another investor) sent Washington to the Forks of the Monongahela in 1753. The two Georges soon formed other ties. They rode and hunted together, served as justices of the peace and vestrymen of Truro parish, and purchased town lots in nearby Alexandria. And both pursued the architectural projects that reveal much about the life of the Virginia gentry.

Today only a fraction of the crowds who tour Mount Vernon go on to Gunston Hall, the small jewel of colonial architecture that Mason was already constructing. The two houses should really be seen together, because they provide visual clues to the changing values of the political elite of the new republic's most populous state. In 1754 Mount Vernon was still a relatively modest clapboard house, with four ground-floor rooms and small bedroom chambers in the half-story above. Its first expansion awaited Washington's marriage to Martha Custis five years later. Its full transformation into the distinctly American house to which we now throng—with its plain bedrooms, its sweeping view of the Potomac, and the key to the Bastille that Lafayette sent its owner as a tribute to his career as national liberator—took place only after Washington retired from military service in 1784.

By contrast, Mason's project better captures the aesthetic values of the late-colonial Virginia gentry. Like Stratford Hall, his friend Richard Henry Lee's birthplace, Gunston Hall took the familiar red-brick Georgian form that is now a cliché of Virginia architecture—a facile link between modern townhouses and office complexes and the lost world of the revolutionary generation. When Mason and his peers built in the Georgian mode, however, they were not seeking to develop a distinctly provincial style. Rather, they were identifying with the English gentry from whom some of them descended and whom they all wished to emulate. Their aspiration was not to become more American—whatever that might mean—but more English.

For Mason this was a conscious choice. Through his younger brother Thomson, who was in London studying law—and befriending John Dickinson—Mason hired William Buckland, a young carpenter-joiner from Oxford, to complete his house. Just starting out, Buckland had to rely on the pattern books he brought with him from England or which his clients already owned. These provided a wealth of models covering every aspect of designing a proper country seat or townhouse. But with Gunston Hall to exhibit his skill and Mason to endorse him, Buckland became a successful architect, earning numerous commissions from the Chesapeake gentry. A portrait by Charles Willson Peale, begun just before Buckland's sudden death in December 1774 (but finished only in 1787), catches the architect smiling, eyes alight, the plans for an Annapolis merchant's house and his pens at his elbow. His expression, though, seems not pensive, like the Boston silversmith Paul Revere's gaze in the famous portrait by John Singleton Copley, but rather expressive of sheer delight in his work and his good fortune. Buckland had good cause to rejoice. Mason's invitation enabled him to escape the long apprenticeship and uncertain prospects that would have been his lot in England.

Buckland's first client soon became a draftsman of another kind. Through the force of a powerful intellect and an exceptional reputation for independence and probity, Mason emerged as the leading framer of the new constitution and the accompanying declaration of rights that the fifth Virginia provincial convention adopted in the spring of 1776. He thereby helped shape the Revolution's most lasting legacy and its most striking departure from British constitutional tradition. This was the great innovation that treated a constitution not as a working description of a government but as a single authoritative document, written at a known moment of historical time, under rules that made it legally superior to all the other acts that the government it created would subsequently adopt. That experiment began a good decade before the framers of the federal Constitution of 1787 met at Philadelphia to produce that one charter of government that we know best and still study and dispute most intensely. But the architecture of the 1787 Constitution took the form it did because its framers could treat the earlier documents written with independence as examples to be improved upon and corrected. No longer would Americans appeal to custom or tradition or the myth of an "ancient constitution" lost in the dim mists of history as the sources of their political values and institutions. Americans had instead become a people who defined themselves by the constitutions they framed and who saw their opportunity to "alter and abolish" constitutions as the great achievement their own struggle for independence should bequeath to the wider world.

That was not how Englishmen or their American cousins originally thought a constitution was properly defined. A decade earlier, while writing in opposition to the Stamp Act, John Adams had struggled to define the "true" or "real constitution" of the British Empire. With no authoritative text to point to, Adams relied on analogy and metaphor to explain how Americans could know what their constitutional rights truly were. First he ruminated about "the constitution of the human body," the treatments a physician would give it, and the "*stamina vitae*, or *essentials and fundamentals of the constitution*," without which the body could not survive. From the organic he moved to the mechanical. "A clock has also a constitution," Adams noted. Clocks must be designed, of course. But Adams was concerned not with how they were created but with how they worked. Whichever analogy he used, for Adams a constitution was known not by its origins but by its function and operation. The challenge in 1766 was to cure or repair its essential parts, not develop a better model or design a new frame.

Not long after Adams published these thoughts, a backbencher in

Parliament went further in rebutting the notion that governments arose from conscious design. "No government ever was built at once or by the rules of architecture, but like an old house at 20 times up & down & irregular," Sir George Saville wrote in a private letter. Indeed, in every form of creative activity, "principles have less to do than we suppose," Saville continued. "The Critics rules were made after the poems. The rules of architecture after ye houses, Grammar after language and governments go *per hookum* & *crookum* & then we demonstrate it *per bookum.*" Governments, in this wonderfully skeptical view, were not really founded or framed; they developed and evolved, or else they were simply imposed by force. Saville was offering only a vulgar version of the position taken by David Hume, the Scots philosopher-historian, two decades earlier. The idea that any known regime had an exact moment of foundation, especially one based on an original contract among its people, was a myth, Hume argued. "Conquest or usurpation, that is, in plain terms, force, by dissolving the ancient governments, is the origin of almost all the new ones which were ever established in the world," he wrote. "And that in the few cases where consent may seem to have taken place, it was commonly so irregular, so confined, or so much intermixed either with fraud or violence, that it cannot have any great authority."

Hume had good reason to be skeptical. Yet as he lay dying in Edinburgh in the summer of 1776, the newly independent American states were putting his doubts to a pragmatic test that the great empiricist would have relished. In this experiment, political philosophy, and even lovers of philosophical wisdom, did take a part. Hume's friend Benjamin Franklin presided over the Pennsylvania convention that wrote the most radical of the new constitutions adopted in 1776. But the spring and summer of independence were hardly a moment when Americans could give their inner philosopher free rein. There was a war to be fought, and countless other urgent tasks undertaken. Constitution making *was* important because Americans were anxious to develop fully legal governments that could replace the extralegal apparatus of committees and conventions that had been governing them since 1774. Still, it was only one revolutionary task among many, to be discharged as quickly as possible, without distracting the provincial conventions from their other duties or fomenting political discord among the people at large.

In practice, then, the business of writing constitutions usually fell to select committees and a handful of widely respected individuals, notably including George Mason in Virginia, John Jay and Robert Livingston in New York, and (later, in 1779) John Adams in Massachusetts. They thought of themselves as "framers," the word we still commonly use, and they described the constitutions they wrote as frames of government. Perhaps that choice of terms was fortuitous, but it was still revealing. Just as a house had its frame, so would a government. The organic and mechanical metaphors of body and watch that Adams had been forced to use a decade earlier no longer fit. A constitution was now to become something visible, a fixed structure with a number of well-designed chambers in which the work of government would be carried on, under the watchful eye of the citizens-passersby who were its ultimate owners. It was, in other words, a work of political architecture and interior design—and George Mason, the master of Gunston Hall, became Virginia's own William Buckland.

The experiment in constitution framing that effectively began in Virginia in May 1776 was soon followed in nearly all the other American provinces that ceased to be colonies in July 1776 and now called themselves states or commonwealths. But first and foremost, they were *republics*, a form of government known to have flourished, for a time, in antiquity, and to have been revived, with dispiriting results, in more recent times in Italy. "Of Republics, there is an inexhaustible variety," John Adams observed in his 1776 tract, *Thoughts on Government*, which was perhaps the best quick guide Americans had to the constitutional project on which they were about to embark. The variations on the republican form that appeared in other states—notably Pennsylvania, Maryland, New York, and Massachusetts—help round out our understanding of what this experiment originally meant, and so it will be helpful to survey how the project unfolded beyond the great commonwealth of Virginia as well.

Anyone who bears a *IV* after his name carries the weight of some kind of tradition on his shoulders. So it was with George Mason of Gunston Hall. Like most members of his class, his family's colonial origins dated to the mid-seventeenth century. The founding George immigrated around 1651, with enough capital to begin purchasing the indentured servants on whose labor the colony depended. Virginia's headright system entitled him to acquire fifty acres of land for every servant transported to the colony. Within a few years, he owned nine hundred acres, several times the average holdings of the region's free planters. Georges II and III improved the family estate in the usual way, by marrying well, cultivating tobacco, purchasing servants and slaves, accruing more headrights to land, and leasing smaller farms to tenants. They also served, quietly, in the House of Burgesses. George Mason III had a seat there from 1720 until 1735, when he drowned in a freak accident while crossing the Potomac in the trading sloop that he and his brother-in-law, John Mercer, had used for the past decade. His son, the future constitution framer, was only nine.

Ann Thomson Mason never remarried, but George had one strongly paternal figure in Mercer, his uncle and legal guardian. Mercer came to Virginia after the death of his own father, a Dublin merchant who left his son no resources other than a sharp mind and an acid tongue. Marrying a Mason was one step toward respectability. Mercer started as a trader but gravitated toward law, which he taught himself through the well-stocked library he built. He learned the subject better than most of the justices who heard his cases and whom he routinely berated with abusive language that incurred frequent warnings and occasional suspensions of his practice. He also opened his library to his young ward. It was there that George Mason did the extensive reading that made him one of the colony's most learned men. Unlike his uncle, Mason never practiced law. He only knew it deeply, as any man of property would be advised to do if he wished to assert his interests and safeguard his rights.

After he came of age in 1746, Mason's political interests appeared modest. He sought election to the House of Burgesses in 1748 but failed to carry the poll. He waited a decade to try again, was elected for Fairfax County, but served only one term. His world remained domestic and parochial. In April 1750 he married sixteen-year-old Ann Eilbeck, the daughter of a Maryland merchant. Three years passed before George V was born, but then Ann bore another eleven children over the next two decades. Four arrived within twenty-one months of the last birth. Three died in infancy, including twin sons born prematurely in December 1772, amid "the long Illness of their Mother," who was thirty-eight. She followed on the afternoon of March 9, 1773, dying "of a slow-fever" following "a painful & tedious illness" endured "with truly Christian Patience & Resignation." In the family Bible, Mason recorded a moving tribute to Ann's "modest Virtues," from "her Complexion remarkably fair and fresh—Lilies and Roses (almost without a Metaphor) were blended there" to the unblemished qualities of mind and temper that esteemed her among women. "Form'd for domestic Happiness, without one jarring Attom in her frame!" he marveled in his grief. "I was scarce able to bear the first Shock," Mason recalled a full five years later. "A Depression of Spirits, & settled Melancholly followed, from which I never expect, or desire to recover."

That vision of "domestic Happiness" captivated Mason and defined the private concerns that he was always loath to set aside. His large brood left little choice. The prospect that his children would know downward mobility was a genuine fear for Mason. Indeed, he regarded it almost as a social law. No matter how great a "Man's Rank or Fortune," he wrote that mournful summer, his "posterity must quickly be distributed among the different Classes of Mankind, and blended with the Mass of the People." Fourteen years later, he expressed the same thought at the Federal Convention. He "often wondered at the indifference of the superior classes of society to this dictate of humanity & policy," Mason observed, in support of establishing a broad electorate, when "the course of a few years, not only might but certainly would, distribute their posterity throughout the lowest classes of Society."

Ann's death left the widower despondent. For days Gunston Hall lapsed into silence, a son later recalled, while "my Father paced the rooms, or from the House to the Grave, for it was close by" (and can still be seen). Yet from mourning Mason soon plunged into a new round of activity that reveal a deepened concern with providing for his family. Within days he drafted a lengthy will that ran to twenty-four pages when copied in the Fairfax County records. It covered all forms of his extensive property: real, personal, and the human chattels whom he devised, by name, among his children. Thus to his namesake George V he left not only Gunston Hall, with its adjoining five thousand acres, and "all of my stock in the Ohio Company," but also "my Gold Watch, which I Commonly wear," the "large Silver Bowl" used to christen his children, and "title to two Negroe Men named Tom and Liberty."

Naming a slave Liberty might seem the cruelest joke a master could play. But Mason was sensitive to the moral questions that slavery raised. In the same passage in which he fretted about downward mobility, he meditated on the evils of slavery. Like Dickinson observing his fellow colonists in London twenty years earlier, or Jefferson writing his *Notes on the State of Virginia* a decade later, Mason was most troubled by the institution's effects on *masters.* "Every Gentlem[a]n here is born a petty Tyrant," he lamented. "Practiced in Acts of Despotism & Cruelty, we become callous to the Dictates of Humanity, & all the finer feelings of the Soul." Mason did not directly address the harm this domination imposed on the slave. Yet he knew that such an evil was daily inflicted. Slaves were "a part of our own Species," possessors "of the Dignity of Man" and "the Rights of Human Nature," even if they also appeared as "Wretches, whom our Injustice hath debased almost to a Level with the Brute Creation."

These remarks quite possibly inspired the famous passage (in his *Notes on the State of Virginia*) in which Jefferson later trembled for his nation when he considered that God is just and His justice would not sleep forever. But for Mason as for Jefferson, this moral insight did not outweigh the need to maintain a slave force, to be conveyed, with other property, to one's heirs. In 1773, acquiring new property remained as important to Mason as arranging its distribution. Three months after Ann's death, he began purchasing additional headrights on a large scale. Over the next year, he bought just over a thousand, entitling him, he thought, to more than fifty thousand acres he hoped to patent in Fincastle County, the vast western county that included the future state of Kentucky. This strategy carried a risk. Just before Mason began these purchases, the king's Privy Council ordered colonial governors to suspend land grants. A year later it adopted a new scheme for auctioning the interior American lands to which George III claimed sovereign title.

Mason learned of this development shortly after he lost Ann. He devoted much time that summer to reviewing the history of Virginia, drafting a lengthy memorandum to which he (or a copyist) gave the title "Extracts from the Virginia Charters with some Remarks on them made in the year 1773." Its purpose was twofold: to validate Virginia's extensive land claims, and with them, the Ohio Company's past purchases; and to emphasize the vital role that the headright system had played in the colony's development by encouraging planters to import servants in exchange for the additional acres they would acquire. But Mason's annotative "Remarks" went beyond the immediate objectives of his research. In addition to his comments on downward mobility and the corrupting effects of slavery, he mused about the origins and character of the colony's political institutions and the dangerous implications of recent British measures, especially those threatening Virginia's territorial integrity. What Mason called "ancient usage & practice"—as applied, for example, to the practice of headrights—were "thereby made a part of the Constitution of Virginia, and cannot be avoided or validated by any Proclamation, Instruction, or other Act of Government."

Mason's equation of customary practice and constitutional legitimacy tapped a powerful strain in Anglo-American thinking. What was customary *was* constitutional, and government acts that threatened settled understandings could be attacked as arbitrary assaults on basic rights. In the new year of 1774, Mason hoped that such reasoning could persuade the assembly, Governor Dunmore, and hopefully the ministry in London to acknowledge the colony's traditional ways of granting land. In May he journeyed to Williamsburg, not as a legislator but a petitioner, intent on securing approval for his Fincastle claims.

Once arrived, Mason found "every body's attention so entirely engrossed by the Boston affair" that he had to postpone the presentation of his "charter-rights." Instead he now joined the handful of leaders who were preparing the assembly's response to the Boston Port Act. Patrick Henry particularly impressed him. "Every word he says not only engages but commands your attention," he marveled; "and your passions are no longer your own when he addresses them." Henry was the "first man upon this continent"—an accolade that would soon fall to his Mount Vernon neighbor—but Mason could imagine the famous orator acting on an even grander stage. Recalling the historical reading he had done in his uncle Mercer's library, Mason imagined that had Henry lived at the time of the First Punic War, "when the Roman people had arrived at their meridian glory, and their virtue not tarnished," his "talents must have placed him at the head of that glorious Commonwealth."

Still hoping to have his headrights confirmed, Mason thought that the assembly would do its normal business before turning to Boston. Governor Dunmore had other ideas and dissolved the assembly before it could do any damage to the empire. Mason was still in town when a rump group of lawmakers met at Raleigh's Tavern to issue their call for an intercolonial Congress. He stayed on into mid-June to present his petition in person to Dunmore and the council. But they declined to act, and soon the arrival of the Quebec Act, extending the boundary of the Canadian province as far south as the Ohio River, made a further decision on the matter even less likely.

So did the deteriorating political situation. As preoccupied as Mason remained with his private affairs, he was a resolute Whig and no less militant in assessing British policy than Henry or Washington. In early July Mason prepared a set of resolutions condemning Parliament's vengeful acts against Boston, which he and Washington proposed to submit to the Fairfax County freeholders. He spent Sunday, July 17, 1774, at Mount Vernon with Washington. The next day the two men took their familiar ride into Alexandria, where Washington chaired the meeting that promptly endorsed the resolutions Mason had drafted. Like other communities meeting across the colonies that summer, his neighbors took a firm stand on the imperial dispute. There was no doubt, their ninth resolution affirmed, "that there is a premeditated Design and System, formed and pursued by the British Ministry, to introduce an arbitrary Government into his Majesty's American Dominions." For Mason, there was nothing new in stating his convictions so boldly. He had actively supported the non-importation agreement of 1769–1770, recommending its continuation even after the Townshend duties were repealed. Even earlier, he had sent an angry letter (signed by "A Virginia Planter") to a committee of London merchants who had denounced colonial opposition to the Stamp Act.

Mason closed that letter by describing himself as someone who "has seldom medled in public Affairs, who enjoys a moderate but independent Fortune, and content with the Blessings of a private Station, equally disregards the Smiles & Frowns of the Great." The reference to "a private Station" invoked a well-known couplet from Joseph Addison's famous (and Washington's favorite) play, *Cato.* "Content thyself to be obscurely good," the virtuous Roman senator advises his ally, the departing prince Juba, as Julius Caesar triumphs and the republic nears dissolution.

> *When vice prevails, and impious men bear sway,*
> *The Post of honour is a private station.*

Being obscurely good in a private station was as much as Mason wanted in 1766 and even in 1774, when he declined Washington's suggestion that he stand for the Virginia assembly. Instead he merely spent a night at Mount Vernon in late August, discussing the agenda for the First Continental Congress as Washington, Henry, and Edmund Pendleton prepared to journey to Philadelphia.

The impious men now bearing sway, however, were in London, not Williamsburg or Philadelphia. And that made the post of duty a public office rather than a private station. Against his inclinations, politics drew Mason into its vortex. In July 1775 his Fairfax neighbors named him to replace Washington as their delegate to the provincial convention. Once there, he found himself "a good deal pressed" to replace Washington again, this time by filling his seat at Congress. By Mason's reckoning, two thirds of the members asked him to serve. Though he persuaded most to relent, a group of twenty, led by Jefferson and Henry, insisted on his nomination. Mason was forced "to make a public Excuse, & give my reasons for refusal, in doing which I felt more distress'd than ever I was in life." His words must have evoked his private grief and burdens, because "I saw tears run down the President's cheeks." His reward for escaping a trip to Philadelphia was election to the provincial Committee of Safety. Again he "beg'd the Convention" for permission "to resign, but was answer'd by an universal NO."

Mason was entitled to a measure of self-pity. His domestic burdens were real enough. At fifty, he was reaching the status of senior citizen in Chesapeake society. None of his namesake ancestors had lived to sixty. No hypochondriac, he suffered from recurring gout that immobilized him for days on end. Nor did he suffer fools lightly—and anyone engaged in public deliberations, even amid the patriotic fervor of 1775, had to endure "the Bablers," who loved the sound of their own voice too well. The sarcasm with which he answered such babbling did not endear Mason to his legislative colleagues. But neither did it diminish the respect he commanded.

That respect had two sources. One was the evident learning and intellectual ability that he had cultivated throughout his life, from his boyhood reading in his uncle's library to his own researches into Virginia history. Like the younger Jefferson, whom he already knew, and the still younger James Madison, whom he would soon meet, Mason read deeply, widely, and thoughtfully. His learning was manifest, and so were his abilities as both orator and writer. But perhaps more important, Mason personified an ideal type of citizen, the independent country gentleman whose opinions were entirely his own, beholden only to an impartial assessment of the public good. Mason's genuine disdain for public office conferred additional authority. No one could accuse him of muffling his voice to curry political favor. Precisely because his inclinations were so private, he symbolized the ideal of republican virtue that permeated the constitutional experiments of 1776. If he was to be a public servant at all, he would act from duty only, not ambition.

Once he accepted office, however, there were ambitions that Mason was happy to pursue. One was the redemption of the headrights that he resumed purchasing even as he set to work on the new state constitution in May 1776. Then too there were the old claims of the Ohio Company, which he still hoped to revive. Little had been done with them since the Proclamation of 1763 confined American settlement to the lands east of the Appalachians. The collapse of British authority, however, meant that Mason no longer had to appeal to an unresponsive royal governor or his superiors in London. A government run solely by Virginians, for Virginians, might be receptive on both fronts, especially if Mason's historical research confirmed just how important headrights and territorial claims had been to their province's fundamental interests all along.

Mason defined those interests much as his forebears had. In their own eyes, the landed gentry who had ruled the colony since the mid-seventeenth century *were* Virginia. They thought of it as their commonwealth to govern, and themselves as its true proprietors. Not that they were oblivious to the concerns of the lesser planters whose votes they routinely courted and to whom they often extended credit to purchase the slaves whose labor all coveted. Yet the social distance separating the gentry from smaller planters working a couple hundred acres alongside one or two slaves was significant. Mason was a patrician, not a plebeian. No passerby could ever confuse Gunston Hall with the ramshackle cabin of an ordinary planter.

Mason thus embodied a tension, or even contradiction, that underlay the republican enthusiasm of 1776. In his politics and professed principles, he was a perfect Whig of a radical stripe. He subscribed to the creed that defined the political faith of a Samuel Adams and their mutual friend Richard Henry Lee. He shared their admiration for the militant virtue of the ancient Roman republic. At the Federal Convention of 1787, he invoked ancient precedent by moving to empower Congress to enact sumptuary laws regulating what people could eat, drink, or wear, in order to promote frugality among the lower ranks of society. Had it ever become fashionable to wear a toga to legislative debates, Mason would have happily donned one to play the role of Roman senator to Patrick Henry's first consul. Yet he was also the owner of the sumptuously decorated Gunston Hall, a man who looked forward to drinking "the first Bottle of good Claret" he expected to obtain with the arrival of "the first Ships from France" after the announcement of the Franco-American alliance. As the active force in both the Ohio Company and the speculative sinkhole of the Dismal Swamp Company, he shared the vision of economic development that animated fellow planters such as Washington and merchants such as Robert Morris. He was, in other words, an acquisitive entrepreneur in practice, yet a frugal republican in principle. Herein lay a tension that the Revolution could only sharpen.

Talk of independence was everywhere in the spring of 1776. With it, inevitably, came talk about the new governments the colonies would need to establish following the break with Britain. That subject had been taboo before 1776, when the colonists loyally insisted that their objective was to restore the rights of their existing governments. But the longer the imperial crisis lasted, the more Americans fretted about the danger of allowing extralegal committees and conventions to be their main instruments of government. The greater the duties these revolutionary authorities assumed, the more they wished to have the full measure of law behind them. Rule by committees and conventions was politically effective but legally defective. If taxes had to be levied, soldiers recruited, dissenters punished, and justice dispensed, patriot authorities wanted to have the authority of law, not mere public opinion, behind them.

Full-throated public discussion of new governments began only after Thomas Paine's *Common Sense* made it safe to debate the larger issue of independence. Privately, however, American leaders began speculating about the subject before Paine published his dual assault on monarchy and empire. One night in November 1775, Richard Henry Lee asked John Adams to jot down his ideas. "It is a curious Problem what form of Government, is most readily & easily adopted by a Colony upon a Emergency," Adams replied the next day. He then offered a matter-of-fact "Sketch" of government, which could be altered in "an infinite Number of Ways, So as to accommodate it to the different Genius, Temper, Principles, and even Prejudices of different People." The work would not prove difficult, he predicted. "A single month" of discussion should be "Sufficient without the least Convulsion or even Animosity to accomplish a total Revolution in the Government of a Colony."

Five months later Adams published an expanded version of this letter under the modest title *Thoughts on Government.* Two exuberant declarations framed his sketch of the new governments the separate American provinces should adopt. Defying "the sneers of modern Englishmen," Adams first affirmed the belief that has defined the American political creed ever after: "that there is no good government but what is Republican." Those English sneers were directed, in part, against the noble (but not aristocratic) principle on which all republics relied: the public-spirited virtue of their citizens. The second statement was even more exultant. "You and I, my dear Friend, have been sent into life, at a time when the greatest law-givers of antiquity would have wished to have lived," Adams wrote, keeping to the epistolary form of his letters to Lee and other delegates. "When! Before the present epocha, had three millions of people full power and a fair opportunity to form and establish the wisest and happiest government that human wisdom can contrive?" The Worcester schoolmaster and Massachusetts circuit rider could now imagine himself occupying the same exalted stage as the great lawgivers of antiquity: Solon of Athens, Lycurgus of Sparta, Moses at Sinai.

Thoughts on Government did not enjoy the smashing literary success of *Common Sense*, but it did provide a focal point of debate for the new subject of how to go about writing constitutions. Thanks to his congressional friends, R. H. Lee and George Wythe, his tract was being read in Virginia well before the fifth provincial convention gave Mason his constitution-making task. That was exactly where Adams wanted his *Thoughts* to be read with greatest care. His "poor Scrap" of a pamphlet, he confessed a few months later, was "callculated for Southern Latitudes," where the "popular Principles and Maxims" of republican government were "so abhorrent to the Inclinations of the Barons" who ruled there. Adams worried that the "barons" of the south would want to follow the forms of British government more closely than they should. That might mean, for example, creating an upper house modeled too closely on the House of Lords, instead of the simpler plan Adams laid out in his *Thoughts* when he proposed that the lower house should simply appoint a senate "from among themselves or their constituents." Adams may also have doubted whether the southern gentry would want to adopt his radical principle of representation. A republican legislature, he wrote, "should be in miniature an exact portrait of the people at large. It should think, feel, reason, and act like them." When one looked at the people's representatives, one should see a rough approximation of the people themselves. It followed, therefore, that the legislature should also be constituted on principles of equality, meaning that an "equal interest among the people should have equal interest" within the assembly. The wording appears clumsy, but the principle it endorsed is the rule of apportionment that we call "one person, one vote."

One friend who shared Adams's enthusiasm for the subject was Thomas Jefferson, who returned to Philadelphia just as Congress was approving a resolution designed to permit patriot leaders in the three proprietary colonies of Pennsylvania, Delaware, and Maryland to replace their old colonial legislatures with new governments. On May 10, 1776, Congress adopted a resolution allowing new governments to be established in colonies "where no government adequate to the exigencies of their affairs have been hitherto established." But that left individual colonies free to decide the matter, and Pennsylvania needed to be pushed. Five days later John Adams brought in a preamble to this resolution that turned political discretion into revolutionary duty. It called for suppressing "the exercise of every kind of authority" under the Crown and the creation of new governments drawing their power directly from the people.

The next day, Jefferson sent the resolution to a Virginia correspondent, adding a revealing hint. If the provincial convention was about to pursue the work of constitution writing, he mused, perhaps it should recall its delegates for a brief period, as other colonies had done. "It is a work of the most interesting nature and such as every individual would wish to have his voice in," Jefferson went on. "In truth it is the whole object of the present controversy; for should a bad government be instituted for us in future it had been as well to have accepted at first the bad one offered to us from beyond the water without the risk and expence of contest."

The whole object of the present controversy: no colonist could have made that claim before 1776, because doing so would have undercut the entire theory and strategy of resistance. But Jefferson, with his visionary gaze already set on the American future, understood that necessity and opportunity had merged. If he had his druthers, he would have hurried back to his old college town, Williamsburg. History had other plans for the lanky, sandy-haired Virginian with the sharp pen; it soon saddled him with the task of drafting the Declaration of Independence in Philadelphia and thereby securing his eternal fame. But even before he began laboring over that text, he worked up several drafts of a constitution for Virginia, which his old professor and congressional colleague George Wythe carried home for him in mid-June. In Jefferson's absence, it fell to George Mason to play the leading role in Virginia.

"A smart fit of the Gout" kept Mason at Gunston Hall as the fifth provincial convention came to order in May 1776. He missed the passage of the fateful resolutions instructing its delegates at Congress to move for a declaration of independence, articles of confederation (which would become the nation's first *federal* constitution), and a plan for making treaties of alliance. But he reached Williamsburg by May 18, to discover that "We are now going upon the most important of all Subjects—Government." He first griped that the drafting committee was "overcharged with useless Members." He expected it to hear "a thousand ridiculous & impracticable proposals," likely to end in "a Plan form'd of hetrogenious, jarring & unintelligible Ingredients." The only solution, he wrote to Richard Henry Lee, was to have "a few Men of Integrity and Abilitys, whose Countrys Interest lies next their Hearts," take charge. He did not mean himself. Mason wanted Lee to return from Philadelphia, much as Jefferson wished to do. But Congress faced equally pressing challenges, and the role that Mason cast for Lee quickly fell to him instead. "The political Cooks are busy in preparing the dish," Edmund Pendleton wrote to Jefferson. "Colo. Mason seems to have the Ascendancy in the great work."

That "great work" proceeded in two stages and produced two distinct documents. One was a declaration of rights, drafted by Mason in late May and finally approved on June 12. By then Mason and his committee were at work on the constitution proper, which the convention adopted on June 29. To our way of thinking, this reverses the proper order. The best-known example of constitution making would occur a decade later, when the new federal Constitution was framed first, and only then amended through the supplemental Bill of Rights. But the course that Virginians took in 1776 better captures the revolutionary logic of their action. In their view, the current crisis had reduced Americans to the condition that John Locke had called "a dissolution of government," the subject of the concluding chapter of the *Second Treatise*, a text that Mason, like Jefferson, knew well. Mason and his colleagues in the convention readily agreed that both George III and his Parliament were fully exhibiting the "Arbitrary Will" and "Arbitrary Power" that Locke made the criteria for judging whether a government was dissolved and its people placed "at liberty to provide for themselves, by erecting a new Legislative." For Locke, as for the revolutionaries of 1776, the legislative was the supreme branch of government, "*the Soul that gives Form, Life, and Unity* to the Commonwealth."

Designing their new legislatures was the most important issue American constitution makers now faced. But a people emerging from a dissolution of government had a prior obligation, or so at least the ideas of compact associated with Locke implied. In the absence of historical evidence, skeptics like Hume and Saville could doubt whether any such "original" compacts had ever existed. But Americans were now living history as well as reading it, and the lack of past examples did not preclude their setting future precedents. A people forming a government had the moral right and duty to affirm the purposes for which they were uniting, to declare the rights they wished to secure. For how else could they judge which form of government—which kind of legislative—best answered their needs?

That agreement logically had to be reached before the deliberations about government proper could begin, and that is why Mason drafted a declaration of rights first. This also explains why that declaration cannot be read simply as a precedent for the amendments that James Madison drafted in 1789. Scholars have faulted the Virginia declaration for falling short of the broader range of civil liberties found in the later Bill of Rights. But that criticism mistakes its very purpose. Madison's project of 1789 was to compile a list of restrictions on the powers of the national government. By contrast, Mason's purpose was to define the duties of the republican citizens of Virginia, and in so doing, to describe who they were as a people.

In part, they were the freeborn heirs of traditions of English liberty that the seventeenth-century struggle against Stuart autocracy had vindicated, but which right-thinking Whigs traced back not only to the Magna Carta of 1215 but to the "ancient constitution" that predated the Norman conquest of 1066. Some of the rights that Mason included thus *were* antecedents for Madison's list of 1789. Articles 8–11 of the final Virginia declaration of 1776 covered basic civil rights such as trial by jury, confrontation of witnesses, and freedom from "excessive bail" and "cruel and unusual punishments." Article 12 endorsed freedom of the press, while Article 13 praised "a well-regulated Militia" as an alternative to a dreaded standing army. Article 16, with a key amendment proposed by the young Madison, recognized "the free exercise of religion, according to the dictates of conscience."

Yet in other articles Mason expressed deeper purposes that would today seem entirely out of place in a constitutional text. Article 15 of the final draft, for example, has no counterpart in a modern American bill of rights: "That no free government, or the blessings of liberty, can be preserved to any people but by a firm adherence to justice, moderation, temperance, frugality, and virtue and by frequent recurrence to fundamental principles." Article 2 instructed Virginians that all officials were their "trustees and servants." Article 3 reminded them that government was instituted for "the common benefit, protection, and security of the people." When government failed to secure those ends, a people had the right "to reform, alter, or abolish it." Article 4 affirmed that just as officeholders did not constitute a separate class, so their offices were neither "descendible" nor "hereditary." Article 5 reinforced this point by stating that those wielding political power "should at fixed periods, be reduced to a private station"—an echo of Addison's *Cato*—lest they forget that they were subject to the very laws they were enacting. Article 6 stated that elections were to be "free" and the suffrage granted to "all men" possessing "a permanent common interest with, and attachment to, the community."

These articles did not state *rights* in our sense of the term. They imposed no firm limits on the power of government, nor was their primary concern to protect subjects against the coercive power of the state. They instead equated the right to enjoy the blessings of liberty with the duty of living as republican citizens. Mason's verb of choice was the monitory *ought*, not the mandatory *shall*. The provision urging officeholders to return to a "private station" had no legal effect; it set no term limits on anyone. Article 6 did not define how "a permanent common interest" was to be measured. Mason cast his articles not as legal commands but as political principles. He used the language of political duty, moral obligation, and civic responsibility, all measured in the personal attributes of individuals. This theory treated liberty not as a libertarian formula promoting personal autonomy, but as the morally superior alternative to its evil twin, licentiousness. In the political language of the era, liberty was best defined in opposition to the vices that licentiousness wantonly indulged.

The moral dimension of Mason's republicanism assumed that a people who would rule themselves collectively also had to govern themselves individually. This was what the historian Joyce Appleby has called "the radical *double entendre* in the right of self-government." It was to inculcate that form of self-government that Article 15 praised the moral attributes of "justice, moderation, temperance, frugality, and virtue." Of these, the last was most important. For the revolutionaries of 1776, virtue meant the ability of citizens to subordinate private interest to public good. The term had an avowedly martial and thus masculine cast. Since Machiavelli, advocates of republicanism had held that the virtue of a citizen inhered in his willingness to take arms in defense of his republic, and not rely on the mercenary standing armies that Article 13 denounced. Yet the other qualities Mason commended were less gender-specific. Indeed, they sounded much like the private traits he admired in his late wife, Ann. They anticipated the new role that American writers would increasingly ascribe to wives and mothers, whose duty would be to sustain and nurture republican citizens.

Mason thus used the declaration of rights to give a republican identity to "the good people of VIRGINIA" in whose name the convention would "ordain and establish the future form of Government" for their commonwealth. In this sense, the constitutional project had profoundly moral dimensions. But the actual business of constitution making was primarily about designing institutions and offices and determining who could join in their appointment and operation.

In drafting a constitution for the "ancient dominion" of Virginia, Mason and his colleagues demonstrated just how widespread republican ideas were. "Land barons" they may have been by the standards of John Adams, whose Braintree farm could well disappear in some small corner of the Mason estate. But in practice the variance between his hopes for the southern colonies and Mason's ideals proved narrow. Such differences as do exist illustrate not disagreement over basic principles but rather the range of experimentation and adaptation that Adams expected would occur.

Mason began his constitution by creating three "separate and distinct" departments and enjoining "that neither exercise the powers properly belonging to the other." This statement paid homage to Montesquieu, the celebrated French philosophe whose great work, *The Spirit of the Laws*, first divided the domestic powers of governance into legislative, executive, and judicial departments. The convention made that prohibition even more explicit by forbidding any *person* from exercising the power of more than one department at any time (except that justices of the peace could serve in the assembly). The constitution then predictably endorsed the bicameral principle. But by making Virginia's House of Delegates and Senate "a complete legislature," it broke with both Montesquieu and Adams by denying the governor a veto over legislation. The only royal prerogatives that Mason proposed allowing the governor to retain were to prorogue or adjourn the legislature and to issue reprieves and pardons. The convention went further, explicitly denying the governor any power to disband the legislature. Lest there be any doubt, the constitution additionally declared that the governor "shall not, under any pretence, exercise any power or prerogative by virtue of any law, statute, or custom of *England*." The governor would be annually elected by the legislature, prohibited from serving more than three consecutive terms without becoming ineligible for another four years, and made subject to impeachment for "mal-administration, corruption, or other means by which the safety of the state may be endangered."

Part of this animus against the executive reflected the anti-monarchical rage that swept America in the spring of 1776. Naively or otherwise, the colonists had asked and expected much of their last king. In 1774 Jefferson reminded George III that he held "the balance of a great, if a well poised empire" in his hands, obliging him "to deal out to all equal and impartial right." His Majesty refused to take the hint. With a boost from *Common Sense*, American disillusionment with the king evolved into a renunciation of monarchy itself. All varieties of royal rule were suspect—Britain's constitutional monarchy, with its reliance on ministers who were accountable to Parliament, as well as the absolutist forms that still flourished on the Continent. But the dangers of unbridled executive power and influence had also been a staple theme of American politics for decades, an obsessive concern reinforced by the disputes between royal prerogative and legislative privilege that repeatedly disrupted colonial politics. In the year of independence, nothing seemed more natural than to make the restriction of executive power a prime object of constitutional reform. So strong was this impulse that it overcame a far more important consideration. Any commonwealth preparing to oppose the dominant empire of the Atlantic world would be well advised to enhance the authority of the executive, not weaken it. It was, after all, the one branch of government whose distinctive virtues of "energy" and "despatch" were most urgent in time of war. But at least in 1776, ideology and memory outweighed prudence and calculation, and reaching agreement on the need to cabin the executive thus proved mistakenly easy.

A second, seemingly more fundamental question also generated little debate: the *extent* of the power to be vested in the legislature. None of the state constitutions bothered to enumerate the specific powers the legislature would exercise. They simply followed the dominant doctrine of parliamentary supremacy and assumed that all forms of human activity were subject to legislative regulation. Power might flow originally from a sovereign people, but once they reconstituted a supreme legislative, its power was plenary in nature. The restraints on its exercise did not depend on the positive enumeration or negative limitation of specific powers. They came instead from the institutional division of authority within the legislature and the vigilance of its constituents, the people themselves. This understanding turned the problem of preserving constitutional balance back to the two great questions with which Adams and other political writers were already wrestling: the character and authority of an upper house, and the fundamental nature of political representation.

Mason had considered both issues well before he took up his new vocation as constitution framer. Like other observers of colonial politics, he was troubled by the conspicuous absence from American governance of any institution that could serve as a counterpart to the House of Lords. The colonial councils were hybrid bodies that acted both as upper legislative houses and as executive councils advising the governor. Worse, councilors were either directly appointed by the Crown or subject to royal veto or removal. How such dependent bodies could fulfill the balancing duty assigned to them under the dominant theory of mixed government* remained an unsolved puzzle before 1776. As Mason noted in his "Extracts and Remarks" of 1773, "no candid Man, well inform'd in the principles of the British Constitution, & acquainted with the tumultuary Nature of Publick Assemblies, will deny" the real value of having "such an intermediate Power between the People and the Crown." But to attain that end, "the members of this intermediate branch should have no precarious Tenure." Ideally they should serve for life, like members of the House of Lords; "be equally independent of the Crown and of the People"; and no longer wield the executive and judicial powers that the councils had also held.

Amid the republican enthusiasm of 1776, the idea of life tenure for any official lay almost beyond contemplation. Almost, but not quite. One notable Virginian *did* think that British practices merited imitation: the congressional delegate Carter Braxton. With one eye on Williamsburg and the other cast askance at John Adams, Braxton published a short *Address* urging the Virginia convention to follow British forms as closely as possible by establishing a governor to serve during good behavior and a senate with life tenure. Braxton disparaged the ideal of individual virtue that Adams had made the fundamental principle of *Thoughts on Government.* Schemes of government based on a self-denying human virtue "may be practicable in countries so sterile by nature as to afford a scanty supply of the necessaries, and none of the conveniences, of life," he warned. But they could never long appeal to a "people who inhabit a country" like America, "to which providence has been more bountiful." Americans were already too eager to consume, too anxious to get ahead, in a word, already too self-interested, to play the ascetic Spartan role that strict republican theory assigned them.

* Under this theory, the balance for which the British constitution was so admired in the eighteenth century depended on the capacity of the House of Lords to mediate between the concentrated authority of the Crown and the dynamic energy of the people, as embodied in the House of Commons. In *The Spirit of the Laws,* for example, Montesquieu made the House of Lords a critical factor in maintaining the liberty-preserving character of the British constitution. Being French, Montesquieu tended to be long on theory and short on empirical accuracy. No scholar today could agree that the aristocratic chamber played the institutional role that Montesquieu assigned to it. But in terms of constitutional theory, his view was the conventional one at the time.

For Braxton, it followed that Americans should replicate, as best they could, the mechanisms of governance that had bestowed stability and prosperity on Britain.

Braxton's doubts about his countrymen's civic virtue earned few plaudits. Richard Henry Lee dismissed his *Address* as a "Contemptible little Tract." Mason's opinion of it was no better. But he was too orthodox a Whig to escape the legacy of mixed government. In his draft constitution, he contrived a new mechanism to give the upper house a distinct social character. Mason proposed dividing the state into twenty-four districts and allowing each county within a district to appoint twelve "deputies, or sub-electors," each "having an estate of inheritance of lands" worth five hundred pounds. The sub-electors of each district would then appoint one senator, possessing an estate worth at least two thousand pounds, twice the value of the comparable qualification Mason proposed for members of the lower house. Senators would serve four years, then be "rendered ineligible for four years," with one quarter of the chamber replaced annually. By Braxton's standards, this would hardly turn the state senate into a chamber of aristocracy. Yet it did imply that those with the greatest holdings of land and slaves constituted a distinct social class whose interests were not identical to those of ordinary planters.

Something about this primitive effort to secure the power of the gentry must have disturbed Mason, for when he turned to the lower house, he adopted a more democratic principle. In the Virginia declaration of rights, he had linked the right to vote with the ability to show "sufficient evidence of permanent common interest with, and attachment to, the community." Conventional views restricted voting to those possessing freehold property. A person owning no real property, whose livelihood depended on the custom and will of another, had no independent judgment to exercise and could thus be denied the highest privilege of citizenship. Mason rejected this view. His draft expanded the idea of "common interest" and "attachment" to cover tenant farmers whose leases had at least seven years to run, as well as "every housekeeper"—that is, homeowner—resident in his county for one year, who "hath been the father of three children."

In effect, Mason was trying to recalibrate the balance of government at both ends of the scale, formally distinguishing the upper house while broadening the electorate. His colleagues in the fifth provincial convention were less inclined to experiment. They eliminated his scheme of sub-electors and allowed senators to be chosen by the same electorate that would choose the lower house. The convention also left the suffrage as it was, under a forty-year-old statute giving the vote to free white males owning twenty-five acres of improved land or a hundred acres of unimproved land. Perhaps because the delegates were confident that the old planter elite would remain in power, they imposed no property requirements for membership in the legislature.

Working out prosaic details like these is always a principal task of constitution making—and often marks the point where great principles and expedient calculations overlap. In June 1776, expediency also limited the attention the convention could devote to the whole enterprise, and thus to resolving the puzzles that their great constitutional experiment was only beginning to pose. Constitution making in 1776 was at once progressive and retrospective, forward-looking yet swayed by traditional principles. Had Americans been governed strictly by logic in 1776, they could have abandoned bicameralism entirely. In a republic there was only one people to represent. If they were represented "in miniature" in one house, why establish a second one based on the irrelevant legacy of aristocracy? Believing as he did that time would carry American families up and down the ladder of social mobility, Mason was an unlikely voice for according additional political rights to some citizens on the basis of property. Yet with his attachment to the ideals of the balanced constitution, he could not escape the orthodox view that a second chamber was essential. And if it was, it had to represent something. Aristocratic wisdom could not be measured or quantified; property could.

Had the convention wished to ponder such puzzles, this one issue could have consumed weeks of debate. But June 1776 was no moment for prolonged haggling. Even if the absent Jefferson was right that securing a free constitution was "the whole object of the present controversy," other tasks commanded the delegates' attention. A scant three weeks passed from the time that Mason prepared his draft (probably June 8–10) until the constitution received final approval on the twenty-ninth; and the entire convention spent only a week on the matter. The speed with which it acted seemingly confirmed that John Adams had been right to think (back in November 1775) that constitution making would not be arduous.

Mason's stature, and the skill and learning he brought to the task, certainly helped ease the process. Then too "the political cooks" at Williamsburg were in general accord about the recipe they should follow. They regarded themselves as the rightful heirs and legatees to a century and a half of self-government, dating to the heralded first meeting of the House of Burgesses in 1619. They did not doubt their ability to maintain control of the legislature. A conservative like Edmund Pendleton could prefer to see the vote "confined to those of fixed Permanent property, who cannot suddenly remove [from the state] without injury to that property." Others, like Jefferson, wanted to extend it "to all who had a permanent intention of living in the country." But their disagreements on this point remained a matter for polite dispute. At least in Virginia, constitution makers could be republicans in principle, devoted to the people's right to rule, because they remained a dominant gentry in practice, confident in their own capacity to govern.

Beyond Virginia, however, the revolutionary act of constitution making did prove more contentious, exposing fissures in the body politic that widened under the grinding pressure of events. The most revealing cases were Pennsylvania, Maryland, New York, and Massachusetts. Like Virginia, each of these states had its own set of political actors and interests. But far more than Virginia, they also contained upstart leaders and groups who saw the Revolution as a moment to assert bolder claims for political recognition.

The most volatile of the new republics was Pennsylvania. In Virginia the adoption of the constitution produced no real change in the structure of power. But in Pennsylvania the collapse of the old proprietary regime created a turbulent situation that allowed a group of political outsiders to seize the constitutional initiative. They acted in a charged atmosphere in which public discussion was already suffused with claims for a radical political equality running beyond Mason's moderately reformist urges. Although Pennsylvania was widely regarded as the most open, fluid, and tolerant society in the Americas, the rhetoric of 1776 suggested that it was rife with unjust inequalities. Philadelphia was a hotbed of agitation, much of it arising from the militia companies that had formed enthusiastically a year earlier, their ranks filled with artisans who called themselves Associators. The initial concerns of these Associators involved military discipline, uniforms, and weapons, and pitted the gentlemen who planned to become officers against the artisans they expected to command. A broader agenda began to emerge, however, with the formation of the Committee of Privates in the fall of 1775, which soon began acting as a forum for airing the grievances of Philadelphia's working population. By the spring of 1776, Philadelphia's newspapers were filled with articles challenging conventional notions of privilege and rank and appealing for a new political equality that did not depend on the traditional association of political rights and property.

The likely authors of these articles and the leading advocates of a new equality were not themselves artisans. Rather, they came from an energized group of activists who welcomed the collapse of the old colonial regime as an opportunity to vie for political leadership. Their social origins and positions were respectable—more so, say, than those of their friend and ally Tom Paine, the corset maker's son and former excise officer. Timothy Matlack was a brewer whose love of horseracing and cards ultimately led the same Quaker brethren who redeemed him from debtors' prison to expel him from Friends' meeting. The Dublin-born George Bryan was a moderately successful merchant, a 1771 bankruptcy notwithstanding. Joseph Cannon studied mathematics at the University of Edinburgh and now taught it at Philadelphia's own college. Yet these men remained upstarts who felt a biting resentment toward both the older proprietary hierarchy and the mercantile elite who looked to Thomas Willing and Robert Morris for leadership.

Had John Dickinson not gutted his own influence by resisting independence, he might have played the same role in Pennsylvania that Mason took in Virginia. But his withdrawal from public life left the field free for the insurgents in the convention that met in Philadelphia in mid-July. It had been summoned in turn by a provincial conference that met briefly (June 18–25) after the legal assembly finally dissolved. The conference expanded the electorate eligible to vote for convention delegates to include all taxpaying Associators who had lived in Pennsylvania for a year, while also enfranchising the province's large German population. With only a fortnight to mobilize for the election of delegates, the Committee of Privates quickly alerted the Associators to the dangers ahead. "Suspicious Characters" would "strive, by every Art in their Power, to recommend themselves to your Favour," its broadside of June 26 warned. Allow such "improper persons" to make your constitution, and "You will have an Aristocracy, or Government of the Great." The committee defined "improper" to include not only "great and overgrown rich Men" (like the well-fed Robert Morris) but also members of "the learned Professions," who "are generally filled with the Quirks and Quibbles of the Schools." Or so at least Professor Cannon, the self-professed author of this broadside, professed to believe.

The convention met on July 15 and turned to constitution making three days later. Like the convention in Virginia, it began by considering a declaration of rights. Though largely copied from Mason's draft, which had been widely reprinted, it reflected Pennsylvania's unique history as a haven for religious dissenters and a magnet for free immigrants from Britain and Germany. Its statement of religious freedom went beyond Virginia's in denying the state any authority over religious belief and exercise. It also contained a wholly new article affirming "that all Men have a natural inherent right to Emigrate from one State to another that will receive them." Most strikingly, the draft declaration advanced a radical notion inspired by the egalitarian talk of the past year: "That an enormous Proportion of Property vested in a few Individuals is dangerous to the Rights, and destructive of the Common Happiness, of Mankind," the final article declared, "and therefore, every free State hath a Right by its Laws to discourage the Possession of such Property."

So bold a vision of republican society was inconceivable in Virginia, where the amassing of land and the bondsmen to work it defined the very essence of economic life. Nor did it survive the scrutiny of the Pennsylvania convention. Even so, it did represent Pennsylvania's refinement on Mason's call for "justice, moderation, temperance, frugality, and virtue." Article 14 of the Pennsylvania declaration borrowed that same republican formula, but with a revealing shift in emphasis. It replaced "virtue" with "industry"—a word that naturally appealed to the artisan community. More provocatively, Article 14 treated these qualities as the criteria to which republican citizens should pay "particular attention . . . in the choice of officials and representatives." Here was a new twist on the double-entendre of self-government, one that shifted the burden from citizens governing themselves as individuals to citizens collectively governing the commonwealth by watchfully monitoring their elected officials.

That too reflected the radical thrust of the constitution that the convention adopted on September 28. It was distinctive in three critical respects. First, Pennsylvania became the only state to vest legislative authority in a unicameral assembly. That decision could be seen as mere inertia, for under William Penn's frame of government, the province had always had a single house. But the upstarts in the convention felt little affection for the Penns or their regime. Their defiance of the best political science of the day was consciously radical. They were pursuing the logic of republicanism itself by supposing that a true republic founded in equality needed only one house because there was only one people to represent. That conviction supported a second noteworthy innovation. Section 15 of the constitution required that "all bills of public nature shall be printed for the consideration of the people" before enactment. Moreover, "except on occasions of sudden necessity," that final action should be delayed "until the next session of assembly." This was republicanism in a pure form. Americans generally agreed that the people themselves were the ultimate check against the abuse of power. But only in Pennsylvania did this belief end in wholesale rejection of the ideal of balanced government. That disdain was reflected in the third innovation that distinguished the Pennsylvania constitution. It eliminated the office of governor entirely, replacing it with a twelve-member rotating council with an appointed president. Rather than treat this weakened executive as the one branch of government where experience should be nurtured, the constitution mandated a rule of rotation. "More men will be trained to public business," the relevant article explained, while "the danger of establishing an inconvenient aristocracy will be effectually prevented."

Constitution making in Pennsylvania was distinctive in one other critical respect. The aspiration for popular consultation expressed in Section 15 carried over into the constitutional deliberations themselves. On September 5, the convention resolved to make the draft constitution available for public comment. It appeared as a pamphlet and was reprinted in two Philadelphia newspapers. Although the convention made no further effort to collect public opinion, this step did enable interested citizens to voice their concerns before a final draft was approved. A flurry of articles in the Philadelphia press—bearing democratic bylines such as "C," "F," "K," and "R," rather than the classical pseudonyms favored by learned gentlemen—discussed not only an array of specific clauses, but also the wisdom of the convention's taking so radical a course without giving the people an opportunity to assess its work. As a result, an initiative mounted in conformity with the convention's populist ethos opened a political fissure that divided the state for years to come. Critics insisted that Pennsylvania would risk too much by ignoring the familiar principles of institutional balance and vesting legislative power in one chamber. But the convention had resolved that issue even before the draft constitution was published, and the victorious majority adamantly opposed its reconsideration. Alerted to the opposition this provision was sparking, the convention insulated the constitution from the further review its critics sought. On September 26 its members passed an ordinance requiring voters to swear an oath of loyalty to the constitution itself.

Such an oath did not divide patriots and loyalists, but rather it split the patriots into two nascent political parties, the Constitutionalists and the Republicans (the former supporting the new constitution, the latter seeking its revision). Within a week, John Adams rightly predicted the result. "We live in the Age of political Experiments," he wrote to Abigail. Some would inevitably succeed, others fail. Adams did not doubt which fate awaited Pennsylvania. It "will be divided and weakend, and rendered much less vigorous in the Cause, by the wretched Ideas of Government, which prevail, in the Minds of many People in it." The turmoil in Pennsylvania confounded his confident prediction of 1775. Now was not the moment, Adams believed, to probe the outer limits of government by consent. He had said as much to Abigail back in the spring, when she wrote her famous letter asking John to "Remember the Ladies" in the new legal code she imagined Congress would draft and warning her absent husband that we women "will not hold ourselves bound by any Laws in which we have no voice, or Representation." Adams the would-be lawgiver did not take his "saucy" wife's plea seriously, replying in mock horror to her audacious request. But he gave a more thoughtful reply to a male correspondent who pressed him to support a significant expansion of the suffrage. To open this question now would prove too "fruitfull a Source of Controversy and Altercation," Adams warned. "There will be no end of it. New Claims will arise. Women will demand a Vote. Lads from 12 to 21 will think their Rights not enough attended to, and every Man, who has not a Farthing, will demand an equal Voice with any other in all Acts." The ferment in Pennsylvania only confirmed Adams in his cautionary opinion. There was a war to win, and too many questions of this nature could only distract attention from that struggle.

The limits of consent were also sharply contested in Maryland, but with very different results. At the outset, constitution making there threatened to grow even more turbulent than in neighboring Pennsylvania. In the summer of 1776 elections for the provincial convention that would draft a constitution were marred by disorder at the polls. In five counties, arms-bearing taxpayers forcibly ejected election judges who tried to limit the poll to legal voters. Remarkably, more than half of the delegates chosen had never served in any of the eight previous conventions. Many of the newcomers seemed ripe to follow the leadership of Rezin Hammond and John Hall, two prominent planters who displayed the demagogic qualities that history indicated would emerge in times of turmoil. Soon after the convention began work on the constitution, the two men presented a petition signed by nearly nine hundred freeholders in Anne Arundel County, demanding that the suffrage be extended to all taxpayers, regardless of their ownership of real property.

The sharpest critic of these developments was Charles Carroll of Carrollton, who had recently returned from Congress, where he had the unique distinction of being the only Catholic signer of the Declaration of Independence (later, he would be the last surviving signer). The Carroll family story is itself a remarkable American saga. Its founder, the first Charles Carroll, came to Maryland in 1688 in the service of its Catholic proprietor, Lord Calvert. Carroll was intent on recouping family fortunes lost in the great confiscation of Catholic lands that followed Oliver Cromwell's conquest of Ireland three decades earlier and further diminished in the 1690s by the Carrolls' continuing support for the deposed James II. The first Carroll's ties to Maryland's proprietors and two strategic marriages launched a successful career that combined commerce, money lending, law, and the ongoing acquisition of land. At his death in 1720, he owned nearly 50,000 acres, a personal estate valued at 7,535 pounds, and 112 slaves. Charles of Annapolis was his eldest surviving son and, within a decade, the only survivor of ten siblings. Being his father's chief heir was an advantage in itself. But so was the steely determination with which the second Carroll pursued his activities in agriculture, iron manufacturing, and money lending. Rather than rely on slaves to toil on his plantations, he imported tenant farmers to work the additional holdings he acquired with each passing decade.

The third Charles Carroll (the signer) thus had the makings of aristocracy about him. So John Adams perceived him when the two met at Philadelphia in September 1774, as the First Continental Congress was getting under way: Carroll was "a very sensible Gentleman, a Roman catholic, and of the first Fortune in America. His Income is Ten thousand Pounds sterling a Year, now, will be fourteen in two or 3 years, they say, besides his father has a vast estate, which will be his, after his father." Adams did not note, and probably did not know, the strange fact that dogged Carroll until he was twenty. He was born out of wedlock, to his father's common-law wife, Elizabeth Brooke, whom the elder Carroll finally married only after his son was near to completing a lengthy European education, which included eleven years of study at Jesuit colleges in Flanders and France and another soul-sapping five reading law at the Inns of Court. As a bastard, Charley could not legally inherit the vast estate his father was grooming him to manage, and the threat of non-inheritance worked to keep him on the path his father assigned. Charley's repeated pleas to be allowed to come home were answered with an endless stream of fresh instructions about his education. Only in 1765 was Charley allowed to end his exile. His adored mother had died four years earlier. But her marriage to Charles of Annapolis made their son a legal heir.

A family as wealthy as the Carrolls should have been a major force in Maryland politics. But since 1718, the colony originally founded as a Catholic refuge had barred Catholics from voting or holding office—largely in reaction to a controversy the first Carroll had sparked. Those legal restrictions did not apply, however, to the extralegal bodies the crisis of 1774 summoned into existence. Charley visited Philadelphia while the First Continental Congress was just getting down to business, meeting delegates and gleaning some news of its proceedings. Returning to Maryland, he was elected to the provincial convention. In 1776, Congress took advantage of his religion and his fluency in French to send him as one of its commissioners to Quebec. On his return, he was elected to Congress in his own right. He was there less than four weeks, time enough to sign the Declaration; then he returned to Maryland in time for the ninth convention, which would draft a constitution.

Carroll was lucky to be able to attend the convention. The voters in his native county of Anne Arundel rejected him, but he managed to obtain a seat from the capital, Annapolis. Once seated, he was appointed as one of the six-member committee that would draft the constitution. Its members were a virtual roll call of the available patriot leadership, suggesting that the newcomers to the convention were willing to defer to their betters. Carroll was not so sanguine. By October, his letters to his father were veering toward despair. Part of his anxiety reflected the worsening military situation in New York. But weighing even more heavily were fears "that no free gover[nmen]ts. can be established during such contention" and that his "poor deluded countrymen" were being gulled by "the detestable villainy of designing men." One had only to look at Pennsylvania to see where it would end. "If they set out thus," he wailed, "how will they finish?" Two weeks later, he still thought that "we shall have a very bad govert. in this State." That fear immediately gave way to a different hope: that "this winter some negotiation will be set on foot, wh will lead to peace on safe terms to the Colonies." If not, Americans would soon face ruin, less from "the calamities of war" than the errors of the "simple Democracies" they seemed bent on creating. Of all forms of government, this was "the worst, and will end as all other Democracies have, in despotism."

Yet Carroll's road to defeatism and constitutional despair took two unexpected turns. One had to await Washington's victories in New Jersey and the new year of campaigning they bought. But the other came much sooner and led him to a more measured view of constitution making. The constitution might not be so bad after all, he wrote to his father on October 20. There would be a bicameral assembly, with senators serving five years, two less than the drafting committee had proposed. Terms for the lower house had also been reduced from three years to one. But that only brought Maryland into line with the republican maxim that made annual elections "a palladium of liberty." Carroll now thought that most of his colleagues "mean to do what is right in their judgments." It was their knowledge that was deficient. With his years of Old World schooling, Carroll had intellectual credentials these untutored rustics lacked: "a knowledge of history, & insight into the passions of the human heart." They knew nothing of "the Governts that have existed in the world!" he marveled, or "the causes wh brought those govts. to destruction! Yet every man thinks himself a judge—and an adept in the great & difficult science of Legislation."

Yet in the end these "adepts" recognized their own limitations and deferred to Carroll and his eminent colleagues on the committee. The reductions in legislative terms were the only real changes made to the committee's draft. Two efforts to broaden the suffrage were defeated, narrowly but decisively. The committee's greatest innovation—the design of a true senate—survived intact. A term of five years would still allow Maryland's fifteen senators to serve longer than their counterparts elsewhere. More than that, they would be safely insulated from the people at large. They would be appointed by a small pool of electors (two per county) owning real or personal estates valued at five hundred pounds. Senators themselves had to own property worth one thousand pounds; members of the lower house, five hundred; the governor, five thousand.

In February 1777 Carroll himself joined the first senate formed under the constitution. Here he learned a profound lesson about the gap between constitutional aspiration and political expediency. Maryland's planter and merchant elite discovered there was a literal price to pay for their success in attaining a balanced constitution. That price was the adoption of a financial program that adversely affected their interests, in two ways. First, it shifted the state's revenue sources from a poll tax to taxes imposed on land and slaves. Second, the assembly enacted a law making both continental and state-issued paper currency legal tender for the payment of debts. Together, these measures reflected a conscious effort to maintain—or perhaps purchase—the political loyalty of small planters, tenants, and artisans, and with it, a willingness on the part of the propertied elite to sacrifice more for the cause.

In principle, the paper money act was exactly the form of unjust legislation the new senate was meant to obstruct. Making legal tender of paper currency that would inevitably depreciate would rob creditors and landlords (and the Carrolls were both) of the fair value of obligations owed them. But when Charles Carroll cast the lone senate vote in opposition, he acted from a different form of family loyalty than mere self-interest. His father, Charles Carroll of Annapolis, had virtually ordered his son to do so. The older man's first comment on the pending legislation was that it "will Surpass in Iniquity all the Acts of the British Parliament agt: America." That was pretty nearly his mildest remark on the subject. Lawmakers who favored such a measure were like "Highway men & pickpockets," acting "unreasonably unjustly Inniquitously & Tyrannically," worthy of the "Brand of Eternall infamy." Should Charley's dissent prove futile, his father counseled, resignation from public life was the only course of action that remained. For "when Vice prevails & wicked Men bear Sway the post of Honour is a Private Station" (quoting, again, Addison's *Cato*, with "wicked" substituting for "impious").

This blast marked the opening of a fascinating exchange that continued for months. The emotional bond between father and son was no less affectionate or intimate than that between husband and wife. Indeed Charley rarely wrote to his own wife directly, but trusted his father to let her read his filial letters instead. The paper money issue forced Charley to issue a declaration of independence of his own. Having obeyed his father and cast his lone dissenting vote against the original legislation, the younger man thereafter declined the path of absolute justice and Cato-like virtue that his father hectored him to take. In "such great revolutions" as ours, the senator replied on March 15, 1777, "partial injustice & sufferings" were unavoidable, and accordingly he had "prepared myself against the worst events." Besides, any man who now chose to withdraw, even on the basis of conscience, risked becoming an enemy to the republic that expected his service. But the elder Carroll was relentless, sorely trying the patience of a son who became ever more exasperated even as he remained devoted. Having worked so hard to expand the estate his father had bequeathed, Charles of Annapolis could not bear the thought of impious men infringing on the basic right of property. Then too he carried more freshly than the senator the historical memory of the political persecutions that had cost his ancestors their Irish estates.

Still, the younger Carroll also recalled that history. If America was conquered, the Carrolls would relive their ancestors' experience. Their estate would be forfeit and America itself "become a second Ireland" ruled "by a military force & corrupt & venal governments." Yet Charley also knew there was no alternative to the sacrifices that political realities imposed. "If we can save a third" of the "monied part" of our estate, "and all our land & negroes," he wrote in April 1777, "I shall think our selves well off." So too the son demurred when his father ranted against lending money to the cause. To avoid the bitter consequences of defeat, was it not better to lend, in the hope that victory would require the public debt to be "redeemed in time," than to pay the inflated prices that the personal purchase of "land, negroes, or any thing useful" now entailed?

Carroll's acquiescence to the democratic will, then, was more calculated than principled. The Revolution allowed him to overcome the legacy of political exclusion that had been the ironic fate of Maryland Catholics. In other respects it was deeply disillusioning. Though both Carrolls were sound Whigs, the challenge of preserving their property in the face of Maryland's financial policies disabused them of the idea that independence had ushered in a reign of republican virtue. The new constitution avoided the obvious errors committed in neighboring Pennsylvania. But a constitution, however carefully drafted, could only set up a framework for deliberation and decision. It could not control their outcome. That was a matter of politics, and Carroll's initial accommodation to the policies his father despised was a mark of how men of his standing responded to the political ferment the Revolution had released.

Still, there were constitutional lessons to learn from the experiments of 1776. The most striking were absorbed in New York. With the British holding New York City, its provincial convention shuffled off to White Plains, Fishkill, and Kingston before finally roosting in Poughkeepsie. The loss of the capital had another effect. Some of the populist sentiments heard in its great rival port of Philadelphia were echoed in New York City. But with its artisan population either quarantined there or dispersed, that voice was muted, though not wholly silenced. As in Maryland, notable patriot leaders dominated the committee appointed in August to draft a constitution, including the moderate core of the New York delegation to Congress: James Duane, John Jay, Robert Livingston, and Gouverneur Morris.

Partly by circumstance, but apparently also by design, constitution making in New York became a protracted process. The circumstance was the war. With part of the state occupied and the rest gravely threatened, the New York convention did not have the luxury of working solely on the constitution. Like conventions in other states, it was also a surrogate legislature whose most urgent priority was mobilizing for defense. But some committee members saw advantages to moving slowly—or so Livingston later claimed. Though the drafting committee made some progress by early fall, it did not report a completed constitution until March 12, 1777, and another six weeks passed before its adoption. The whole process, Livingston observed, required "well timed delays, indefatigable industry, & a minute attendance to every favorable circumstance," all designed to avoid the gross errors of Pennsylvania.

Whatever its cause, this lag allowed the convention to grasp a key truth that the republican enthusiasm of 1776 had too easily overlooked. War was no time to experiment with a radical purging of executive power. Too much was at stake to hobble the one department on which the daily operations of government depended. The New York constitution marked the point where the flood of anti-monarchical feeling that swept the country after the publication of *Common Sense* began to recede. The governor would serve for three years, not one, and be elected by the people, not the legislature. Equally noteworthy, the constitution stipulated that the governor would again become a third branch of the legislature through the exercise of a veto over legislation. True, the constitution vested this prerogative not in the governor alone, but in a joint executive-judicial Council of Revision, in which the governor would cast only a single vote. The expanded appointment powers of the executive were given to a second council; here again the governor could cast only one vote. But a governor elected by the people at large had the potential to become a dominant political force, not only on these councils, but within the state as a whole.

Only one state took longer to adopt a constitution than New York: its eastern neighbor, Massachusetts. In June 1775, responding to the urgent situation in that colony, Congress authorized Massachusetts to resume a semblance of legal government under the royal charter of 1691, with the provincial council acting in place of the governor. In September 1776, the house of representatives asked the town meetings to authorize the General Court (that is, the house and the council) to draft a constitution. Three fourths of the towns responding approved the idea. But twenty-three towns did not, and some raised an objection that pointed to a fundamental problem: what steps had to be taken to make a constitution truly constitutional—that is, superior to the ordinary acts of the legislature? It was simply improper, these towns objected, for an existing legislature like the General Court to write a constitution. "The same body that forms a constitution have of Consequence a power to alter it," Concord protested. Such a constitution, being "alterable by the Supreme Legislative is no Security at all to the Subject against any Encroachment of the Governing part on any, or on all of their Rights." The same criticism could be levied against the constitutions adopted in other states. Their conventions too had acted legislatively, doing the ordinary business of running the revolution. But those states had no alternative. Massachusetts, with a legal government (of sorts) in place, did.

The General Court tried to answer this objection in an ingenious way. In May 1777 it asked the towns to empower the representatives chosen at the spring elections to sit as both legal assembly and constitutional convention. When the towns apparently complied, the newly elected representatives appointed a committee to draft a constitution while they conducted normal business. But when the resulting draft was sent to the towns, it too met significant opposition, much of it addressed to the *merits* of the proposed constitution, but some of it again protesting the *mode* of its framing. In February 1779 the assembly made one last try, asking the towns whether a separate convention should be called "for the sole purpose of forming a new Constitution." When the response proved overwhelmingly positive, the assembly set new elections, with an enlarged electorate embracing all free males over age twenty-one. The first true constitutional convention in American history came to order in Cambridge on September 1.

Its three hundred members were far too many to work as a committee of the whole. They gave the task to a drafting committee of thirty, who in turn passed it on to a subcommittee of three notables: John Adams, Samuel Adams, and James Bowdoin. The first of those so honored was right to predict that "I shall have a laborious Piece of Business of it." After an eighteen-month absence on his first diplomatic posting to France, John had been home a week when his Braintree neighbors elected him to the convention. He welcomed their familiar approval after suffering the slings and arrows of his former employers in Congress. Adams took the project back to Braintree, where he could work surrounded by his children as he labored to restore himself to Abigail's good graces. That last duty grew no easier when word came that Congress wished him to return to Europe as its sole peace negotiator. The lone vote against him had come, he learned, from Dickinson, the opponent whom Adams had disparaged in his "innocently intended and so unlukily divulged" letter of 1775. With an eye on his new posting, Adams could ply his "new Trade as a Constitution monger" with rare equanimity, knowing he would not have to defend his draft against the comments of several hundred querulous delegates.

There was one point, though, to which Adams would have clung regardless. This was to have a legislature of "Three Branches"—meaning that the governor should have an absolute veto over legislation. That had also been his view three years earlier. But since then his rationale for the veto had evolved. In his 1776 tract, *Thoughts on Government*, Adams followed conventional wisdom in treating the upper house as the chief balancing mechanism in the government. Now he believed that a veto would enable the governor to act as a "Reservoir of Wisdom" against the legislature, or between its contending houses. "We have so many Men of Wealth, of Ambitious Spirits, of Intrigue, of Luxury and Corruption," he wrote to Elbridge Gerry, "that incessant Factions will disturb our Peace, without it." That conviction supported a second shift in his thinking. In 1776 Adams favored allowing the legislature to elect the governor. Now he followed New York—never an easy thing for a Massachusetts man to do—by proposing that the people should make the appointment (though for a suitably republican term of one year, not three).

Adams sailed for France in mid-November, taking with him his two elder sons, John Quincy and Charles, on the perilous voyage. Even after a difficult and dangerous passage and an unplanned trek through northern Spain, they arrived in Paris well before Massachusetts completed the constitution he had left in its care. The heaviest snows in sixty years had delayed the convention from reassembling. Not until March 2, 1780, was the constitution ready to go to the towns. As Adams predicted, his absolute veto did not withstand the convention's scrutiny. But a limited veto survived, as did the idea of popular election of the governor, and with them the movement toward reinvigorating executive power that began in New York. The convention tinkered with his draft in other ways. It eliminated the qualification that legislators "be of the christian religion" and revised the modest property requirements for election that Adams had proposed. But the final draft was recognizably his, not only in substance but also in form. For Adams brought to his composition a lawyer's eye for structure and coherence missing from the other state constitutions. By comparison, they were jumbles of separate articles, arrayed in almost pell-mell fashion. Adams set his constitution in neatly captioned chapters, with subordinate sections and articles. The greater dignity of his text suggested that a constitution was more than a device for restoring legal government. It was law itself—indeed, supreme law.

In history, or the myths about it, such laws were given by men of exceptional wisdom, like "the greatest lawgivers of antiquity" who Adams thought would envy his generation. In June 1777, when the General Court was acting as a constitutional convention, he had asked Abigail, "Who will be the Moses, the Lycurgus, the Solon" of Massachusetts? "Or have you a score or two of such?" Ever mindful of "the fatal Experience" of Pennsylvania, he thought it would be "a Pity" if the constitution had to be laid before the people for their approval. Constitutional debate could only "divide and distract them," he warned James Warren. "However their Will be done."

So it was in the spring of 1780—but in terms less clear than a philosophical lawgiver might desire. In setting the constitution before the towns, the convention encouraged them to review it by sections, while requiring its approval by "two thirds of the male Inhabitants of the age of twenty one years and upwards." When the delegates reconvened in June 1780, they faced a perplexing task of sorting an untidy mass of disparately organized returns. They had no handy formula to reduce these unwieldy materials to one simple result. In the end, the convention threw up its hands and declared the constitution ratified.

The adoption of this new "frame of government" closed the first chapter in the formation of a distinctive constitutional tradition. In 1774 Massachusetts had been the first colony whose "ancient government" was "wholly abrogated" by imperial fiat. As the last revolutionary commonwealth to form a constitution, it also became the first to perfect a workable concept of popular sovereignty. Elsewhere popular assent meant only the *prior* election of delegates to a convention that would include constitution making among its tasks. In Massachusetts popular consent meant *subsequent* ratification of a document framed by a body called for that purpose alone. This innovation had profound consequences. It made it possible to distinguish the supreme law of a constitution from all the other legal actions the government it created would henceforth take. Turning a constitution into an authoritative text would also spare interpreters the task of rummaging among organic and mechanical metaphors, as John Adams had done back in 1765, to explain what a constitution was or how its guarantees could be preserved.

The precedent set in Massachusetts had one further effect. Building upon the unexpected circumstances that made constitution writing both an act of revolutionary necessity and "the whole object of the present controversy," it helped seal a defining aspect of American nationhood and citizenship. Americans were a people who wrote constitutions, and the constitutions they wrote defined their character as a people. That is why the early declarations of rights remain so revealing, however we grade the care with which they enumerated civil liberties. They made acceptance of republican principles the principal obligation of self-governing citizens. That is also why the debates over suffrage could never be conducted with the prudent restraint that John Adams deemed essential. Adams himself helped set a new standard of accuracy by adding the image of the "miniature" to the familiar language of political representation. This metaphor (or variations such as "mirror" or "portrait") so captured the political imagination that it soon became a cliché. Yet even had he written less boldly, the same clamor for political recognition would have erupted. Revolutionary constitutions premised on government by consent would have to confront tough questions as to exactly whose approval was required, not only for acceptance of the constitutions themselves but also the new laws they would authorize. If America was not yet ready to require the consent of "saucy" women like Abigail, it still had to listen to vocal artisans like Philadelphia's Associators and the small planters in Maryland whose populist sentiments so alarmed the two Charles Carrolls.

One significant segment of the American population, however, could claim no place in the miniature. The figure of a solitary African was a familiar element in eighteenth-century art. One of the most famous was painted by John Singleton Copley, the distinguished Boston-born painter who had left Massachusetts in 1774, not out of disaffection with the American cause but to pursue greater artistic horizons in London. (Aspiring English artisans such as William Buckland might better their lot by immigrating to America. But serious American artists, such as Copley or Benjamin West, found greater inspiration in Europe.) The central figure in Copley's famous painting *Watson and the Shark* (1778) is an African seaman, standing erect in the center of a longboat as his mates rescue Watson, the maimed youth. The African's left hand holds the rope that Watson is too weak to grasp; his right arm extends to his side but does not stretch outward to pull in the victim. That is the work of the other boatmen, who kneel outboard between him and Watson. The African watches the rescue as intently as they do. But where their gazes convey the anguish of the attack and the anxiety of the rescue, his tends toward the mute and impassive. Copley's lone African is a central element of the painting and the historical moment in Havana Harbor that it portrays. Watson had Copley paint the African from life, because such a sailor had indeed figured prominently in his rescue. But to the casual visitor to the National Gallery in Washington, where the painting now prominently hangs, the African seems isolated within the scene in which he is a dominant figure.

African and African American slaves, like George Mason's cruelly named Liberty, were part of the people whom the new constitutions would govern. Indeed, no other members of American society felt the full weight of law more oppressively. But whether they could ever number among the people who figured in Adams's miniature, those whose authority made constitutions supreme law, was a question that the revolutionaries were only beginning to pose—and one it would take their descendants scores of years to answer it.

5

Vain Liberators

THE VAST MAJORITY of Africans who sailed the Atlantic in the eighteenth century were not able seamen, like the solitary figure in Copley's painting *Watson and the Shark*, but rather desperate and despairing captives who crossed the pitching ocean chained below decks, wallowing in filth and bilge, on the Middle Passage to permanent enslavement. Few colonial merchants who participated in this wretched trade profited more than Henry Laurens of Charleston.* From 1748 to 1762, his firm, Austin and Laurens, handled the sales of over ten thousand slaves, the labor on which South Carolina's profitable rice and indigo plantations depended. Then they dissolved the firm. George Austin and his nephew, George Appleby, who had joined the partnership, returned to England with their profits. Laurens invested his in eight plantations in Carolina and its frontier outpost, Georgia. All were worked by slaves.

Henry Laurens was a widower of fifty when the Boston Tea Party crisis broke early in 1774. He observed it not from Charleston but from London, where he was supervising the European education of his three sons while seeking treatment for his painful gout. Laurens was present in "the Cockpit" on January 29 when Solicitor General Wedderburn ripped a stoic Benjamin Franklin for his role in sending the purloined letters of Thomas Hutchinson to the governor's enemies in Boston. Two months later he hastened to Westminster to hear the House of Commons debate the news from America, only to be sorely distressed

* Until 1783, the city was known as Charles Town, but I have followed the later usage.

when the speaker insisted that the galleries be cleared. Like other men of property, Laurens could not shake the belief that British policy rested on a terrible misjudgment of American sentiments and a foolish commitment to repression rather than diplomacy. After sailing home to Charleston later that year, he immediately plunged into revolutionary activities, serving first as a member of the provincial convention and the colony's committee of safety, then as a delegate to Congress and eventually its fourth president. In 1779 Congress ordered him back to Europe to join its fledgling diplomatic corps. Captured at sea, he was interned in the Tower of London, uncertain whether he would be tried for treason as the most important political leader the British managed to snatch.

Compared to the mortal risks his son John was fond of taking, however, the elder Laurens may have considered his imprisonment almost a lark. When war erupted in April 1775, Jack (as he was called) was yet another American student at the Inns of Court, dutifully slogging through the same dreary legal treatises that had taxed Charley Carroll a decade earlier. Jack was a sensitive youth with artistic talents, naturalist interests, and a vocation for medicine. Those were fine subjects to study, his father grudgingly conceded, but they did not offer the knowledge or skills he wanted Jack to acquire so he could ably manage the property he would one day inherit and watch over the four younger siblings for whom he would be responsible. Though Henry was not as imperious or manipulative a father as the elder Charles Carroll of Annapolis, he was just as intent on keeping his son on his destined path, and Jack had learned to curb his own desires to satisfy his father's wishes—until the war at last enabled him to plead a nobler calling. Though it took months of importuning, Henry finally yielded to his son's determination to abandon his studies and join the Continental army, where his education quickly brought him to Washington's attention. Jack soon formed close friendships with Alexander Hamilton and the marquis de Lafayette. Like them, he welcomed the opportunity to demonstrate his physical bravery. Indeed, Jack bowed to no one when it came to putting his life in danger, and both Hamilton and Lafayette took to warning the elder Laurens that his son was taking needless risks.

Jack Laurens exhibited a different kind of courage—moral rather than physical—when he began challenging the peculiar institution that was the basis of his family's fortune. Well before 1776, his father had also begun to regret his past activity in the Atlantic slave trade, and even, perhaps, his continuing involvement in plantation slavery. But like most planters, Henry simply could not imagine how the southern economy could survive without it. Jack went further, expressing his opposition to slavery with a moral clarity and youthful fervor that his father admired yet also regarded as naive and impractical. As a member of Washington's staff, the younger Laurens developed a scheme to recruit regiments of slave soldiers, to be drawn originally from his father's own chattel property. Doing so would enable Jack to gain a combat command and thus pursue the military glory he craved. But Jack intended his scheme to serve a second, nobler purpose. Slaves who risked their lives to secure American independence would demonstrate their own capacity to live in freedom, not subjection, as potential citizens and not the mere objects of legal oppression.

Jack's opposition to slavery can be traced to those few years he spent in London, dutifully reading cases and memorizing causes of action while his mind wandered elsewhere. His circle of London friends included John Bicknell and Thomas Day, authors of *The Dying Negro*, a poetic if melodramatic assault on the injustice of slavery. Its composition was occasioned, the authors wrote, by the real suicide of a slave who was seeking baptism in order "to marry a white woman, his fellow-servant." His master, the captain of a West Indian merchantman, discovered the plan and arranged to ship his defiant property off to America. But the slave got hold of a gun and, rather than use it in a futile act of resistance, took his own life. The poem imagines the epistle the slave would have written to his intended bride before his last desperate deed.

First published in 1773, *The Dying Negro* had its third printing in 1775, with a new preface aimed directly at the American colonists. "Let the wild inconsistent claims of America prevail," Bicknell and Day wrote in conclusion, but only "when they shall be unmixed with the clank of chains and the groan of anguish. Let her aim a dagger at the breast of her milder parent, if she can advance a step without trampling on the dead and dying carcasses of her slaves." It takes little imagination (especially for any American who has ever studied in Europe!) to conjure a sheepish Jack Laurens, sitting mutely in a Holborn coffeehouse as his English friends rail against the rank hypocrisy of his countrymen, or gamely trying to justify plantation slavery with the argument that Europeans were unsuited to work in tropical climates. Neither Laurens, father or son, could ignore how fashionable the link between liberty and slavery had become. It was no less a figure than Samuel Johnson, the eminently quotable man of letters, who laid down the most devastating retort against American pretensions. In his 1775 tract, *Taxation No Tyranny*, Johnson rebutted the charge that a ministry bent on enslaving liberty-loving Americans would soon turn upon liberty-loving Britons. "If slavery be thus fatally contagious," he asked, "how is it that we hear the loudest yelps for liberty among the drivers of negroes?" The great dictionary maker was connecting two distinct definitions of *slavery*. One was legal and embraced the chattel labor system of the southern colonies. The other was political and described the condition to which Americans would sink if ruled by parliamentary laws made without their consent. It was this second definition that Henry Laurens invoked when he wrote home in 1774, warning his South Carolina countrymen that if they did not show "True sincere love of Country & Constitutional Liberty," they must "submit to wear the Badge of Slavery." He was not alluding to the thousands of Africans his firm once sold or the hundreds he still owned.

An authority even greater than Johnson raised an even more profound question about slavery in 1772. In the famous *Somerset* case, Lord Mansfield, chief justice of the Court of King's Bench, had rendered a decision that quickly reverberated across the Atlantic. James Somerset was a Virginia slave whose master, the Scots merchant (and Boston customs official) Charles Steuart, brought him to England in 1769. After Somerset escaped and was recaptured, Steuart decided to export him to Jamaica for sale (just as the model for "the dying Negro" was to be transported). With aid from Granville Sharp, a leading English opponent of slavery, Somerset sued for his freedom under the great writ of habeas corpus. The *Somerset* case was widely reported in newspapers on both sides of the Atlantic. (Henry Laurens was in London while it was decided, and may well have attended the proceedings.) In his ruling barring Somerset's deportation, Mansfield called "the state of slavery" a condition "so odious" by nature "that nothing can be suffered to support it, but positive law"—that is, legislation.

As a matter of law, the *Somerset* decision had no effect in the plantation colonies of America. There, a veritable hogshead of positive laws made slavery a wholly legal institution. Yet as a *moral* matter, Mansfield's pronouncement seemed significant. Its impact was evident in Virginia, for example, when the delegates to the provincial convention of May 1776 debated the declaration of rights. As drafted, George Mason's first article declared "That all men are born equally free and independent." This statement was pure Locke, and uncontroversial—until the delegates considered one implication: that no one could be born into a state of slavery. Virginia planters knew that they no longer needed to import slaves from Africa. Chesapeake slaves were reproducing in numbers capable of creating a self-sustaining labor force. Hapless victims ensnared in Africa might still endure lifetime labor under the conventional legal fiction that they were captives in a just war. But what about their Virginia-born children? Locke had argued, after all, that the just rights that a conqueror gained over his captives did not extend to their families or their descendants. How then could slavery become a perpetual status, to be passed from one generation to another, world without end?

Even so, the notion that "all men are born equally free and independent" was troubling enough to merit amendment. Article 1 was accordingly revised to state "That all men are *by nature* equally free and independent, and have certain inherent rights, of which, *when they enter into a state of society*, they cannot, by compact, deprive or divest their posterity." Whatever natural rights slaves could claim did not alter the legal status that gave them no rights at all. Slaves had never voluntarily entered society, and the revised language of the declaration left *their* posterity bound to endless servitude. Republican Virginia might proscribe an aristocratic right to rule, "descendible" from one generation of officials to another. But the yoke of slavery remained hereditary, with no jubilee of liberation provided or promised.

Had he been more committed to his legal studies, Jack Laurens might have admired the sort of verbal dexterity the Virginians demonstrated in tamping down Mason's language. But in the spring of 1776 it was the moral argument against slavery that registered. "I think we Americans at least in the Southern Colonies, cannot contend with *a good grace*, for Liberty, until we shall have enfranchised our Slaves," he wrote to a South Carolina countryman of loyalist leanings. "How can we whose Jealousy has been alarm'd more at the Name of Oppression sometimes than at the Reality, reconcile to our spirited Assertions of the Rights of Mankind, the galling abject Slavery of our negroes?" Here was another echo of Edmund Burke's March 1775 speech promoting reconciliation. Americans "augur misgovernment at a distance," Burke reminded the Commons, "and snuff the approach of tyranny in every tainted breeze." Jack was old enough to remember the "noisome smell" that plagued Charleston in June 1769. Its source was soon traced to the "putrefaction" of rotting bodies, the human refuse of slave ships thronging the harbor where the Ashley and Cooper rivers meet, as residents say, to form the Atlantic Ocean. But the tyranny slaves endured was no distant breeze or passing foul odor. It was perpetual and physical, measured in the beatings they endured for goading their masters too much. To his moral credit, Jack Laurens, as privileged a son as America could produce, came to confront this central aspect of slavery directly. His concern went beyond its corrupting effects upon masters—the issue that troubled his forerunners at the Inns of Court, John Dickinson and Thomson Mason—and the moderate qualms felt by his father, Henry. The war he was so eager to join might provide slaves with more than an escape from their toilsome burdens. It might also enable them to prove that they deserved the active rights of citizenship. Just as Jack longed to exchange the tedium of legal study for the glory of war, so he came to think that military service could transform African Americans from bondsmen to citizens.

Henry Laurens was born in 1724, making him, like Samuel Adams and George Mason, part of an older cohort of leaders whose age could have limited their involvement in the Revolution. As a member of the South Carolina political elite, he sat in the provincial Commons House of Assembly from 1757 to 1771, when, recently widowed, he carried his sons Jack, Harry, and Jemmy to Europe in search of a better education than Charleston offered. Two daughters remained at home—until 1775, when he sent Martha and Polly to England in the company of his brother James. Seven other children had died, and in May 1770 so had their mother, Eleanor Ball Laurens—better known as Helen or Helena—after twenty years of marriage. "The bare mention" of her name, Laurens sighed a year later, "hurries reflexions upon my mind which betrays a weakness that I can not yet conquer." For Laurens as for Mason, widower's grief made family a far greater priority than politics. But then, after returning to America, with all his children in Europe, it freed him for the duties the Revolution thrust upon him.

The Laurens family, like John Jay's, was part of the great Huguenot diaspora. Henry's grandfather André, the founder, settled first in London in 1682, tried his hand in Ireland, and then carried his family to New York City in the 1690s before removing to Charleston in 1715. André died soon afterward, leaving his "four Sons & one Daughter with Such portions as put them above low dependance." So Henry recalled in 1774, writing (in English) to a family in Poitiers who might be distant relations. "Some of them retained the French pride of Family," he added rather gracelessly, considering his correspondents, "& were content to die poor." Not so Henry's father, né Jean but known as John. "He learned a Trade, & by great Industry acquired an Estate with a good Character & Reestablished the Name of his Family."

That trade was saddler. The fact that it marked the first step on the Laurenses' path to wealth is a reminder that early-eighteenth-century South Carolina was still a frontier society. Indian traders who dealt for deerskins needed saddles, as did the hands who herded the free-ranging livestock that formed a major part of South Carolina's early economy. Many of these hands were slaves, carried to the mainland by the Barbados planters who formed South Carolina's first elite. The arrival of the Laurens family roughly coincided with the critical transformation that made rice the colony's dominant crop. By then slaves formed a demographic majority of the province, making South Carolina the only mainland British colony where they did so. The rapid expansion of rice cultivation and the development of indigo as an increasingly profitable crop guaranteed that African slaves would continue to be imported into Charleston until the Revolution.

In 1744, when Henry was twenty, his father apprenticed him to James Crokatt, a Scots merchant who had long resided in Charleston before resettling in London. It was a promising placement. Crokatt had actively promoted the growing of indigo in Carolina, even securing a bounty for its cultivation from Parliament. As a reward the assembly made him its agent in London. In June 1747 Laurens sailed home, only to learn that his father had just died. He returned to England in 1748, expecting to rejoin Crokatt's firm. To his surprise, Crokatt produced copies of letters he had sent Laurens, "charging me heavily with Cruelty & ingratitude." Though they soon patched up their misunderstanding, Crokatt had taken a new partner and had no room for Laurens. Twenty-five years later, Crokatt was still lamenting his mistake, while Laurens chortled over the mortification that those who "through Sheer Envy" had misled the elder merchant now felt at "having been the Instruments of my prosperity."

For Laurens already had a contingency plan in place to form a partnership with George Austin, a Shropshire merchant recently arrived in Charleston. Before returning home, Laurens visited Liverpool and Bristol to contact several houses actively involved in "the African trade." He and Austin had clearly planned this beforehand. Most slave ships were owned by British firms. But once they reached Charleston, local factors handled the sale of their cargo. In his initial approach, Laurens offered to perform an array of useful tasks, from provisioning ships for their onward voyages to remitting payment to England. Soon the firm was conducting slave sales. In the *South Carolina Gazette* for July 29, 1751, for example, Austin and Laurens advertised "a Cargo of healthy fine SLAVES," imported "directly from GAMBIA, in a Passage of *Seven Weeks.*" A notice from January 1752 announced the sale of another three hundred, also arrived "directly from *Africa.*" A 1755 letter to a Barbados correspondent described the latest cargo from Angola: "They were in very pretty order & well assorted (vizt.) 116 Men, 45 Women, 49 Boys, & 33 Girls, total 243 Slaves & they averag'd £33.15.6 Sterling." This sale occurred amid feverish speculation about how slave prices would be affected by the expected outbreak of the war that young George Washington's trips to the Forks of the Ohio helped trigger. "Very contrary to the opinion of most of us there was such pulling & hauling who should get the good Slaves that some of them came to collaring & very nearly to Blows." The impetus for this "pulling & hauling" was the high price for indigo. Austin and Laurens dealt in it too, as well as in rice, deerskins, and naval stores, the colony's other main exports. The partnership, extended to Austin's nephew, George Appleby, flourished until its dissolution in 1762. In the decade after 1751, they were the largest handler of slave cargoes in British North America, responsible for the sale of more than sixty shiploads—and thus thousands of Africans.

As a merchant Henry generally wrote about slaves in matter-of-fact terms. They were simply another commodity, subject to laws of supply and demand and the news and rumors that shaped all prices. Slave trading required the coldest of all the cold calculations any merchant could make. "Never put your Life in their power a moment," he warned the young captain of his sloop *Montagu*, prior to a Jamaica voyage on which slaves were a possible purchase. A single incautious moment could lead to "the destruction of all your Men & yet you may treat such Negroes with great humanity." His meaning was not that they *should* be treated so, only that humane treatment offered no assurance against rebellion.

When the firm dissolved and its two English partners returned home, Laurens ended his active involvement in the slave trade. That meant forgoing the hefty profits that Charleston factors continued to earn until the Revolution. For the great demographic transformation of American slavery took hold more slowly in the palmetto land of rice and indigo than it did in the tobacco provinces northward. Chesapeake slaves were already forming a creole community, with native African Americans outnumbering new Africans bearing the "country marks"—the ritual scars—of their origins. In 1760 Africans still constituted two fifths of the roughly 60,000 slaves resident in South Carolina, but only a seventh of Virginia's 140,000 slaves. Even though his preferred method of remitting payment to English merchants gave him a typical profit of 8.5 percent, rather than the 14 percent reaped by merchants using another mode, Laurens would have grown wealthier still by remaining in the trade.

His decision to retire had multiple sources. He was reluctant to carry on without his partners and eager to devote his attention to his children. There was also a nagging moral concern, which Laurens mentioned occasionally. One such occasion came in 1768, when as a legislator he opposed petitions seeking a bounty for the importation of a group of "poor Protestant Christians from Ireland" who had been exposed to greater "Cruelty" at sea than he had ever seen in his dozen years "in the African Trade. I quitted the Profits arising from that gainful branch," he recalled, "principally because of many acts, from the Masters & others concerned, toward the wretched Negroes from the time of purchasing to that of selling them again." *Wretched* was a word that Laurens often applied to slaves, not out of contempt for Africans but pity for their condition. The same sentiment briefly appeared in 1769, when he sold seven slaves for a Bristol firm. One more died at sea, and another "poor pining creature hanged herself with a piece of a small Vine which shews that her carcase was not very weighty." Then Laurens added a brief postscript: "Who that views the above Picture can love the Affrican trade?" But an inability to love the trade was not the same as actively hating it. Only months later, he informed another

English firm that he would "do you a piece of service by employing a friend" to manage any slave cargo they chose to sell. It would be a good market, he promised. When he sent his own ships to the West Indies, slaves were one item he authorized his captains to purchase.

Unease over slavery thus had its limits. When his friend the Moravian minister John Ettwein expressed his opinion "that often where a Man has Slaves his Children become lazy & indolent," Laurens concurred. He had often "wishd that our oeconomy & government differ'd from the present system," Laurens replied. "But alass, since our constitution is as it is, what can individuals do?" Nor did withdrawal from the slave trade end his involvement in slavery. Like prospering merchants elsewhere, Laurens invested his profits in land. In 1756, with a brother-in-law, he bought his first rice plantation. He bought another, Mepkin, along the Cooper River, in 1762, and additional land near Charleston, which he laid out as a suburb with a fine English name, Hampstead. These marked only the start of a series of investments in land that Washington, Mason, and the Carrolls, no slackers in this field, could envy. Soon Laurens owned other plantations in South Carolina and Georgia, a day's sail away. By his own account he had not set out to become a great planter. "I have been insensibly drawn into such an extent" of land holding, he noted in 1768, "by means & circumstances quite adventitious." But no merchant as diligent as Laurens could truly ascribe such opportunities solely to the good luck of timing and location. His transformation into a major planter was part of his plan to take his sons to England for schooling. "My plan of life lies within a very narrow compass," he wrote to the governor of East Florida in 1769, "& I hope soon to live quietly at Macedon." This was not the name of a favorite plantation, but an allusion to Alexander the Great's lamenting that he had no worlds left to conquer.

Plantation ownership made Laurens the master of hundreds of bondsmen, whom he had to regard differently from the thousands he sold. Unless disease or a hard crossing left slaves too feeble for quick sale, a trader never concerned himself with their welfare for long. Arriving slaves were anonymous beings, interesting only for the qualities—appearance, demeanor, age, gender, strength, country of origin—that affected their price. Not so plantation slaves. Laurens's papers routinely discuss the best way to put them to use, a task combining managerial calculation with psychological strategy. A 1774 letter from London illustrated these dual aspects of slaveholding. Laurens first instructed his manager at Mepkin to conduct a planned experiment. "Put two careful midling hands" to work "transplanting" mature indigo seeds into "an exact Quarter of an Acre in a well prepared piece of Ground," he ordered. Using "midling hands" would produce greater "accuracy" in gauging slave productivity than one could gain "if two best or two bad hands were employed." But Laurens gave a further instruction. "Tell the Negroes that I hope to meet them when they are beginning to thrash Rice, & to find at least as many in number of my Cattle, Sheep, Hogs as I left," he wrote, "& that I Shall be provided with proper rewards for Such of them as have behaved well." Like other masters, he understood that an owner's use of his human resources involved a daily calculus of punishments and rewards, threats and incentives.

As a slave owner, Laurens consciously played the role of patriarch. "The reflection is comfortable, that my Servants are as happy as Slavery will admit of, none run away, the greatest punishment to a defaulter is to sell him," he wrote in 1768. He fired one overseer for keeping a slave mistress, in part for the "jealousy & disquiet" it would sow, but also because "your familiarity with Hagar" was "wrong & unwarrantable in itself." This same Hagar (he owned two with that name) may have been one of the three "old Domestics" who welcomed him home in 1774: "my Knees were clasped, my hands kissed my very feet embraced," he wrote to his son, describing his property as "humble sincere friends." But his dispersed holdings required Laurens to trust his overseers to act with the humane attitude he idealized—always a risky proposition for patriarchal masters. It is revealing that he let his Mepkin overseer pick the two hands to perform his indigo experiment. A genuine patriarch would have known which ones to choose himself.

He had to rely on his overseers even more as he prepared to take his three sons to Europe. The trip was long delayed, but by the late summer of 1771 Harry, Jack, and Jemmy (just turning six) were ready to sail. Eight-year-old Harry had been sent ahead because "he bears the hot weather badly." The crossing took twenty-nine days. No sooner had they arrived than Henry wrote home, advising Carolina friends that "Rice will bear a very high Price at all the Markets in Europe." On whichever side of the Atlantic he found himself, Laurens could simply not slough off his merchant's habits. But his main task was his sons' education. He first placed Jack and Harry in a small school kept by

Richard Clarke, a former Carolina clergyman. But he was quickly dismayed after Harry had "a Candle, wantonly and maliciously thrust into his Face, by a Vagabond Brother of Mr. Clarke's." After finding a new school near Birmingham for Jemmy, Laurens and his two older sons set out in late May 1772 for Geneva, the free republic where John Calvin once held sway. Here he hoped to find better educational prospects than England offered.

Their journey was something less than the Grand Tour that English gentlemen of Jack's age were expected to make. In Calais they purchased a "clumsy close Chaise, two wheels, drawn by a large Shaft Horse, and two Bidets"—nags, in the lingua franca. Laurens allotted Paris all of two days, part of it spent visiting Gobelins, the great tapestry manufacturer. His sons took greater pleasure in "The Droll Scenes" of the street: "Bag Wigs on Rags and Dirty, embroider'd wooden Shoes, Ladies astride on Mules and Asses, Powder'd Postilions in tatter'd Vestes." But Jack also appreciated nobler things and "feasted on several Pieces of fine Painting, some of Carving, and a few in Sculpture." He had been drawing for years; his sketch of a soft-shelled Carolina turtle, forwarded to a prominent English naturalist, had even been published in London.

Jack's artistic talents and naturalist interests were related to the questions he and his father now faced. Finding acceptable schooling in Geneva proved easy. Henry returned to London confident that his boys would have "more friendly attention paid to them in all respects than we could hope for in a Kingdom overwhelm'd by Luxury and Vice, where Interest is almost universally the main spring of Action." But in fact this was only the beginning of an ongoing struggle over Jack's education and vocation. In August Jack wrote that he would honor his father's wishes and study law. But he cast his decision in terms that would unsettle any progressive eighteenth-century parent eager to rule by reason and trust rather than guilt or fiat. "I ought not to abandon myself wholly to my own inclinations," Jack wrote. "I leave my favorite Physick"—medicine—"grieved to the heart," in order "to embrace that which I know would give my Dear Papa the most pleasure," law.

These were not the words Henry longed to hear. "If you really feel grief at heart," he hurriedly replied, some other decision must be made. In all his business dealings, Henry was a stickler for duty, obligation, and honor. With Jack, he followed John Locke, the reigning educational authority, who instructed parents to rear children to choose for themselves. Within a few days, his old business partner, George Appleby, convinced Laurens that he had read Jack's letter too severely. The potential quarrel evaporated, and Henry enrolled his son as a future student at the Inns of Court. A second trip to Geneva in the fall found Jack "Satisfied & in a regular plan of Education," freeing Henry to plan a spring return to Carolina.

In fact two years passed before he sailed home. As a sole parent deeply invested in his sons' education, he found it difficult to leave. The sorry plight of his niece, Mary (or Molsy) Bremar, also delayed him; she had been seduced and left pregnant by her brother-in-law, Sir Egerton Leigh, the president of the South Carolina council and thus a royal appointee. Just days before she was due, Leigh packed his ruined sister-in-law on a ship bound for England. Their child was born, and quickly died, at sea. Once arrived, Molsy in effect became Laurens's ward. It did not help his view of the situation that he and Leigh had tangled bitterly back in 1768–1769, after a customs official seized two of Laurens's vessels, and Leigh, as a judge of vice-admiralty, failed to grant the relief Laurens sought. Their spat escalated into a nasty pamphlet war, which even ensnared Jack when his father asked the young Latinist to translate relevant passages from the writings of the great jurist Samuel Pufendorf. Henry also made fruitless efforts to mediate the quarrel between his other former partner, George Austin, and Austin's debt-ridden alcoholic son—more lessons in the science of parenting and the art of teaching children to avoid vice.

Until news of the Boston Tea Party broke upon London early in 1774, Henry's letters betray few political concerns. They offer no hint that he was bursting with anger at the empire or that he foresaw a revolutionary upheaval. On those rare occasions when politics reared its head, Laurens spoke as a patriot, but not a militant one. Early in his stay he found himself "attacked by a Talkative Gentleman at his own Table, on the ingratitude of America." After heated discussion, his "violent Friend" at last conceded that Britain's policies toward America "were *inexpedient*, not *wise* nor *politic*." But such talk exhausted him. "This kind of wrangling politics is the least agreeable Business," he noted. He far preferred topics like "Contrivances for pounding Rice, felling Trees, improving Lands and Marshes, building Wharves, and many other Things which eventually may produce some Benefits to my Carolina Friends, and to my Country." Political quarrels struck him as a game not worth the candle. "These disputes are abundantly more alarming than fatal," he observed in November 1773. "The merest trifle in politics may in one year bury the Remembrance of all that passed with you in September last as if the transaction had happened in the last century."

Two months later Laurens was caught up in the far from trifling crisis over Boston. From the outset he sensed that American patriotism "will be put to trial & possibly by a Severe test." After watching Wedderburn's caustic Cockpit examination of Benjamin Franklin, Laurens wrote to his brother that the Crown lawyer "went out of his way & far overshot the Line of Truth to abuse Doctor Franklin." Wedderburn's show of anger revealed how "really Seriously alarmed" the North ministry was. Americans should "seize Time by the forelock & prepare a Shield against the Evil Day which I apprehend we are advancing to." Accommodation was possible if the duty was repealed and Boston proved "So wise & So honest as to pay for the Tea Soaked in Salt Water." But he doubted that the government was capable of "treating with propriety such a bagatelle, as the Recovery or loss of the affections of three Millions of Subjects." When a well-placed acquaintance asked, "What then would you have us do?" Laurens displayed a merchant's sense of how to weigh outsize risks against modest benefits. The taxes Britain wished to impose "are Galling to the Americans" yet "Yield no benefit to the Mother Country," he replied. "What then are we contending for, Imaginary emolument, at the risque of Thousands of Lives & Millions of Pounds, possibly of the Dignity of the British Empire"? There was simply "no medium between Compulsory measures by Fleets & Armies & a Wise Retraction on this Side."

On March 11 Laurens hastened to St. Stephen's Chapel at Westminster to hear the Commons debate "the American papers." He had three members ready to introduce him but was distressed when the galleries were closed to onlookers. "This resolution of the Speaker gave us American out of door folks much umbrage," he wrote to Jack. But the texts of Lord North's punitive measures were soon available, reinforced by the threatening "Language of many Men above the lower Rank & of almost every one about the Court." Your "harbours shall be blocked up," they boasted, "& if those steps prove inefficacious, your Houses shall be lain in Ashes, & then see what you'll do!" With other "forlorn Americans," Laurens signed petitions of protest to both houses of Parliament and personally waited upon Lord Dartmouth, the secretary of state for America, to deliver a third petition, this one addressed to the king.

The ease with which Laurens used the term *American* exposes the basic miscalculation shaping British policy. Well before the Port Act reached the colonies, Laurens knew it would become a continental affair. "Boston is pitched upon by the King who is prime Minister & his little Select Cabinet to bear & be punished for the imputed Sins of all the American Continent," he wrote to a Charleston friend. Their object was "to terrify the other Colonies into a Compliance with Measures which nothing but Power can support." He readily linked the pending measures to a decade-old scheme to subject America to parliamentary jurisdiction. Now those claims were to "make a good platform for the invincible reasoning from the Mouths of four and twenty pounders." But even if British arms managed to "fix the Badge of Slavery, upon the Sea Coast," that would only "hasten the beginning of Independence out of the reach of Fleets & British Troops." Of course, Laurens had already done his share to affix a different and more durable badge of slavery along the Carolina coast. But when he dashed off this observation in March 1774, the association between political and chattel slavery had not yet been made.

Henry was generally pleased with his sons' two years in Geneva. Even so, his letters to Jack were filled with the moral admonitions and prudent advice that flowed naturally from a loving parent who made mercantile discipline and Christianity his own guides. When he spent the New Year of 1774 at Bath, taking "Upwards of 3,000 Strokes from the hot Pump" for his gout-ridden left leg, he passed on the lessons he drew from the "Scenes of Folly & Ruin" he observed. "Folks from 7 to 70 of both Sexes & all Ranks" could be found "working at the Card Table 12 Sometimes 18 hours in 24," Laurens noted, shocked that gambling was going on around him. "My Ideas traced consequences from the Acts & the whole view was horrible." Laurens prayed that his sons may "dare to be Singular or among the unfashionable few, who are Still the Cement of Society & the Witness of Truth & Virtue." Even Geneva had temptations, and any intimation that Jack had indulged in distractions alarmed Laurens.

As he was preparing to depart, Henry learned that Jack hoped to spend another year in Geneva. Without questioning his son's motives or reasons, Henry put his foot down—metaphorically, since his gout-inflamed legs were "so stiff as to seem jointless & so very feeble as renders me incapable of moving even upon crutches." Pleading concerns of health, family, and the situation at home, he ordered his sons back to England. By late August 1774 the family was reunited in London, and by the morning of November 7, Henry was at last ready to sail from Bristol on the packet *Despencer*, bound for Charleston. At age fifty he was about to become a revolutionary.

His last advice for Jemmy and Harry was to "wash your Mouth every Morning, Noon & Night with Cold Water." Jack received a sterner message: "do not stir from your Studies unless you perceive that such an excursion may be made without any prejudice to them." Henry was not pleased when Jack's first letters revealed that he had already been out and about, escorting two homeward-bound friends to their departure from Gravesend and planning a holiday trip to Cambridge. A paternal lecture followed. Were you only awaiting my departure to act "like a Bird after long confinement" fleeing its cage? Henry asked. But he had a more serious concern. Jack had joined "a Croud of Anxious Politicians" attempting to hear the Commons debate America, only to be dispersed "with the dreadful Cry of Clear the Galleries." His political enthusiasm remained unabated. If war came, Jack wrote in early December, "there is no Man I hope would more gladly expose himself, or hold his Life more cheap." His father found this declaration more alarming than admirable. "Reserve your Life for your Country's call," he replied, "but wait the Call" and "mind your chosen business."

On neither side of the Atlantic did this advice prove practical. For his part Henry plunged immediately into revolutionary politics. In January 1775 he served in South Carolina's first provincial congress. When a second congress met in June, he was made its president. When it adjourned and left a committee of safety to oversee the province's preparations for war, he became its president as well. He filled letters to his sons with political news and reports of his own distress. Writing to ten-year-old Jemmy in late May 1775, Laurens despondently wondered whether "I shall ever again have the pleasure of embracing you or even of writing to you." Who knew what fate awaited him? "I may be a prisoner, in Chains & under Sentence of Death," all for "Striving to transmit that Liberty to my Children, which was mine by birthright & compact." Writing this one letter drained him emotionally; he shed tears as he imagined Jemmy "cast upon the inhospitable Shoar of this wide world, pennyless & friendless."

Just after New Year 1776 he wept even more bitter tears when he learned that it was instead Jemmy who had died, fracturing his skull by jumping from a ledge at the house where the brothers lodged. Signing himself "the most unhappy of Sons," Jack wrote home to relay the sad news, while exhorting his "Dear Father, not to abandon yourself to Grief as if all your hopes were buried here." Patriotism could be a refuge for the mourner. "You have great and important duties to perform upon the Earth," Jack wrote. "Your family your Country looks to you with confidence."

The morning of the accident, Jack had sealed another letter to Henry, expressing his desire to return home and bear arms in defense of his country. Even had Jack taken to the study of the law, he could not have resisted the call of patriotism. After being chastised for keeping bad company, he fell dutifully in line and used his letters home to sound suitable notes of contrition. But the gloomy reality was that the nuances of legal doctrine and practice held no allure for him. The only reason to pursue this career was his father's insistence upon it. Yet even as he repented, Jack made his own feelings plain. "Yes my Dear Father I will obey you fully in all that you require," he submissively wrote in November 1775. Then came the telling counterpunch: "I feel like a Man avoiding the Service of his Country because his Father tenderly commands him to be out of Danger." To his uncle James, just arrived in England with Jack's sisters, he spoke even more plainly of the "dishonour" and the "humiliating Situation" in which filial duty had placed him.

Henry responded by cautioning his son against dropping one course of action to pursue another for which he was not prepared. "Be not ambitious of being half a Soldier half a Lawyer & good for nothing," he wrote. "Aim at Character, which you could not expect in any high style if you were to commence Soldier tomorrow." Jack was reluctant to press his grieving father. Then for some months the ordinary difficulty of conducting a transatlantic correspondence prevented the issue from coming to a head. Finally, in mid-August 1776, Henry sent off a long letter that signaled a change of heart. Overwhelmed with public duties, he had let nearly four months pass since his last letter, and he began with a lengthy review of events. As he often did, Laurens portrayed the political crisis as a family tragedy, invoking the image of imperial parent and colonial child that writers on both sides often used as a metaphor for the Anglo-American bond. "Even at this Moment my heart is full of the lively sensations of a dutiful Son," Henry wrote to his own dutiful son, "thrust by the hand of violence out of a Father's House into the wide World." For most writers the parent-child metaphor of empire was simply a handy cliché. For Henry, who had now outlived eight of his children while worrying about the remaining four, it cut deeper. Now he extended his hand to invite Jack to return to his father's house. Finding himself "brought into a new World & God only knows what sort of a World it will be," Henry reminded Jack, just turned twenty-one, that "you are of full Age entitled to judge for your self." With his entire family in England, Henry felt lonely in his private life and morose when he considered public events, "more than ever anxious to see you" while beset by the "Clouds & Darkness [that] are before me."

These thoughts ended a letter in which Henry also denounced slavery and indicated that he was "devising means for manumitting many of them & cutting off the entail of Slavery"—meaning the hereditary transfer of slave status from one generation to another. He was not responding to any fresh statement of Jack's on the subject. The stimulus came instead from a May trip to his Georgia plantations, and even more from outrage that "many hundreds" of slaves "have been stolen & decoyed by the Servants of King George the third" in acts of "inglorious pilferage to the disgrace of their Master & disgrace of their Cause." As Henry saw it, Britain was the true author of the slave trade and its joint beneficiary. British merchants, aided by their government, still controlled the trade. "Kings & Parliaments" once had a joint share with colonists in establishing the law of slavery. Now they "employ their Men of War to steal those Negroes from the Americans to whom they had sold them," with the likely intent of reselling "them into ten fold worse Slavery in the West Indies."

"I am not the man who enslaved them," Laurens continued. "They are indebted to English men for that favour." In a literal sense this was true. Legally enslavement began in Africa with an initial capture and abduction and the coastal sales that enchained captives below decks on ships like those whose cargoes Henry once sold. Even so, a modern reader can only consider his disclaimer as either hypocritical or disingenuous. When Jefferson proposed similar language in the Declaration of Independence only weeks earlier, his colleagues had the good sense or moral embarrassment to delete it. Yet before condemning Laurens too quickly, we should note that he did not limit his objections to the slave trade. That was the easier target of early antislavery sentiment, perhaps because the pathos of individuals being plucked from their native country and shipped away engaged the moral and literary imagination more readily than the brute tedium of plantation labor. But as the owner of eight plantations, Laurens knew more about the final destination of the slave trade than its places of origin. He trusted the loyalty of his own slaves. In both Georgia and Carolina, they "to a Man are strongly attached to me," he noted. In fact, even as he wrote, five were fleeing to Florida from one of his Georgia plantations. He had nonetheless come to believe that he should now act as "a promoter not only of strange but of dangerous doctrines" and in some unspecified way help his countrymen "to comply with the Golden Rule" by freeing their slaves.

How Laurens came to this point is not easy to determine. Nor is it easy to measure the depth of his feelings. In England he certainly encountered the emerging antislavery sentiments of reformers like Granville Sharp. He was in London when the *Somerset* case was argued, yet his two brief references to it suggest that it hardly disturbed him. "I was going to tell a long and comical Story" about the case, he wrote to a Charleston friend in late May 1772, a month before Mansfield rendered his decision, but "they say supper is ready." In another letter three months later, he noted that he would avoid commenting on Mansfield's judgment "until we meet, save only that his Lordship's administration was suitable to the times." This was not meant as praise. Laurens had sharper words for "the able" John Dunning, the attorney for the defendant slave owner, Steuart, who opened "by declaring that he was no advocate for Slavery, & in my humble opinion he was not an Advocate for his Client nor was there a word said to the purpose on either side." If *Somerset* marked a milestone in legal thinking—and it clearly did—the passing comments of this leading trader and owner of slaves hardly treated it as such.

Henry's August letter expressing his antislavery ideas reached Jack in late October. The reply his son dashed off was enthusiastic on two key points and sheepish on a third. Jack was overjoyed at gaining permission to return home. He also greeted Henry's emancipation plans "with rapture," thereby assuaging the worries his father had expressed about the impact of his decision on his children's fortune. With his heart set on military glory, wealth was the least of Jack's ambitions. It was also the one argument for slavery he found most disturbing. Whenever he discussed its injustice with southern or West Indian friends, Jack wrote, they were ultimately "driven" to rejoin, "Without Slaves, how is it possible for us to be rich?" The more difficult challenge was to imagine how to prepare slaves to live in freedom. "We have sunk the Africans & their descendants below the Standard of Humanity," Jack observed. "By what Shades and Degrees they are to be brought to the happy State which you propose for them, is not to be determined in a moment."

Only after this extended discussion did Jack offhandedly mention his own arrival at a different "happy State"—matrimony. Henry had just become both a father-in-law and an expectant grandfather. Jack had seduced Martha Manning, at nineteen the youngest of three daughters of a London merchant who once traded in the West Indies. Well into her sixth month, Martha had begun to show, and out of "pity" Jack felt obliged to marry her. But as he made clear to his new father-in-law, he was bent on "fulfilling the more important Engagements" he owed "to my Country." Jack did not abandon Martha immediately. They spent two months in a Chelsea house that William Manning had rented. In late December Jack left for Dover, intent on crossing the Channel to secure passage to America on a French ship. His daughter, Frances, survived a difficult birth a month later. Her father sailed from Bordeaux a few days after that, bound for St. Domingue, and thence to the mainland.

Jack welcomed his father's permission to answer the call to duty as a release from domestic obligations he had no wish to assume. Yet with the shameful case of cousin Molsy fresh in his memory, he knew that his father would not easily withhold judgment on his conduct. Jack's ship arrived at St. Domingue just as his father was absorbing the news conveyed in his October letter. When he reached Charleston in mid-April, he learned that Henry was off visiting his Georgia plantations prior to taking his new seat in Congress. Jack hurried south for the reunion.

Not for the first time, Henry was disturbed by how casually Jack conveyed personal news. Writing to William Manning in early June 1777, as he and Jack were starting for Philadelphia, he called Martha "my Dear Daughter" and promised that "she and her dear little Girl shall not be unnoticed in my last Will and Testament." As for Martha's errant husband, Henry still hoped that Jack's "Talents & his diligence" could make him "much more extensively & essentially useful to his Country" in some line other than the military. "As a Soldier One Man can act only as one (very few Cases excepted) & in point of real usefulness will often be excelled by Men of moderate abilities & better nerves." Education and knowledge of languages, Henry felt, qualified Jack for a political or diplomatic post, where he might make a difference. Such a choice would allow Jack to act as "the promised Support of mine & the builder of his own family." Alas, none of these appeals swayed his son. "He heard them & pursued the dictates of his own mind," Henry wrote to Manning.

En route to Philadelphia, father and son would have had time to consider Henry's half-formed ideas of emancipation—a safer topic than Jack's affairs and ambitions. But it is also possible that Henry's reforming enthusiasm of 1776 had already cooled. The references to slavery in his letters during this period concern ordinary questions of plantation management and problems related to the relaxed oversight to which war and the master's absence inevitably led. In the coming months the initiative to think boldly about slavery came from the son, not the father. And if Jack's ideas built upon the earnest discussions of the subject he had joined as a student in London, they now gained force from his experiences as a Continental officer.

Within two weeks of his arrival at Philadelphia in July 1777, John Laurens became a member of General Washington's staff. He clearly possessed the "character" and "property" that the commander sought in his officers. For his part, Henry was immersed in the doings of Congress. Like many newcomers after 1776, he found a troubling gap between its inherited reputation and current condition. A "Short three weeks" of attendance convinced him "that a great Assembly is in its dotage" and made him regret his election. Within three months he succeeded John Hancock as president. His new position partly eased the feeling that he was engaged in "useless Service" at enormous expense, particularly with the boomtown prices delegates paid during their winter 1778 exile at York. Laurens became the main conduit for correspondence between Congress and its commander. As president he was really the agent of Congress, not an independent actor with great discretionary power. But with his son so closely attached to Washington, the relation between president and commander inevitably deepened. Not that Laurens was entirely immune to the criticisms of Washington that others were voicing. Shortly after the battle of Germantown, Henry wrote to Jack a short note recording what the "loose Tongues" chattering around him were saying—a litany ("buz! says one") about Washington's failings. Though he regarded Washington as "the first of the Age," he worried that one criticism was justified. "A good Heart may be too diffident, too apprehensive of doing right righteous proper Acts, lest such should be interpreted arbitrary," Laurens noted. In other words, Washington could be faulted for not fully exercising the authority that Congress had given him to commandeer the supplies his soldiers lacked.

When it came to his son, however, a nearly opposite concern proved far more disturbing. Jack usually had time to dash off only short letters to his father. But from Jack's new friend, the marquis de Lafayette, Henry had already received an unsettling account of his son's first combat. When Washington and his staff came under fire at Brandywine, it was Jack who took the greatest risks. "It was not his fault that he was not killed or wounded," Lafayette, who was also hit, told Henry Laurens. "He did every thing that was necessary to procure one or t'other." Jack performed even more valiantly at Germantown three weeks later. As the American assault began, a musket ball passed through his shoulder. Ignoring the pain, he joined the misguided effort to roust the British force holed up in Benjamin Chew's stone mansion, this time taking a bruising "Blow in his Side from a spent Ball." At York, Henry waited days before a letter dispelled the "suspicion" that "tonguetied" friends were withholding fatal information about Jack. The mere sight of the familiar handwriting on the envelope "cost me a few tears which probably would not have been so soon started if I had heard of his death." As a father he still resented Jack "for the Robberies he has committed" by taking "a husband & father from his young family, a Guardian from his Brother & Sister, a Son & friend from a dependent Father." Any new battle could bring crushing news. But Laurens had to admit that Jack was acquiring "qualities more valuable than Courage, he understands the Science of War, in theory & is getting fast into practicable knowledge." Writing to Jack, he had to content himself with a gentle reminder about "the distinction between ge[nuin]e Courage & temerity."

That was a lesson Jack never learned. Washington's own indifference to danger required his aides to be just as bold when balls and bullets were flying. So did the currents of affection and emulation that gave young officers their wartime camaraderie. Jack was already forming friendships with Lafayette and Alexander Hamilton, who had also narrowly escaped death in September on his mission to destroy flour mills along the Schuylkill. But what bound them together was less the intensity of combat and more their devotion to Washington. That loyalty far exceeded any commitment that Henry Laurens could feel to the institution that he nominally led. At the same time, he intuitively grasped that Jack's personal attachments bound him as much to the commander's "family" as his own. Jack still addressed Henry as "my dear father and friend." But as Henry ruefully noted, "If I may judge from his conduct he had forsaken Father & Wife & Child Houses & Beds & all for the sake of the Country & the Cause which possesses his whole Heart." Jack expressed no longing to be reunited with his wife and daughter. He did not oppose his father-in-law's opinion that a wartime ocean crossing was too dangerous for Martha and Fanny to risk. On the rare occasions when he heard from Martha, the spare accounts of her letters that he relayed could only have irritated his father. Henry admitted that "my mind is [much] exercised by this subject" of family reunification. "Often do I wish for that aid, or at least that alleviation, which might be derived from your conversation," he wrote to Jack, "but you seem to be absorbed by only one passion."

To offset the pathos of the patriarch, Henry had the consolations of the patriot. Jack's unstinting devotion to his duties contrasted sharply with the ignoble attitudes and behavior he sensed everywhere around him. Like so many others, Henry was distressed by the difficulties the army faced in keeping fed and clothed. On his way north in July, taking "the upper Road 300 Miles from the Sea," he had been amazed by the "fine Farms" he observed, "abounding with Crops of Grain, Cotton Flax & Hemp the Spinning Wheel & Loom are Seen in every House & very few with less than eight Children." Britain had to be mad "to attempt to restrain & check this growth." They might as well "attempt to build a Bridge from Plymouth to Philadelphia."

Yet amid this plenty the army was manifestly suffering. Henry found no shortage of explanations for its woes. These began with Congress itself. Not only was it undermanned and overworked. Its ingrained inefficiency, its propensity for "running whole days into weeds of unmatured conversations," could only drive a merchant like Laurens to distraction. Most of its decisions on matters of finance and logistics seemed badly misguided. It was a mistake, he feared, to rely as much on French loans as Congress was disposed to do. Worse still were the lax terms on which it spent public funds. The "diabolical motto—'He to be accountable'" that it routinely applied to its purchasing agents "maddens me every day," he complained. Three decades of conducting business inclined Laurens to condemn those who did not match his strict sense of commercial rectitude. As the army prepared for the desperate winter of 1778, he grew increasingly critical of those responsible for its provisioning. His suspicions soon implicated his predecessor, John Hancock, as well as Robert Morris, whom he feared Washington trusted too much. They and lesser figures made up a cast of "prompters & Actors, accomodators, Candle Snuffers, Shifters of Scenes & Mutes" who "taken together, form a Club whose demands upon the Treasury & War Office never go away ungratified."

Thus dismayed by politics and purchases, Henry proudly idealized Jack's selfless devotion to duty. "You are one of those, who are honestly running head long, hazarding life & forfieting the sweets of domestic life," Henry wrote to Jack soon after the army encamped at Valley Forge, "duped by a parcel of fellows who are picking our Pockets & who for pelf I verily believe [would] sell us tomorrow." But Jack's desire for domesticity existed only in his father's imagination. It was Henry whose frequent hints about retirement drew stern rejoinders from Jack. The letters between them mirrored those between the two Charles Carrolls, only here it was the son who chastised the father. "No man has so much influence over me as my worthy friend my Virtuous Son John Laurens," Henry wrote in late March 1778. "If after half an hours conversation he will confirm his present advice," he added, "I will be governed by it," indicating that his scruples about remaining at Congress were abating, if not quite yet removed.

Far more than a half-hour would be needed to resolve their differences on another matter. Writing from Valley Forge in mid-January, Jack made his radical proposal "to augment the Continental Forces from an untried Source." Instead of waiting for his inheritance, Jack asked his "dear father" to "cede me a number of your able bodied men Slaves" now, and he would train and outfit them as soldiers. Since they had "the habit of Subordination almost indelibly impress'd on them," they already possessed "one essential qualification of Soldiers." The Laurens family could thereby bring "those who are unjustly deprived of the Rights of Mankind to a State which would be a proper Gradation between abject Slavery and perfect Liberty."

Jack's proposal was inspired by many sources. The most immediate lay all around him as he wrote: the bedraggled soldiers of the Continental army a month into their winter encampment. He also drew on the high principles he imbibed as a student in London: the antislavery sentiments of his English friends. His father had endorsed those convictions in the letter permitting Jack to return home. Their subsequent conversations on "the means of restoring [slaves] to their rights" suggested that he was open to the idea. Then there were more personal motives. An officer who raised a unit of soldiers could expect to command them as well. While selfless service to Washington might prove his devotion to the cause, no staff officer could ever gain the glory that came by leading men in battle. That glory would be greater and more lasting still if earned with soldiers who had once been wretched slaves.

Henry's initial reply struck a note familiar to any parent blessed with children prone to sudden and earnest enthusiasms. "I will not refuse this, if after mature deliberation you will say it is reasonable." But he doubted that Jack had thought his proposal through or accounted for the "Caution & great circumspection" it would need. Before Jack could develop his ideas, Henry wrote again to express his reservations. To avoid "the reproach of Quixotism," Jack would need to find at least twenty allies to support his scheme. A deeper objection concerned the slaves themselves. "Have you considered that your kind intentions towards your Negroes would be deemed by them the highest cruelty," Henry asked, "& that to escape from it they would flee into the Woods"? Believing that his own slaves appreciated their kind treatment, Henry thought that they would reject Jack's offer. They would "interpret your humanity to be an Exchange of Slavery[,] a State & circumstances not only tolerable but comfortable from habit," for a terrifying condition "where Loss of Life & Loss of Limbs must be expected by every one every day" (a situation they were bound to face, given Jack's penchant for risking the glorious death he had told his father he was willing to accept).

Thus challenged, Jack defended his scheme on its merits. The main point at issue was not "that monster popular Prejudice," potent as it was. It was rather the capacity of the enslaved to accept the bargain: liberty for danger. He conceded that slaves were an "unhappy species" enduring a state of "perpetual humiliation" where they were "debased by a Servitude" that only death relieved. Yet if they were a "species," it was their condition that formed them, not their nature. Jack spoke of slaves in the language of sympathy, one of the great subjects of eighteenth-century moral philosophy, not race. Though a "trampled people," they still "have so much human left in them, as to be capable of aspiring to the rights of men." They were also capable of moral improvement, for "like other men, they are the Creatures of habit, their Cowardly Ideas will be gradually effaced, and they will be modified anew." Military training would impart other habits. As individuals, slaves would feel the same "hope that will spring in each mans mind respecting his own escape"—the same irrational belief that encourages soldiers to think they will avoid the fate of those falling around them.

Beyond the question of the slaves' capacity for service lay a simpler matter of principle. All of twenty-three years of age, Jack could still claim that he had "long deplored the wretched State of these men" and ever abhorred "the bloody wars excited in Africa to furnish America with Slaves" and "the Groans of despairing multitudes toiling for the Luxuries of Merciless Tyrants." Unlike Henry, Jack did not deflect moral responsibility for slavery from its American beneficiaries to British officials and traders. Neither did he directly accuse his "dearest Friend and Father" of complicity with evil. Still, the underlying moral question remained inescapable, and so did its association with the American cause. When might the restoration of rights "be better done," Jack asked, "than when their enfranchisement may be made conducive to the Public Good"?

Henry wasted no time answering. He parried Jack's implicit moral accusation by applying the principle of liberty to the objects of his son's concern. If Jack indeed denied that slaves could be held as property, he should first "set them at full liberty & then address them in the Language of a recruiting Officer to any other free Men." Given this choice, Jack would be lucky if "four in forty" accepted his offer. And of these four, three would likely turn "Deserters in a short time." Henry would not pretend that he was "an Advocate for Slavery—you know I am not." But moral conviction did not override his sense that this "Negro scheme" was wholly impractical. If Jack's real purpose was to obtain a combat command, he could easily return to Carolina and raise "a Regiment of White Men." But should Jack remain intent on defying "the opinions of whole Nations," Henry virtually dared him to pursue his plan.

Rather than take that challenge, Jack apologized "for the trouble which I have given you on this excentric Scheme," which he renounced "as a thing which cannot be sanctified by your approbation." He was troubled only over his father's "imputing my Plan in so large a degree to ambition." There the matter rested. For all the candor with which they had discussed it, the project was still only a passing diversion amid the urgent issues that preoccupied Congress and the army during the Valley Forge winter. Against this background, it seems remarkable that Jack's visionary scheme could be considered at all and unsurprising that it was shunted aside so quickly.

Yet impractical as Henry made Jack's enthusiasm seem, the "Negro scheme" involved more than the naive vision of an idealistic young officer. "Have you consulted your Gen[eral] on this head," Henry had asked, presuming that Washington would second his objections. Jack's response likely took him aback. "He is convinced that the numerous tribes of blacks in the Southern parts of the Continent offer a resource to us that should not be neglected," Jack replied. "He only objects to it with the arguments of Pity, for a man who would be less rich than he might be" should his private property be lost to soldiering. The enlistment of African Americans, whether free or slave, was a question Washington had occasionally faced since taking command. Free blacks were already part of the provisional army that awaited him in Cambridge in 1775. But should they form part of the national army he intended to fashion? His initial answer was no. A preliminary order of July 10 prohibited the enlistment of "any stroller, negro, or vagabond." On October 8, 1775, Washington had his council of war review the issue, explicitly asking "whether there be a Distinction between such as are Slaves & those that are free." The council unanimously rejected using slaves, while "a great majority" voted "to reject Negroes altogether." They probably acted with only minimal debate. This was the last of ten items on the agenda, and no discussion was recorded in the minutes.

The distinction between free and enslaved mattered. As property, slaves had no legal will or personality of their own. Their recruitment would need the assent of their owners, who would presumably deserve compensation. To enlist *slaves* would also "dishonor" the cause by implying that liberty-seeking Americans required aid from the most "wretched" elements of their society. The question of recruiting *freemen* was less easily answered. When Edward Rutledge of South Carolina urged Congress to order Washington to "discharge all the Negroes as well Slaves as Freemen in his Army," he "was strongly supported by many of the Southern Delegates but so powerfully opposed that he lost the Point." For his part, Washington soon modified his position. On December 30, responding to protests from free black soldiers upset "at being discarded," he revoked his order prohibiting their reenlistment. His new order implied that additional free blacks could be recruited. But Congress clarified the point by limiting eligibility to those who had already "served faithfully in the army at Cambridge . . . but no others."

Yet if Washington shared the prejudices of his class and region, he was neither close-minded nor dogmatic. Before leaving Boston, he invited Phillis Wheatley, the young and already celebrated African writer who had honored him with a worshipful poem, to visit his headquarters at the future Longfellow house in Cambridge. Her poem and the accompanying letter praising the "Generalissimo of the armies of North America" got buried on his paper-strewn desk for four months. But his reply, when it came, was gallant, apologetic, respectful, and personable. Being gracious to an admirer, however, was easier than weighing the costs and benefits of recruiting black soldiers. That raised political and legal questions that even an imperious "dictator"—rather than the cautious one that Washington proved to be—would hesitate to push. The growing numbers of black soldiers in the army owed more to the initiative of civilian governments, especially in the north, than to the enterprise of the commander in chief, whose own view was closer to the calculated reservations of Henry Laurens than the youthful enthusiasms of his son.

Yet calculated reservations can always be recalculated. In January 1778 Washington promptly approved Brigadier General James Mitchell Varnum's request to be authorized to raise "a Battalion of Negroes" from his native Rhode Island. That state had a large concentration of Quakers and Baptists whose conscience told them that slavery was wrong from the standpoint of religion and humanity. Washington conveyed his approval of Varnum's request to the governor of Rhode Island without comment.

Still, the use of black soldiers remained a minor question as Washington and his staff waited to see whether the British would abandon Philadelphia. Like everyone else, Jack Laurens welcomed the change that came when the army began tracking Sir Henry Clinton across New Jersey in mid-June 1778. It brought relief from the stink of hundreds of horse carcasses rotting in shallow graves around Valley Forge and also promised a chance for the action that the young Laurens coveted. When the armies clashed at Monmouth in the stifling heat on June 28, Jack's wishes for active service were fulfilled. Sent by Washington to reconnoiter the ground between the American forces and the enemy, he became "impatient and uneasy" over Charles Lee's failure to press the attack he had asked to lead. Lee's orders to retreat came "with a rapidity and indecision calculated to ruin us," Jack wrote to his father two days later. He was present as Washington rode up, dressed down Lee for this "unaccountable Retreat," and rallied the Americans. As Jack saw it, the engagement ended with the Americans "masters of the ground—the Standards of Liberty were planted in Triumph on the field of battle." Jack let two days pass before informing his father that he and Baron von Steuben had narrowly avoided capture. Left unmentioned was the fact that Jack had again been slightly wounded and that his horse had been killed beneath him. Henry got that news from a New Jersey delegate whom Hamilton favored with a report of the battle. Henry waited a fortnight before reminding his son that he did not have "to tempt the fates" on every possible occasion.

Monmouth was the last significant engagement fought in the northern states, and it effectively ended the second major phase of the Revolutionary War, the two years of conventional maneuvers that began with the British landings near New York City in July 1776 and then extended into the hard fighting back and forth across New Jersey. A good three months before Monmouth, the British government had fixed upon a new strategy. On March 8, 1778, Lord Germain dispatched instructions ordering Clinton to carry the war to Georgia and South Carolina. If the American union had a soft underbelly, this was it. Former royal officials and American loyalists had insisted that there were significant pockets of Tory strength in the south, particularly among Highland Scots in the backcountry. Because southern states were less densely settled, their militia would be more difficult to mobilize. Once mobilized they would face double duty as the fresh threat from British arms joined the customary fear of slave insurrection. A southern campaign would also enable the British to exploit their naval supremacy, which allowed them to move hundreds of miles by sea while the Continental army slogged overland across a country where few good roads existed.

Within a week, however, news of the Franco-American alliance led to a second, more profound shift in British strategy. Imminent war with Albion's ancient enemy across the English Channel forced the government to consider how it would protect all the global pressure points where vital imperial interests were at risk. The threat began at home, where there was a genuine fear of a French invasion. Should Spain enter the war on behalf of its French ally, it would surely try to regain the great fortress on the Gibraltar promontory. Combat was also likely in distant India, where British and French interests were already competing for commercial supremacy. Above all, Britain would need to protect the sugar-producing islands of the West Indies. In the Atlantic economy, sugar remained a far more lucrative crop than Chesapeake tobacco or Carolina rice. The North American theater of operations would have to compete for strategic attention and scarce resources with all these other interests and commitments.

Once his army regrouped around New York City, Clinton accordingly found that his first task was to launch an expedition against the island of St. Lucia. Not until late November 1778 did he dispatch a much smaller force to Georgia. Its success was dramatic. Savannah fell quickly, and through the support of troops that had marched north from East Florida, the British gained control of most of Georgia by the spring of 1779. This initial success suggested that, with reinforcements, a southern strategy might indeed work. Restoring civil government in one or more provinces might then inspire loyalists in the south to act, as Tory refugees repeatedly claimed they would.

This turn of events did not surprise Henry Laurens. Back in September, intelligence from home indicated that the British would soon make South Carolina their objective. Why shouldn't they, when its occupation could bring "the expected plunder of an abundance of Provisions, Merchandize, many thousands of Negroes, great quantities of Cannon and warlike stores, Horned Cattle, Sheep, Hogs, and Horses, an immense value of Indigo, and upwards of 200 sail of Ships"? At his behest Congress asked Virginia and North Carolina to send a combined force of four thousand militia and Continentals southward. That was only a stopgap resolution. As the situation deteriorated at year's end, he understood that the first challenge remained to raise forces for defense of the lower south. With adequate warning South Carolina might mobilize "10,000 Men, one half of them badly clad & badly armed." But if the British summoned their Indian allies, "the Inhabitants of the back Country will not leave their families exposed on that frontier."

These developments finally united father and son in their thinking about Jack's "black project." The younger Laurens had spent the late summer in Rhode Island, acting as liaison to the French forces under Admiral d'Estaing. In a failed attempt to liberate Newport, he commanded a rear-guard detachment that came under heavy fire as the Americans withdrew. Washington copied the strong praise Jack's conduct earned from Nathanael Greene into a personal letter to Henry Laurens. In late October Jack rejoined his father in Philadelphia. While there, Congress recognized "his brave conduct in several actions" by voting him a permanent commission as lieutenant colonel and urging Washington to give him a command "whenever an opportunity shall offer." In doing so Congress deviated from the rule of promotion, which relied on strict seniority to minimize the disputes over rank that frequently rolled through the officer corps. Embarrassed by the imputation that Congress was favoring a member's son, Jack declined the promotion. Congress then adopted another resolution praising him for his "disinterested and patriotic principles."

On December 23 Washington himself arrived in Philadelphia to confer with Congress. While the general was paying an informal call on Congress, Jack Laurens and Hamilton were making their way to a rendezvous with Charles Lee outside the city. This private meeting was the climax of a dispute that had escalated over the summer. Washington's two young aides had charged Lee with cowardice and misjudgment for his actions at Monmouth. Lee in turn insisted upon a court-martial, which predictably ended with his conviction. Devoid of political sense, Lee then appealed to Congress. In early December, with Congress set to uphold the verdict, Lee published a defense of his conduct that criticized Washington's generalship. He also besmirched his commander's character in private conversations. This was too much for the "family," and Jack challenged Lee to a duel. The two met in the wintry dusk on the twenty-third. Walking toward each other, Laurens grazed Lee with the first exchange. The two prepared to reload, but at the insistence of their seconds, the duel ended with each man's honor satisfied.

What either Henry or Washington—Jack's two fathers, as it were—thought of this affair is not recorded. Dueling was a practice that Washington had sought to suppress, and Henry did not admire his son's fondness for needless risks. Yet it is difficult to imagine Jack hiding the matter from his father while they were lodging together in Philadelphia. Moreover, Henry was engaged in a matter of honor himself, arising from allegations of commercial misconduct he had directed against Robert Morris; this would soon lead him to fight a duel with the North Carolina delegate John Penn. Equally noteworthy, when Jack returned from his encounter with Lee, he was almost certainly greeted by the man whose honor he had just defended. For Washington, joined by Martha, lodged with the Laurenses during his six-week stay in Philadelphia.

On Christmas Eve, Washington returned to the State House for a more formal meeting with Congress, then withdrew as it appointed a five-member committee, including Henry, to consult with him over the next campaign. They were still in discussion when reports of the reverses in Georgia arrived on January 20, 1779, and were promptly referred to the committee. Henry immediately jotted down a rough estimate of the resources available to both sides in this new theater of war. In his account of the enemy's strength, Laurens wrote *800* next to the notation "Negroes sufficient for Pioneers, servants & all fatique [*sic*] Duty by Land & Water [and] Negroes who may be armed." A second entry placed the number of potential "Rebel Negroes" as high as thirty-five hundred. Laurens then reckoned the forces available to the Americans at seven thousand "auxiliaries" from North Carolina and Virginia. From his own state "when beset by Britons, Indians, Negroes & Tories, on all sides in her bosom," no more than five thousand might be raised. His calculations gave the enemy the advantage in both the quantity and quality of troops available and indicated that the Americans would need new sources of manpower. He also had ample time to discuss with Washington and Jack where these men might be found, for the commander and his young aide tarried in town another fortnight before returning to the new headquarters at Middlebrook, New Jersey.

The fear that the British might draw significant strength by arming American slaves was not new. It first surfaced with Governor Dunmore's proclamation of November 1775, offering freedom to any Virginia slave or indentured servant who rallied to the British standard. But Dunmore had been operating from a ship in Chesapeake Bay, and until the British could establish a permanent presence in the south, their ability to exploit such offers was limited. Nor were British officers and officials of one mind about recruiting slaves. The new southern strategy rested on appealing to loyalists, not slaves, and Tories were slaveholders too. When the news from Georgia reached Congress, some delegates assumed that the British would naturally arm slaves, just as Laurens calculated. "The greatest source of Danger, is the accession of strength they will probably receive, from the black Inhabitants," one Maryland delegate wrote, and "if they are resolved to prosecute the Measure, and to break through every tie of honor and Humanity, they will gain considerable Strength." It was an open question, however, whether the British would act more as liberators or as plunderers or new exploiters. Henry's quick estimate of late January allowed for this uncertainty, adding "Negroes not less than 5000" as likely plunder from Georgia to the others whom he counted as potential military laborers or rebels.

The news from the south gave Jack a new opportunity to seek martial glory. Once back at camp, he wrote to his father in mid-February, announcing his intention of reviving "my black project" and asking whether "the force of example, Argument, and above all that of impending Calamity will determine our countrymen to embrace the salutary measure which I propose." For the project to succeed Jack would have to teach by example—meaning that Henry needed to allow his own slaves to enlist. His reward would be "the glory of triumphing over deep rooted national prejudices" and enjoyment of "the most delicious and enviable feelings" to which his "sacrifice" would give rise. "It will be my duty and my pride," Jack concluded, "to transform the timid Slave into a firm defender of Liberty and render him worthy to enjoy it himself."

By early March Jack was chafing to depart. He knew he would need "to spend as much time as possible in disciplining and instructing my soldiers before I introduced them to the enemy." Washington granted his request for leave to go to South Carolina, assuring Jack that he could always "resume his place in my family" and providing a laudatory recommendation that attested to "the very distinguishing proofs of his bravery." Jack was all the more "anxious" to be on his way because his "reputation" would depend on the fate of his project.

Henry agreed and sprang into action when Jack asked to have his plan endorsed by Congress. Stopping in Philadelphia on his way home, Jack presented John Jay, who had succeeded Henry as president, with a letter from Hamilton supporting his scheme. Hamilton candidly defended the idea that "the negroes will make very proper soldiers, with proper management" of the kind Jack could offer. Hamilton first toyed with the military "maxim" which held that "with sensible officers soldiers can hardly be too stupid," which he cited to counter the common objection that black slaves "are too stupid to be soldiers." He did not accept this objection. While noting that "their natural faculties are probably as good as ours," Hamilton joined his friend in thinking that "the habit of subordination which they acquire from a life of servitude," along with "their want of cultivation," might actually "make them sooner become soldiers than our White inhabitants." The greater obstacle to overcome was the "prejudice and self-interest" of whites. That it was prejudice—"the contempt we have been taught to entertain"—Hamilton did not doubt. The decisive fact, though, was that slaves' loyalty was now up for grabs. If the Americans "do not make use of them in this way, the enemy probably will." By giving "them their freedom with their muskets," the Laurens plan would "secure their fidelity, animate their courage," and even exert "a good influence upon those who remain, by opening a door to their emancipation."

This same logic appealed to the new committee that Henry asked Congress to appoint on March 18, to consider the situation of the southern states, and on which he again served. The committee met with Daniel Huger, a special emissary sent by Governor Rutledge to describe the dire situation of South Carolina. In remarkably candid language, the committee stipulated that the state was "unable to make any effectual efforts with militia, by reason of the great proportion of citizens necessary to remain at home to prevent insurrections among the negroes, and to prevent the desertion of them to the enemy," who could be expected "to excite them either to revolt or to desert." The South Carolina delegates and Huger agreed, the report continued, "that a force might be raised in the said State from among the negroes which would not only be formidable to the enemy from their numbers and the discipline of which they would very readily admit, but would also lessen the danger from revolts and desertions by detaching the most vigorous and enterprizing from among the negroes." What the committee delicately called the "inconveniences" of this proposal required that it "be submitted to the governing powers" of South Carolina and Georgia. If they accepted it, the United States would "defray the expence" of recruiting three thousand slaves, to be formed into separate battalions, commanded by white officers appointed by the states (and therefore respectful of their customs). Owners would receive up to one thousand dollars for each slave "of standard size, not exceeding thirty five years of age" who served throughout the war. The recruits would be "cloathed and subsisted" at federal expense but not paid. Those who served "well and faithfully" until war's end would be emancipated and paid fifty dollars. With some debate and further amendments, Congress approved this proposal on May 29.

This set of resolutions captured the political, legal, and moral predicament in which its patrons—Henry Laurens foremost—found themselves. South Carolina and Georgia were virtually insulted for their internal weakness, yet allowed to decide whether to save themselves through the measure proposed. Slaves whom the law treated as passive, unthinking objects of its regulation—or whom masters like Laurens regarded as faithful bondsmen—could now be viewed as moral agents capable of choosing their allegiance. If they went over to the enemy, they could be accused of "deserting" the same side that feared their revolt. If they enlisted, their masters would receive most of the compensation they would earn, at great risk. If lucky enough to survive the war, they would receive only a marginal wage. Emancipation might be priceless, but it would also be nearly payless. Yet emancipation *was* a recognized aim of this policy, and what is most remarkable about its adoption. This measure "will I suppose lay a foundation for the emancipation of those poor wreches in that Country," wrote William Whipple, a New Hampshire delegate and Henry's friend, "& I hope be the means of dispensing the Blessings of freedom to all the Human Race in America."

Congress confirmed that this was John Laurens's project by again voting him a line commission as lieutenant colonel. This time Jack accepted promotion but again worried about the resentment of his fellow officers. He was relieved when Hamilton assured him that he had acted just as properly in accepting this offer as he had in rejecting the earlier one. But what one Pennsylvania delegate called "a noble proposal in that Young Gentleman" remained the visionary project of a young man whose own political capital in his native state was inherited rather than earned. Even with the blessing of his father and the approval of Congress, Jack was undertaking an uphill struggle whose outcome was doubtful. Its acceptance in South Carolina would likely require the success of British arms to persuade individual planters to part with the slaves they valued most. But British advances were just as likely to encourage slaves to "desert" or masters to hustle their labor force off to safer parts—as Georgia planters were already doing.

Jack also had to wrestle with the knowledge that no one was more skeptical of his scheme than Washington. If he did not know this before he left camp in mid-March, his father would have told him before he headed south at month's end. "Had we arms for 3000," Henry privately wrote to Washington on March 16, and "such black Men as I could select in Carolina, I should have no doubt of success in driving the British out of Georgia & subduing East Florida before the end of July." Washington disagreed. "The policy of our arming Slaves is, in my opinion, a moot point, unless the enemy set the example," he replied four days later. If we do it, the enemy will only follow suit, and then the question would be "who can arm fastest, and where are our Arms?" Assume, that is, that both sides tried to mobilize and arm slaves whose allegiance was up for grabs. Resources might give the British the edge. This cautious response ignored the obvious rejoinder that any decision the British took would be made without worrying about how the Americans would act. But thinking like the slave owner he was, Washington readily identified other costs. Giving some slaves the chance to enlist "will be productive of much discontent" among those left "in servitude," he predicted. That would presumably lead to either decreased productivity or increased turmoil on plantations. Either way, Washington was unwilling to take the idea seriously. Four years into the war, he had to confess that "this is a subject that has never employed much of my thoughts." Whether from prejudice or calculation, he did not regard it as a workable solution to the defense of the south. This was a private letter, however. Just as Laurens did not relay it to Congress, neither did Congress solicit Washington's thoughts on the "black project."

His role in this discussion was limited to confirming that he could not spare troops from his own northern command to reinforce the American position in the lower south.

Jack left Philadelphia in early April 1780. Four weeks of travel overland brought him home just as a British assault force was crossing the Savannah River and heading toward Charleston. With his new commission and Washington's recommendation in hand, Jack promptly asked the local commander, William Moultrie, to let him lead a Carolina force sent to withdraw a rear guard posted along the Coosawhatchie River. Moultrie granted the request and quickly regretted it. Ignoring his orders, Laurens forded the river in search of action, exposed his men needlessly, and was forced to retreat as British troops, safely lodged in houses he had neglected to clear, fired upon his men. Two (including a free black soldier) died and seven were wounded. Jack was one of them, with another musket ball through his arm and another horse shot from beneath him. Moultrie held his tongue when Laurens reported the action but believed that the morale of his untested troops suffered badly from the young man's rash leadership.

When Henry heard of the action four weeks later, he was less critical than he might have been, had he known more about Jack's conduct. Writing his "Dear son & beloved fellow Citizen," he wryly noted that his trip home was evidently "not calculated for pleasureable amusement." He would "not ask the insulting question," he added, "were your party surprized?" That was as close as he came to expressing his usual concern over Jack's foolish bravery. He closed instead on a literary note, quoting Brutus's farewell to Cassius from Shakespeare's *Julius Caesar:* "'If we do meet again, why, we shall smile; if not, why, then our parting was well made.'" Then he affirmed his trust in "that Holy God" of Whom neither of the pagan republicans "had just conceptions."

Henry had to be philosophical about events. Intelligence from his state was scant, and letters typically took a month or more to reach Philadelphia. Not until August did Congress learn that the British had lifted their siege of Charleston and withdrawn to Georgia. The city had been close enough to falling for Governor Rutledge and his council to discuss terms of surrender. Not for the last time in the state's history, members of its ruling elite were disposed to act as if the Union existed primarily for their benefit. If other states failed to give Carolina the support it needed, then it might be absolved of its federal obligations.

Jack was part of these discussions, adamantly opposing any idea of surrender while bridling at the request that he personally carry its terms to the enemy. He was a South Carolinian by birth but an American by military vocation, and he had not left the service of the nation's leading patriot to play so ignominious a role. Henry proudly wrote to Governor William Livingston of New Jersey, who had become something of a pen pal, that his son "when solicited refused at all hazards 'to go out with a flag on any such errand.'"

The British retreat did nothing to alleviate either the city's or the state's vulnerability. Whatever security they enjoyed owed more to the strategic uncertainties of 1779 and the emergence of the West Indies as a theater of combat than to any improved capacity for defense. In New York, Sir Henry Clinton wallowed in self-pity over his dispersed forces but still hoped to mount active operations of his own. Whenever troops became available, his orders and the early success in Georgia indicated that the lower south would witness renewed fighting.

Jack Laurens thus had every reason to pursue his primary mission, and he did so in late May, while British forces had yet to leave the state. While recuperating from his wound, Jack presented his "plan of black levies" to Rutledge. It was discussed in council without his presence and promptly rejected. The case for enlisting slaves was renewed by General Benjamin Lincoln, the commander of Continental forces in the state, with the same result. Not yet daunted, Jack planned to push his project in the legislature, to which he had been elected. Writing to Hamilton, he wished "that I were a Demosthenes—the Athenians never deserved more bitter exprobration [reproach] than my countrymen." If the Greek and Latin aspects of this sentence attest to his classical learning, his further lament that "I am doing daily penance here, and making successless harangues" indicates that he was gaining a political education as well. In the assembly he mustered barely a dozen supporters.

None of this surprised the elder Laurens. Henry supported his son morally and was prepared to do so financially. But having steered the proposal through Congress, he did not try to advance it politically in Carolina. Instead he assuaged Jack's frustration by gently recalling how "I long since foresaw & foretold you the almost insurmountable difficulties which wou'd obstruct your liberal ideas." There was "nothing wonderful in all this," for "it is certainly a great task effectually to persuade Rich Men to part willingly with the very source of their wealth &, as they suppose, tranquility," or to overcome "rooted habits & prejudices." Jack should take the long view. If he succeeded, his name would be "honorably written & transmitted to posterity." If he failed, he could still enjoy "unspeakable self satisfaction. The work will at a future day be efficaciously taken up & then it will be remembred who began it in South Carolina." But Henry sounded a different note of "self-satisfaction" two weeks later. "I learn your black Air Castle is blown up with contemptuous huzzas," he wrote, but "a man of your reading & of your philosophy will require no consolatory reasonings for reconciling him to disappointment." His friend Hamilton struck a similar note. "Even the animated and persuasive eloquence of my young Demosthenes will not be able to r[ouse] his countrymen from the lethargy of volup[tuous] indolence," he wrote.

Nothing could relieve Jack's disappointment more quickly than the promise of action, and one arose even before his father's consolations reached him. In late August Count d'Estaing appeared off Georgia with a force of twenty-five French warships and five thousand soldiers. Though homeward bound, d'Estaing couldtarry long enough to permit a Franco-American assault on occupied Savannah. Benjamin Lincoln soon took a mixed force of fifteen hundred Continentals and militia southward. In the attack of October 9, Jack led a unit of Charleston militia, which reached the city ramparts. There they came under heavy musket fire, followed by the dreaded bayonet charge and the vicious hand-to-hand fighting in which so much eighteenth-century combat culminated. After an hour of bloody mayhem, Laurens ordered his troops to withdraw. For once he emerged unwounded, but many of his men lay behind, gashed, beaten senseless, or dead. The attack failed, and with the French preparing to depart, the siege had to be lifted as well.

When Laurens returned to Charleston, he learned that Congress wanted him to serve as secretary to Benjamin Franklin, its minister to France. For nearly a year, Congress had been waging a divisive debate over its diplomatic establishment and foreign policy. Henry was, to put it mildly, an active force in these disputes, deepening friendships and enmities alike. His son's appointment was a minor part of the resolution of these quarrels; so was Henry's own election in early November 1779 as a commissioner to negotiate a loan from Holland. Anxious to leave Congress, Henry accepted his appointment willingly but doubted that Jack would do the same. That was in any case for his son to decide for himself. Though the appointment appealed to Jack, his duty as an officer weighed against it. "I might have accepted," he wrote to Henry, "had our siege been successful." But since it had not, he felt honor-bound to defend his native province, the more so since it now lay "reduced to our original State of weakness."

That same condition soon sent Jack back to Philadelphia, deputed by General Lincoln to urge Congress to send reinforcements. He expected to be reunited with his father, but Henry had left for Carolina a week before Jack arrived. Once there, Jack wavered in his decision against going to France, even returning to Philadelphia for one last discussion after setting out homeward in mid-December. "After undergoing the severest conflict that ever I experienced," Jack finally declined again (thereby avoiding a possible reunion with the wife and child he had abandoned with no regret at all). He was back in Charleston by January 11, 1780, and promptly took a post commanding marines aboard the frigate *Providence.* His father, seemingly more bemused than anxious, quipped that his son seemed "determined to pursue the Enemies of his Country in every element." Writing to his father while still at sea looking for prizes, Jack appealed to Henry to use his seat in the state legislature "to procure the levy of a few black battalions," citing "the renewed recommendation of Congress on the subject." This may have been a slightly generous reading of congressional intent. But Congress had indeed renewed its May resolutions relating to troop recruitment in South Carolina and Georgia, and that presumably covered Jack's favorite proposal as well. Moreover, General Lincoln supported the idea and urged Governor Rutledge to reconsider his opposition to it. Rutledge transmitted the request to the assembly, where it was referred to a committee that Henry Laurens chaired during the few days of legislative service he spared from preparing for his departure. Rather than endorse Jack's broad plan, the committee merely recommended the use of a thousand slaves for artillery crews and garrison duty. But Jack was also a legislator, and once returned to port, he took his seat in time to make his case. He must have been the unidentified lawmaker who moved that slaves who performed well, even in the limited capacity proposed, would be "enfranchised at the expiration of their term of service." Perhaps too much should not be read into that single word, *enfranchised.* But its use suggests that Jack was thinking of citizenship, and not merely emancipation, as the reward for service.

The motion was rejected, and the assembly adjourned without acting on the committee report. While it was deliberating, Sir Henry Clinton had been landing an invasion force only thirty miles away. Thanks to what Jack called their "natural tardiness & excessive caution," the British did not besiege Charleston until April 1, and the capital, defended only by "simple field intrenchments," held out another six weeks. Its surrender on May 12 made Jack a prisoner of war. But he spent his actual captivity in Pennsylvania, to which he was paroled under the conventions that prevented his return to active service until exchanged for a British officer of equal rank. The sole consolation was to be reunited with his father, who had come north to Philadelphia after failing to find passage to Europe from a southern harbor. On August 13 the two sailed down the Delaware aboard the packet *Mercury;* then they parted. The ship did not live up to its name. Two weeks at sea, it was taken by the HMS *Vestal.* Like his son, Henry was now a captive. Two months later, after a brief sojourn in Scotland Yard, he was confined in the Tower of London and charged with high treason.

Were it not for the commanding place that slavery occupies in the modern understanding of American history, Jack's "black project" would merit only passing mention in the narrative of the Revolution. Even then it might seem more interesting as a family drama than a serious political issue—an American forerunner of Turgenev's *Fathers and Sons.* In the pull and tug between them, the realism of the father, a man troubled by slavery but also inured to it, appears more sensible, though not more admirable, than the romanticism of the son. For was not Jack's commitment to his personal cause consistent with his battlefield propensity to make vainglorious valor the better part of discretion? As Henry knew all along, their Carolina countrymen would never risk their most valuable property in this way. They were even less likely to agree that military service could offer slaves an avenue to emancipation and citizenship.

Nor was emancipation the only link that Americans could find between slavery and military service. Late in 1780 the Virginia assembly faced the same test that South Carolina had already failed, needing to raise troops to augment Continental forces in the region and to ready itself for an anticipated British invasion. After taking Charleston, the British had gained a second striking victory at Camden, South Carolina, in early August, routing an army commanded by Horatio Gates. The hero of Saratoga leapt on a horse and was next heard from two hundred miles away in North Carolina. The American position improved in early October, when patriot militia destroyed a loyalist column at King's Mountain, just north of the line dividing the two Carolinas. That victory was a harbinger of the problems the British faced in implementing their southern strategy. British soldiers were the best troops in the region, but they lacked the numbers to pacify the backcountry. In the vicious guerrilla warfare that erupted between militias, the patriots ultimately gained the advantage. And when Congress replaced the discredited Gates with Washington's great protégé, Nathanael Greene, the Continentals finally had a commander who was up to the challenge that the British were mounting.

That development was not certain, however, when the Virginia legislature met for its fall session. To meet its quota of Continental soldiers, the lower house took up a bill to give "a bounty of a Negro not younger than ten or older than 40 years for each recruit." Rather than free slaves in exchange for *their* service, this proposal would convert white soldiers drawn from the lower ranks of the free population into slaveholders. When he learned of this proposal, the young Virginia delegate James Madison wrote home from Congress to propose an obvious alternative. "Would it not be as well to liberate and make soldiers at once of the blacks," Madison asked Joseph Jones, one of his political mentors, "as to make them instruments for enlisting white Soldiers? It would certainly be more consonant to the principles of liberty," he continued, "which ought never to be lost sight of in a contest for liberty." Virginians need not fear that this policy would create a dangerous class of freedmen who would sow unrest among their former fellow bondsmen, "experience having shown that a freedman immediately loses all attachment & sympathy with his former fellow slaves." Madison did not identify the source of this "experience." But we can plausibly speculate whether his objection to the pending Virginia measure was shaped by contact with John Laurens. The two could easily have met in Philadelphia that fall. Three years apart in age, the well-educated sons of wealthy planters, fearful for the security of their region, they would have readily agreed that principle and circumstance supported the same conclusion. Slavery was a moral disgrace to the American cause, and granting slaves the opportunity to enlist would enable them to demonstrate their capacity for freedom.

Madison's casual confidence that freed slaves would not endanger the stability of a slave society was not shared by the other famous Virginian with whom he is regularly linked: Thomas Jefferson, who was now beginning his second term as governor. Before then his chief political work was to serve on the distinguished committee charged with the comprehensive revision of the state's legal code. Among the bills Jefferson drafted was number 51, "concerning Slaves." In it he proposed to restrict that status to those already enslaved "and the descendants of the females of them." Any slaves brought into Virginia thereafter would become free after a year's residence. They would then have another year to "depart the commonwealth." But those not doing so "shall be out of the protection of the laws"—denied legal recognition or protection of any kind, and thus reduced to a state of nature that owed more to Hobbesian fear than the Lockean duty of mutual preservation. Jefferson applied the same draconian proscription to two other categories of persons: "Negroes and mulattoes" who came to Virginia "of their own accord" and "any white woman [who] shall have a child by a negro or mulatto." It is deeply disturbing that he used racial terms rather than the legal definition of a slave to describe the persons whose lives and liberty he would allow to be pursued so unhappily. The same troubling assumption guided the amendment to this bill that Jefferson later claimed to have prepared, laying out a scheme to "emancipate all slaves born" after its passage, but then to colonize them as a free people to some other region or country whose location he never managed to identify.

John Laurens grounded his project on different assumptions. Like Jefferson, he too had naturalist interests, and perhaps even keener powers of observation from his youthful training as an artist. But his sense of what was morally necessary and politically desirable rejected the pessimistic observations that led Jefferson to suppose that racial differences were unbridgeable. True, Laurens and Hamilton treated the seeming "stupidity" of slaves as a serious question deserving explanation. But they preferred to think it was the *condition* of servitude, rather than an *aptitude* for it, that best explained all those facets of slave behavior—their careless use of tools, their maddening defiance of the simplest instructions—that aggravated active planters like Henry Laurens and the overseers on whom they relied. Allowing slaves to become soldiers would fairly test their capacity for citizenship in a way that the mind-numbing and soul-crushing routines of the plantation never could. This was a truly visionary position, and for all his youthful naïveté and his student's enthusiasm, Jack Laurens deserves moral credit that few of his fellow Carolinians ever earned. But it was Jefferson's fumbling effort to state a theory of racial difference, rather than Jack's notion of potential civic equality, that better predicted the American future.

6

The Diplomats

WHEN HENRY LAURENS finally sailed for Europe in August 1780, he was embarking on the last of three diplomatic missions that Congress had planned a year earlier. Laurens was bound for Holland, where he was instructed to seek diplomatic recognition from the Dutch confederation and, just as important, to secure whatever loans he could negotiate in the great credit market at Amsterdam. His successor as president of the Continental Congress, John Jay, had been the first to depart. In late October 1779, Jay boarded the U.S. frigate *Confederacy*, along with his wife, Sally; her brother Henry Brockholst Livingston; and Conrad Alexandre Gérard, the first French minister to America. Sally did not yet know she was pregnant; her thoughts were directed instead to three-year-old Peter, left in the care of her parents, the first family of New Jersey. They were bound for Madrid, where Jay would be minister to Spain—if the Spanish court received him. But first he planned to consult the doyen of American diplomats, Benjamin Franklin, in Paris.

John Adams sailed a few weeks after Jay. In the deepening gloom of a mid-November afternoon, he and his older sons, John Quincy and Charles, rowed out to the French frigate *Le Sensible*, anchored in Boston Harbor. The father and the precocious Quincy knew the ship well, for it had carried them home only five months earlier, after Congress terminated Adams's first appointment in France by making Benjamin Franklin its sole minister to the court at Versailles. Adams hoped to steal on board "as silently as possible," but his sons were probably not upset when the crews of *Sensible* and another French ship "came upon deck and huzza'd" them. On their first crossing to France, John Adams and Quincy were saved from British warships by a three-day storm, which kept them clutching for handholds and praying for safety below deck. Perhaps that was why the future sixth president of the United States began the diary he would keep for the next seven decades by noting that *Le Sensible* packed an armament of "28 twelve Pounders." Now his father would be the nation's sole negotiator in the unlikely event that peace talks broke out in Europe. Adams had been sorely miffed when he mistakenly thought that Congress had made him a mere commissioner while giving Jay the higher rank of minister plenipotentiary. "Mr. Jays name was never heard till 1774," he grumbled to Elbridge Gerry. "Mine was well known in Politicks in 1764." A similar slight had marred his previous trip, when his name appeared below that of Arthur Lee in the commission for France even though Adams held weightier credentials as lawyer and public servant.

With his capture and confinement, Laurens could claim bragging rights to undertaking the most dangerous mission. He could have made an even greater adventure of it after his escorting officer disappeared for two days while they stopped in Exeter en route to London. In his absence a mysterious gentleman offered to hide Laurens in his "very private & retired" house until "the bustle of enquiry & pursuit should be over." Then he could proceed safely to Holland. The visitor was "amazed" when Laurens declined, even though he had sworn no parole to the absent lieutenant. "Why Sir, Kings & princes in your circumstances have made escapes," he protested. But royal escapades were the last examples the fiercely republican Laurens was likely to emulate.

The Jays' and the Adamses' voyages of 1779 were almost as eventful as Laurens's ill-fated trip the following year. Once at sea, after a slow crawl down the Delaware, the Jays encountered "a brisk gale" that left both retching. Sally "soon recovered," she noted in the running letter she was writing to her mother, "but my dear Mr. Jay suffered exceedingly at least five weeks and was surprisingly reduced." Given that the tall, hollow-cheeked Jay always had a lean and hungry look about him, he must have been verging on the cadaverous. That did not stop him from racing topside when "an unusual noise upon deck" and "the lamentations of persons in distress" roused the entire ship before dawn on November 7. In quiet seas a freak shift of wind had completely dismasted the *Confederacy*. A storm that night left the rudder useless as well. It took two weeks of desperate repairs to make the ship navigable. At this point Jay had to choose their destination. They were in mid-ocean with winter approaching. Gérard amazingly preferred continuing toward Europe. The ship's officers thought they should divert to the West Indies, and Jay concurred. Three weeks' sail brought them to Martinique, which Sally, a child of the Hudson and Delaware valleys, called the most "verdant, romantic country I ever beheld." Their tropical stay was brief. Three days after Christmas, they resumed their voyage aboard the frigate *Aurora*, bound for the Mediterranean harbor of Toulon.

When they reached the Spanish port of Cádiz on January 22, 1780, reports of British naval activity in the Mediterranean led Jay to abandon his plan to visit Paris first. But he could not proceed to Madrid at once. Spain had yet to recognize American independence, and he could not appear unheralded at the Spanish court. Jay instead sent his secretary, William Carmichael, ahead to Madrid to announce his mission. Not until early March did permission to proceed arrive, and even then Jay was told that he would not be received in a public capacity. In late March they began their trek in a six-mule carriage whose "peculiarities" of design fascinated Jay. The mules had more charm than the surly innkeepers they met en route. Jay was most irked by one tavern keeper who charged for the use of more beds than their party numbered. "It was impossible for Eight persons to use fourteen Beds," Jay protested. "They replied that was not their fault." Their route took them across La Mancha, "Don Quixote country," as their guides proudly informed them. A time would come when Jay might have preferred having Sancho Panza for an assistant, rather than his surly brother-in-law or Carmichael, who seemed to think he was Jay's diplomatic equal. They reached Madrid in early April.

John Adams also detoured through Spain. While the *Confederacy* was making its mid-ocean repairs, *Sensible* was wallowing in heavy seas, having sprung a leak two days out, which gave the passengers steady work on the ship's two pumps. The captain changed course for Ferrol, a naval port in the northwest corner of Spain. From there the Adams party went by mule, carriage, and foot to Paris. Still feeling abashed for abandoning Abigail so quickly, John used his first letter home to offer the kind of extenuating apology that husbands traditionally plead. Just imagine how much worse off we would be, he asked, "if You and all the family had been with me." Still, such thoughts did not prevent the observant traveler and serious eater from making the best of the journey. En route they enjoyed "Eels, Sardines, and other fish" as well as "tolerable Oysters, but not like ours." Adams also relished "the most excellent and delicious" pork, "fatted upon Chestnuts and much more upon Indian Corn." Regrettably his route bypassed other regions where diets of sweet acorns or beheaded vipers produced even tastier meat. Passing "thro several, very little villages" east of León, they saw "the young People" doing "a Dance that they call Fandango," each holding a wooden "Clacker" and wearing "wooden shoes, in the Spanish fashion, which is mounted on stilts." Adams finally reached Paris in February, relieved to trade smoky cottages and filthy inns for the Hôtel de Valois, rue de Richelieu.

Amid their travels and travails, whether retching in their cabins, tossing and turning in a Spanish parador while fleas gorged on their ankles, or gazing at passersby from a street-level room in the Tower of London, the diplomats had ample time to reflect on how far the Revolution had carried them. They could measure that distance in nautical leagues and land miles traveled, in months and years separated from children, and above all in the gap between the private pursuits they had laid aside in 1774 and the public duties they took on thereafter. These new missions were their most important tasks yet, the occasions when they would have the greatest opportunity to serve their republic. So too their appointments marked a critical transition in the conduct of American diplomacy. The first commission to Paris—Franklin, Silas Deane, and Arthur Lee—had attained the signal triumph of completing dual treaties of amity and commerce with France in February 1778. But the disorder in the commission's affairs and the animosity between Deane (supported by Franklin) and Lee also made them the laughingstock of Paris, a city that lived on gossip fed by the endless intrigues of the court of the young Louis XVI, barely two years on the throne. Congress had been drawn into examining their charges and countercharges, and soon it too came under attack from its own citizenry as Deane, Lee, and their partisans took to the newspapers to vent their accusations and grievances.

Compared to the team Congress had hastily thrown together in 1776, a diplomatic corps headed by the venerable Franklin and reinforced by two of Congress's former presidents and one of its leading original members would obviously be a marked improvement. Just as important, the United States needed even stronger representation in 1780 than it had required before 1778. Essential as it had been to have Franklin available to press the case for alliance, the treaties of 1778 were much more a response to events than a tribute to diplomatic ingenuity. France had its own *raisons d'état* for supporting American independence: the victory at Saratoga demonstrated that Americans had the staying power to resist conquest; the French navy was again ready to challenge British maritime superiority; and the alliance was a natural, even logical consequence of these facts. The situation the United States faced in 1779 was far more perilous and uncertain. Part of the south was occupied by the enemy, and the rest exposed to invasion. The whole system for financing the war and keeping the army fed and clothed was tottering. And Americans faced the uncertain prospect that European powers might have their own reasons to conclude a peace before the independence of all thirteen states could be firmly assured. Under these circumstances, having a diplomatic team in Europe that Congress could actually trust was a matter of crucial importance. Hence the appointments of 1779.

A year after Adams and Jay made their wintry Spanish landfalls, Congress expanded its peace mission by making Franklin, Jay, and the captive Laurens joint ministers with Adams. Laurens would play only a cameo role in those negotiations, but his confinement in the Tower had at least one other notable effect: it freed John Adams to shift his own base of operations from Paris to Holland, and thus to escape the unique purgatory he would have felt had he remained in the French capital and continued to work in Doctor Franklin's shadow. From their respective bases in Paris, Madrid, and Amsterdam, Franklin, Jay, and Adams thus had the opportunity to pursue and refine their own conceptions of the proper relations between America and the powers of Europe. Yet when Franklin summoned Jay and Adams to Paris in the summer of 1782 as peace talks finally got under way, these three independent-minded gentlemen, representing distinct points on a broader spectrum of American political attitudes, had to determine what their common positions would be. They could answer that question only after they first asked themselves whether they had to adhere to their formal instructions from Congress, which they knew had been drafted under heavy, perhaps improper pressure from France, and which were meant to limit the very judgment and discretion for which Congress had presumably appointed them. In effect, for a few months the three diplomats were responsible for determining the national interest of the United States. In exercising that trust, they also illustrated a recurring dilemma in the conduct of American foreign relations: the tension between the idea that the foreign policy of a republic should rest on the collective choices of the people's representatives, and the reality that there are tides in the affairs even of republics when the national interest must be left to the decisions of a handful of well-situated actors. Whether any other set of American diplomats has matched the distinctive qualities and idiosyncrasies of this early team of blessed peacemakers is a question to ponder.

The diplomacy of the Revolution is essentially a tale of three treaties, three cities, and (on the American side) three men. The three treaties are the complementary treaties of alliance and commerce of 1778 and the provisional peace agreement of 1782, signed with minimal revisions the next year. The three cities are the capitals of Paris, London, and Philadelphia. Each had a distinct political culture, with its own form of intrigue and maneuver, reflecting the differences between the absolutist monarchy of France, the constitutional monarchy of Britain, and the novel American confederation of republican states, which treated the Continental Congress as "a deliberating Executive assembly"—a body that resembled a legislature but wielded powers over war and diplomacy customarily associated with the Crown. The three men are Franklin, Jay, and Adams—the troika who negotiated peace in the summer and fall of 1782.

The treaty of alliance that the first American commission signed at Paris on February 6, 1778, had this simple but decisive effect: it altered the strategic equation of the American war by making it impossible for Britain to deploy the full complement of regiments and naval squadrons required to bring its rebellious subjects to submission. After 1778 the American rebellion became a global war for empire, and Britain had valuable assets to protect around the world, from the great peninsular fortress at Gibraltar, to the sugar-producing islands of the West Indies, to the trading posts of the Indian subcontinent. True, in North America the British retained the strategic initiative well into 1781. But with ships and men diverted elsewhere, this advantage could not be translated into a lasting solution to the conflict so long as the British lacked the means to reoccupy American territory and govern its defiant people.

The peace treaty of 1782–1783 had a more complicated set of consequences. One was obvious: it ended the war and brought formal, if grudging, British recognition of American independence. Two other results merit greater emphasis. By making the Mississippi River the western boundary of the United States, the treaty ensured that this cluster of seaboard settlements, with the bulk of their population still huddled east of the Appalachians, would instead become an empire in the making. That presumptuous adjective *continental* that was attached to the First Congress of 1774 now became the providential augur of a manifest destiny. Yet the opportunities the peace treaty created also brought obligations that Congress proved ill equipped to fulfill. Its inability either to realize those opportunities or meet those obligations became a principal occasion for the great discussions that led to the drafting of a new federal Constitution in 1787.

Both treaties were negotiated and signed in Paris, the city that always takes a star turn in the standard histories of Revolutionary diplomacy, along with the magnificent palace at nearby Versailles. With its six hundred thousand residents, Paris in the mid-1770s was roughly thirty times as populous as Franklin's Philadelphia. Its residents lived within a short walk of the Seine, mostly on narrow streets that bore little resemblance to the neat grid on which William Penn had first mapped his provincial town. The hulking prison of the Bastille lay at the eastern end of the city, just within the walls that still separated Paris from the fields outside. Similarly, to stroll west from the Louvre, through the Tuileries, and on to the Champs Élysées was to leave the city and again enter the countryside. The rooms that Franklin and Deane occupied at Passy were part of the Hôtel de Valentinois, an estate on a bluff overlooking the Seine owned by Jacques-Donatien Le Ray de Chaumont, a well-connected and wealthy supporter of the American cause who was part of the network of suppliers of arms and goods that the Secret Committee of Congress had sent Deane to assemble. Today the site lies west of the Eiffel Tower, on the Right Bank, but in 1778 it was a suburban setting nicely located on the same road to Versailles down which the commissioners traveled to court—and along which the urban crowd would flock eleven years later to command their doomed sovereign to return to his capital. Here Franklin was spared the din, filth, and stink of Paris, a city that pleased the eye and assaulted every other sense. Arthur Lee took rooms three miles away at Chaillot, from which he repeatedly hectored his colleagues over sins of omission and commission.

Paris in the 1770s was as full of spies as Cold War Berlin. The most effective was Edward Bancroft, the American commission's secretary. A former pupil of Silas Deane and one of Franklin's many protégés, Bancroft was a talented naturalist and chemist, with expertise in plant dyes and tropical poisons. He also believed, correctly, that the shocks delivered by certain South American eels were electrical in nature—a hypothesis that Franklin naturally welcomed. As a youth in Connecticut, he had been a physician's apprentice before making his way to Surinam, where he found work tending to the health of a local planter and his slaves. By 1767 he was in London, studying medicine at St. Bartholomew's hospital and publishing the results of his tropical researches. Bancroft became a novelist as well. The commercial success of his quasi-biographical *History of Charles Wentworth*, John Adams later complained, was due to "the plentifull Abuse and vilification of Christianity which he had taken care to insert in it." That did not bother Franklin, who promoted Bancroft's election to the Royal Society of Fellows in 1773. Bancroft was then only twenty-nine: self-educated, freethinking, and gregarious. At dinner he was extraordinarily loquacious, Adams recalled, inspired by "a little generous Burgundy" and the "enormous quantities of Chayan Pepper" he heaped upon his victuals. He was also ambitious for wealth, fond of descending "into the deepest and darkest recesses of the Brokers and Jobbers, Jews as well as Christians," there to gather "the News and Anecdotes true as well as false" that might help his stock speculations. Playing the London exchange remained his chief private preoccupation after Franklin and Deane made him their secretary. Deane was more than happy to join him in speculation.

Bancroft had another profitable sideline that may account for his loquacity. In December 1776, he entered into a contract with Paul Wentworth, a member of the family that long dominated New Hampshire politics and possibly both the same planter he had served in Surinam and the model for the central character of his novel. Wentworth was also the leading recruiter of American spies for the British secret service—Arthur Lee had already spurned one approach—and Bancroft was his great catch. The naturalist's propensity for "Tittle Tattle," which so annoyed Adams, may have been contrived as a cover: no one who blabbed so much could ever be a spy. Once a week Bancroft strolled through the Tuileries gardens adjoining the Louvre and left in the hollow of a tree copies of papers that Wentworth might wish to read. Thanks to Bancroft and the success of British interdiction, King George often knew far more about the work of the American legation than did Congress.

Even so, there is little reason to think that the success of British espionage or the tiffs between the paranoiac Lee and his two sane colleagues affected the course of diplomacy. France was an absolutist state, and the individuals whom the Americans needed to influence numbered only a few: the comte de Vergennes, the foreign minister; his master, the young King Louis; and a few other ministers and aides whose counsel would be taken in any decision about how far France could go in aiding the rebellious colonies or risking war with Britain. The allure of recounting the complicated details of revolutionary diplomacy in a Parisian setting may explain why so many popular writers have been drawn to this seemingly prosaic subject, just as the titles of their works almost tell the story in itself: *A Great Improvisation* by *The Virgin Diplomats* that led to *Triumph in Paris.* But the critical fact about American diplomacy in Paris was that key decisions were taken by a few men who needed to consult no one but one another.

Very different conditions prevailed in the rival monarchy across the Channel. In London, as in Paris, many decisions could be taken by the king and a small circle of advisers, most notably his chief minister, Lord North, and Lord George Germain, who replaced North's stepbrother, the religiously devout Earl of Dartmouth, as secretary of state for America in 1775. But the British monarchy was constitutional, not absolutist, and its decisions could be freely debated and criticized in Parliament and the press. Parliament had played an active role in military affairs since 1689. It approved "supplies" for the army one year at a time, and it adopted the Mutiny Act (the statute empowering the Crown to issue articles of war regulating military discipline) on the same basis. But this kind of active involvement in war making did *not* render Parliament politically superior to the government. So long as he enjoyed the personal confidence of the king—and he always did—Lord North could maintain a healthy majority of support in the Commons (the House of Lords was wholly docile). Yet the Commons had enough independent members to produce notable swings and tightening majorities on particular issues, and those swings offered a rough index to the confidence that the Commons—and the "political nation" it represented—felt in the ministry's policies.

The sharpest parliamentary critic of the American war was Charles James Fox, the brilliant orator and dissolute, shaggy-eyed gambler ("the Eyebrow," as his friend, the glamorous Georgiana, Duchess of Devonshire, called him) who liked to sport the blue-and-buff colors of the Continental army for his appearances in the Commons. Like Edmund Burke, who was something of a mentor, Fox had a rhetorical flair no American—with the possible exception of Patrick Henry—could match; and Henry honed his oratorical skills playing to Virginia juries, an audience rather less sophisticated than the House of Commons. As a study in dissolution, Fox had no American peers either. He was one of those legendary figures who could go straight from Parliament to Brooks Club, the gambling den where he often lost thousands of pounds in a night's wagers, then spend the next day pursuing one of his many amours before returning to Westminster to rip North or Germain over the latest setback in the American war—and still have backbenchers hanging on his every word and snickering over every rumor.

With critics like Fox and Burke ready to pounce, the North ministry faced a form of accountability unknown in Paris. Yet until 1782 the periodic need to answer probing questions about the lack of a strategy for victory and the mounting costs of the war never outweighed one fundamental fact about British politics. To acknowledge that America *was* lost required an intellectual and emotional leap that remained inconceivable, not only for the king, who was the fiercest supporter of his government's policy, but also for a majority of members of Parliament who privately relished Fox's rhetoric but rejected the defeatist conclusion to which it led. After 1778, North himself sank into a despair over the war that hovered between political despondency and psychological depression. Repeatedly he begged his sovereign to let him retire and give way to someone better able to bear the burden of office. Repeatedly the king refused—a refusal North had to honor, not merely from duty, but because George had rescued him from serious financial debts. North remained loyal to the policy whose success he now doubted. But it was also a policy to which he, like most members of Parliament, could see no alternative. To renounce British rule over America remained just as unthinkable in 1780 as abandoning parliamentary supremacy over the colonies had been in 1776.

So long as London held to that view, neither John Adams nor the enlarged delegation of 1781 would have any prospect of pursuing peace. But the diplomacy of the Revolution could have taken a very different course had the king and his ministers ever swallowed their qualms and sent peace commissioners to America with real authority to negotiate. Not only would that have enabled Philadelphia, the third city in this triangle of capitals, to displace Paris as the real site of diplomacy. It might also have allowed it to supplant London as the city where the maneuvers of political factions would have the greatest impact on negotiations. For Congress was its own executive arm—or rather, a deliberative body that possessed the powers over war and diplomacy that European monarchs wielded as a matter of course. Had the Howe commission of 1776 been allowed to do more than issue pardons of grace to repentant Americans, moderates such as Dickinson, Morris, and Jay would have jumped at any serious offer of negotiations, forcing Congress to revisit all the hard questions that the North ministry's intransigence allowed it to duck, and to do so under intense public scrutiny. Two years later, when word of the Franco-American alliance reached London, North hastily appointed a second commission and sped a bill through Parliament, proposing terms that Congress might have accepted in 1776 but that now fell far short of the full independence Americans believed the French alliance would guarantee.

On both occasions, then, Britain's inability to perceive that real offers of negotiation might divide Americans over their basic war aims instead worked to unify opinion both within Congress and "out-of-doors." But that unity was badly threatened in February 1779, when Conrad Alexandre Gérard, the first French minister to the United States, urged Congress to define its terms of peace. France's chief ally, Spain, would soon offer to mediate an end to the conflict, he reported. Should Britain accept that invitation, Congress would need to state its conditions of peace. Gérard hoped Congress would respond quickly. Instead it plunged into months of partisan debates, which soon spilled into the newspapers. New England delegates clamored to ensure that their fishermen would gain access to the rich banks off Newfoundland. Southern delegates worried about American navigation rights on the

Mississippi River. But with the enemy shifting its military operations southward, they were inclined to appease Gérard and not take a hard stance on the Mississippi. When it came to "cod and haddoc," the New Englanders were made of sterner stuff. The dispute raged until September and soon became entwined with the question of who could best represent American interests overseas—and in which positions. It ended only when Jay, Adams, and Laurens received their new assignments and instructions.

From this tortuous and politically embarrassing affair, informed observers could draw two disturbing conclusions. One was that a deliberative body like Congress was not the ideal forum for framing foreign policy. True, it had performed that function well enough in the first years of the Revolution, when British obstinacy and the need for consensus enabled its members to plot a path toward independence and a French alliance. Then there had been only one overarching national interest to pursue. But would that national interest remain so obvious when other issues had to be resolved, or when clusters of states or whole regions recalled that they had particular interests to protect and promote? If the states did find themselves jockeying to advance their special aims, could a politically accountable body like Congress avoid the partisan bickering that marked the prolonged debates of 1779? In this sense, the making of foreign policy in Philadelphia differed radically from that of Paris, where the king and his key advisers were all that mattered, and from that of London as well, where royal prerogative was balanced by parliamentary accountability.

Yet Philadelphia differed from the other capitals in one further respect. Of necessity, it could be only a distant observer of negotiations taking place across the Atlantic, perpetually awaiting information that could be months out of date before Congress could receive, absorb, and respond to it. Without fresh or reliable intelligence, Congress became increasingly vulnerable to French manipulation. The disputes of 1779 took months to resolve, not only because the questions facing Congress were difficult in their own right, but also because Gérard continuously pressed the delegates for their response, even when the Spanish offer of mediation had obviously come to naught. Without French meddling, in other words, Congress might have avoided much of the acrimony of 1779.

That awareness of the heavy-handed French efforts to manipulate

Congress played a key role in shaping the behavior of the American diplomats who negotiated the treaty of peace. It would have been difficult to assemble a more independent-minded team of negotiators than the one Congress had in place by 1781 when, again under French prodding, it enlarged its peace commission. Though formally respectful of Congress, the diplomats all knew enough about its divisions and internal pressures to make them wonder how closely they should follow its instructions. Even had they not felt such qualms, they had enough confidence to act with the independence that Congress valued in them. Why else, after all, had they been chosen, if not for their patriotism, probity, and judgment?

The story of Revolutionary diplomacy, then, is not only about the treaties of 1778 and 1782 and the ways in which three capitals—two great metropolises and one provincial port—made foreign policy. It is also (from an American perspective) a tale of three diplomats, each acting on a distinct conception of the national interest yet ultimately discovering a greater harmony of sentiments than they might have supposed possible.

Benjamin Franklin was the most acclaimed American to visit Paris until President John F. Kennedy accompanied his wife, Jackie, there almost two centuries later. But it was not only his international celebrity that led Congress to make him its leading representative in Europe. In 1776 Franklin was also the only American with any diplomatic experience worth noting. He had been the leading spokesman for American interests in imperial London since the late 1750s, when he launched his futile campaign to make Pennsylvania a royal colony. As agent for the Pennsylvania and Massachusetts assemblies, he represented two of His Majesty's most important provinces, one a magnet for the free British immigration he first championed in his 1751 pamphlet, *Observations on the Increase of Mankind*—a landmark in demographic thinking—the other the great political nuisance that imperial officials longed to chasten. The old newspaperman also applied his polemical genius in the press, publishing numerous essays (all pseudonymously, of course) vindicating American claims.

Never one to miss a quip, Franklin brought these efforts to a hilarious conclusion in his September 1773 essay, *Rules Whereby a Great Empire May Be Reduced to a Small One.* Its barbs and jests notwithstanding, in its analysis of imperial missteps it was just as acute as Edmund Burke's brilliant speech on conciliation two years later, a running catalog of everything Britain could do to alienate American affections. "Suppose all *their* Complaints to be invented and promoted by a few factious Demagogues, whom if you could catch and hang, all would be quiet," Franklin counseled the blundering minister for whom *Rules* was purportedly meant. The same logic sent Gage's men to Concord in 1775, in the naive belief that detaining Samuel Adams and John Hancock would help restore order in Massachusetts. Behind the sarcasm lay a keen insight into the illogical nature of imperial policy. No colonist of his era knew Britain (or its rulers) better than Franklin, and before 1774 few were more committed to preserving the bonds of empire—"that fine and noble China Vase," as he called it in 1776. Yet as Franklin well understood, the deepest bonds were of the kind that drew him to London: language and culture, history and affection. The idea that American loyalties depended on the blanket assertion and raw imposition of imperial law and authority struck him as absurd.

That conviction led him into the misconceived affair of sending the damning letters of Thomas Hutchinson back to Massachusetts in 1772. Remove this loose cannon of imperial politics from that volatile province, Franklin calculated, and the empire might be saved. Instead, the affair of the letters stiffened Hutchinson's resolve to strike back at his critics, and first the Tea Party and then the Coercive Acts were at least partial results of Franklin's failed epistolary coup. So was the vicious dressing-down in the Cockpit that he endured in January 1774, standing stoically in his new coat of Manchester velvet as Wedderburn ripped into him, while lords and ladies snorted and Americans such as Henry Laurens looked on in dismay. Franklin stayed in London for another year, first trying to discourage the punitive measures against his native Massachusetts, then conducting private talks with well-placed English contacts: William Pitt (the Earl of Chatham); Lord Howe, the future naval commander, who acted as a go-between with North and Dartmouth; and two well-meaning Quakers. Franklin was present, "leaning on the Bar" in the gallery, when Chatham presented a conciliatory plan to the Lords on February 1, 1775, but to no avail. Franklin kept the same impassive mien he had shown a year earlier, as "so many of these *Hereditary* Legislators declaim[ed] so vehemently against" Chatham's plan and Franklin as its supposed surreptitious author. But he left Westminster quietly seething, as he would again after a later debate in the Lords featured "many base Reflections on American Courage, Religion, Understanding, &c." The deep irritation he felt as the Lords ranted on, treating Americans "with the utmost Contempt, as the lowest of Mankind, and almost of a different Species from the English of Britain," was more than he could stomach.

Through these intense discussions of early 1775, Franklin nonetheless preserved the mentality of a diplomat. He believed the imperial controversy could easily become negotiable, and he would do anything in his power to further an accommodation—so long as he held to the positions his countrymen had laid down. That the colonists remained his true countrymen, however, was one of the great discoveries that Franklin carried home in late March 1775. That discovery made him no less fond of Europe, no less willing to recross the Atlantic a year and a half later, no less charmed to find that Paris could be as diverting as London. So it was that Franklin combined the temperament of a diplomat with the passions of a patriot. He could mask his feelings to further a discussion, and retreat into that elusive personality that attracts and puzzles biographers. Yet he was also capable and at times compelled to speak directly—perhaps not with the unobstructed candor of John Adams—but with heartfelt conviction still.

Had Franklin been the most polished diplomat who ever lived, and not merely the most experienced and thoughtful American available, his skills alone could not have persuaded Vergennes to offer an alliance that France was unprepared to make. Events would determine whether and when that alliance should be made. For his part, Franklin accepted the basic position that John Adams had expressed in drafting the "model treaty" that the commissioners were instructed to seek. Americans had no interest in forming permanent political alliances with any European powers. They should presume instead that France would be satisfied to see America independent, its markets open to the world, its commodities flowing in new channels, and its vanquished British masters no longer capable of drawing strength and sustenance from their kinsmen across the water. The commissioners should seek a treaty of "amity and commerce," not a lasting alliance. And in the meantime, it would be extremely helpful if France continued to permit the shipment of badly needed arms and munitions to America.

Vergennes was willing to cooperate on this last point, extending to the Americans generous credit to cover the purchase of supplies. But there were other sources of friction. Vergennes grew upset when the commissioners ignored his pointed reminders that American privateers could not use French ports to stage raids on British shipping. France was not yet ready to risk the naval war that formal recognition of independence or overt military support could entail. These considerations kept Franco-American diplomacy in an uncertain state well into 1777, leaving everyone anxious to learn the results of the military campaign that was so slow to get under way.

Under these circumstances Franklin's philosophical equanimity served him well. He was enjoying Paris too much to allow his diplomatic frustrations to burden him. Entering his eighth decade, he still relished the company and kisses of the beautiful and cultured women whom he met for tea or dinner or who accosted him on his walks. His favorite companion and the object of his first flirtation was Madame Brillon de Jouy, his thirty-something married neighbor in Passy. Her charms went well beyond her exceptional musical talent or her newfound interest in chess, two diversions they shared during his frequent visits. Though she addressed Franklin as *mon cher* or *mon bon papa*, his desires were not limited to quasi-paternal affection—at least until Madame Brillon insisted that a surrogate father-daughter relationship was what it had to be. In 1777 Franklin was waging two campaigns—one targeting Vergennes, the other Madame Brillon—and both called for similar strategies of tact, patience, and verbal dexterity.

One other dimension of Franklin's diplomacy bears mention: his realization that his own reputation was the best asset Americans enjoyed in France. He happily abetted the popular craze that began with the June 1777 engraving of his casual portrait wearing the fur cap he had picked up on a congressional mission to Canada in early 1776. Soon that image and others were everywhere: on medallions, busts, tea services, statuettes, a cornucopia of Frankliniana ranging from fine *objets* fit for aristocratic admiration to tchotchkes for bourgeois consumption. In this way, Franklin became a practitioner of cultural diplomacy *avant le fait*, the first to grasp that the American idea (or ideal) that he personified could exert its own political allure. Although France was still an absolutist monarchy, it was not an authoritarian dictatorship or a closed society. Public opinion was an active ingredient in the rich political culture of the ancien régime, equally expressed in the underground literature that made Marie Antoinette, the young queen from Austria, into an imagined pornographic star, and the bric-a-brac that converted Franklin into the symbol of American nationality.

Here was the obverse of the affection that Britain had alienated after 1774. Even if France would support America only for *raisons d'état*, not affection, new bonds might be established to supplant the old ones that the Revolution had sundered. That was an assurance that Franklin felt he could extend once news of Burgoyne's surrender at Saratoga reached Paris in early December 1777. After a year of impasse, the crucial discussions took barely two months to complete. Vergennes now had the proof of American staying power he needed to press the case for alliance and intervention with his king. He did so knowing that the program of naval refitting was far enough along to make an open break with Britain acceptable. When the Americans renewed their suit, Vergennes answered promptly. At Versailles on December 12, the foreign minister explained the generous terms his sovereign would offer. Beyond accepting the treaties of amity and commerce that the envoys had proposed a year earlier, France would not prevent the United States from making "a separate Peace" with Britain, "whenever good and advantageous Terms were offered," so long as one condition was met: that the Americans would never "give up our Independency, and return to the Obedience of that Government."

Vergennes had to insist on this point because both he and the commissioners expected that "probably Britain would be making some Propositions of Accommodation." When those offers did come, they were more remarkable for Franklin's rebuff than the diplomatic openings they predictably failed to provide. There were two approaches. One came from Edward Bancroft's spymaster, Paul Wentworth, who met with Deane in mid-December and Franklin on January 6. The second came in letters from David Hartley, an old friend of Franklin's and one of the few members of Parliament still identified as pro-American. Hartley was also assisting the efforts of the commissioners to aid the hundreds of American seamen held in wretched captivity in British "gaols" and prison ships. Their plight disturbed Franklin deeply, and he was genuinely grateful to Hartley for his efforts.

Yet in both cases Franklin flatly rejected the idea of any accommodation that did not begin with a permanent recognition of American independence. The unfeigned passion that he displayed in their two-hour conversation at his lodgings above the Seine stunned Wentworth. "I never knew him so excentric," he reported. Ordinarily Franklin was the most focused speaker he knew, "but He was diffuse and unmethodical to day." Yet Franklin was nothing if not methodical even as "he lost Breath in relating the burning of Towns, the neglect or ill treatment of Prisoners," and other "Barbarities inflicted on His Country" by an enemy pursuing a "regular system of devastation and Cruelty." Franklin checked his emotions when he wrote to Hartley five weeks later, but his sentiments were unchanged. How could any sensible Briton expect Americans to spurn the aid of France or any other country, he asked, "when your Nation is hiring all the Cut Throats it can collect of all Countries and Colours to destroy us"? Franklin then gave the familiar metaphor of empire a new twist. America "was a dutiful and virtuous Daughter" until a "Cruel Mother-in-Law turn'd her out of Doors, defamed her, and sought her life." Now "her friends hope soon to see her honourably Married" and never again forced into a "Return and Submission to the Will of so Barbarous an Enemy."

Barbarous and *barbarities* were not words that Franklin used thoughtlessly. Britons who naively hoped that the memory of ancient ties could outweigh three years of brutal war had to realize that the empire could never be restored. As a voice for moderation before 1774, Franklin had argued that trust and affection offered greater security to British rule than formal claims of parliamentary supremacy. He could only scoff when Hartley predicted that Americans could again feel at home in Britain while they would long remain strangers in France. "Americans are received and treated here with a Cordiality, a Respect and Affection, they never experienced in England when they most deserved it," Franklin replied. Here was yet one more allusion to his humiliation in the Cockpit four years earlier, with members of the Privy Council "chuckling, laughing, and sometimes loudly applauding" in response to Wedderburn's taunts, or to the abuse heaped on Americans, which he found so rankling a year later. It was with added pleasure that Franklin attended the treaty ceremony of February 6, 1778, in the same velvet suit he had worn in the Cockpit.

If Franklin founded the cultural approach to diplomacy, Jay and Adams can be similarly hailed as pioneers of another essential aspect of the American diplomatic tradition: the belief that the conduct of foreign relations is properly the province not of grand strategic thinkers but of lawyers, with their ingrained concern to ensure that contracts are enforced, contingencies foreseen, and rules laid down to adjudicate all disputes. Franklin was open to the possibility that diplomacy could involve a measure of moral trust, even among nation-states. Jay and Adams regarded trust as something that had to be carefully constructed, provision by provision. But when Congress called Jay and recalled Adams into diplomatic service in the fall of 1779, it was not trying to establish a diplomatic tradition of any kind, but simply turning to the two ablest men it could recruit at a moment of crisis.

After leaving Congress in the spring of 1776 and drafting the New York constitution a year later, Jay had accepted appointment as the chief justice of the state's Court of Appeals. When he returned to Congress in December 1778, he came on something of a legal mission as well, charged with pressing New York's claims to the quasi-independent republic of Vermont. Jay meant to stay only for a few months. But shortly before his arrival, the *Pennsylvania Packet* printed an angry manifesto from Silas Deane, blasting Congress for having recalled him from his position in France. Four days later, irked that Congress had *not* taken umbrage at this insult to its authority, Henry Laurens opened the morning session by resigning the presidency. Jay was elected in his place.

As president, Jay had extensive dealings with Conrad Gérard, the French diplomat who pressed Congress to define its terms of peace. Gérard was an active force in congressional politics, wining and dining delegates and courting their support. Jay was an early and favored object of his attention. Their conversations included a two-hour pipe-smoking sequel to a dinner that Gérard hosted in December. The talk turned to the need to fix a lasting boundary between the United States and the neighboring possessions of Spain. Jay concurred when Gérard opined that the new nation "was already too large to hope to be well governed." It was Jay who asked Gérard to present the prospect of a Spanish mediation in a noon audience on February 15, 1779, with the delegates circled around him. Jay was a known supporter of Deane, whom the French still backed, and he impressed Gérard as a "man of spirit," a close student of human nature, and by reputation "one of the best orators in America." Gérard was pleased when Jay spoke of "the French blood flowing in his veins." (Whether Jay also described how his

Huguenot forebears had fled the persecutions of the 1680s—a story he loved to tell his family—is not known.)

As the overlapping controversies over peace terms and diplomatic appointments dragged on, Gérard regularly received delegates wishing to "speak confidentially about present affairs." When his supporters wavered, he fortified their courage. One critical interview came in early July, when Gérard arranged a meeting with Jay and "two other well-intentioned delegates." As he explained at length in a typically self-serving letter to Vergennes, his friends seemed dispirited by months of debate and "the threats and audacity" of their opponents, whom Gérard strangely labeled *le parti Anglois*—the English party—as if Samuel Adams and Richard Henry Lee secretly hankered for reunion with Britain. Eventually France would have to support New England's access to the fisheries, Gérard's guests argued, and if then, why not now? But Gérard was not biting. The conversation dragged on past midnight, with the diplomat urging the delegates to stand firm. Finally they relented, and Jay asked Gérard to return to Congress to present his latest reports from Europe as a way of advancing their common aims.

Jay's status as president clearly did not deter him from taking sides in this most partisan phase of congressional politics. "There is as much Intrigue in this State House as in the Vatican," he wrote to Washington, "but as little secrecy as in a boarding school." Yet Jay did not use his office to chastise those delegates who were leaking reports of this intrigue to the press. Jay was now France's preferred candidate for peace negotiator. But because Gérard's active politicking alarmed New England delegates, they wanted the position to go to someone they could trust to protect their regional stake in the fisheries. The obvious choice was John Adams, recently returned from his frustrating first tour in Paris, playing unhappy understudy to Franklin's command performance. On the last weekend of September 1779, Congress settled the question by giving the peace mission to Adams and making Jay minister to Madrid.

Jay may have had a few qualms about his appointment. He had just resigned his position as chief justice, pleading family duty but noting that he would stay in Congress if his brothers would "undertake to attend constantly to our good old Father and his unfortunate Family" (meaning their disabled siblings). Yet he would not have left his name in contention if he was unwilling to serve, and Sally seemed ready to accompany him wherever he was sent. Adams accepted just as quickly. But his decision came at a greater personal cost: another lengthy separation from Abigail and his knowledge of the wound it would impart. The very name of the ship on which they would cross the Atlantic seemed to trigger a mood of self-pity in Abigail. "O Why was I born with so much Sensibility," she wrote to John while *Le Sensible* was still anchored in Boston Harbor, "and why possessing it have I so often been call'd to struggle with it?" Was this a charge against her character or the husband who was testing it?

John was far from an unfeeling husband who simply presumed that his wife existed to support his career. But he did expect her to soldier on, with a sense of patriotic duty to match his own. The greater test awaiting him was to master the emotions, which were never buried very deeply beneath the surface of his personality, that a return to Europe would release. His first tour in Paris had been a sore trial. He arrived too late to assist in the treaty he was sent to help negotiate, and instead found himself immersed in the mutual recriminations between Franklin and Deane, on the one hand, and Arthur Lee. Mounting torment over a situation that would defy the "Wisdom of Solomon, the Meekness of Moses, and the Patience of Job" ended only when word arrived that Congress had made Franklin sole minister in Paris. Though Adams still anguished over being set adrift without an assignment, the news came as "the greatest Relief to my Mind." He could look forward to returning to Abigail and resuming the practice of law.

Now he was en route to Europe, embarked on "the most important Commission for ought I know" that Congress "ever issued," yet painfully aware that months or years might pass before peace talks occurred. If that was the case, he wrote to his friend Elbridge Gerry at Congress, he hoped to be useful in either Holland or Prussia—he would ask no additional salary. All Adams wanted was the opportunity to perform the useful service that the circumstances of his previous mission prevented. As to working with Doctor Franklin, Adams had mixed feelings. He understood Franklin's vital importance to the American cause in Europe. One could argue for his recall, as Gerry himself did, "but his Name is so great in Europe and America and the People have rested upon him in their own Minds so long, however erroneously, it would take so much Time and Pains to let the People into the Grounds, Reasons and Motives of it, that I have ever hitherto hesitated at it." But in his heart, Adams knew what those grounds were. Franklin lacked Adams's driving determination to seek the public good for its own sake. "He has a Passion for Reputation and Fame, as strong as you can imagine, and his Time and Thoughts are chiefly employed to obtain it, and to set Tongues and Pens male and female, to celebrating him," Adams had recorded in his diary in May 1779. Franklin was vain and indolent, and Adams sailed for Europe intent on proving that his own character was made of nobler stuff.

By the time he reached Paris on February 9, 1780, Adams had determined a course of action. Franklin would have no part in it at all, for Adams refused to tell him anything about his mission. When they rode to Versailles to call on Vergennes, Adams balked at openly telling the foreign minister why he had returned to Paris, lest Franklin learn the secret as well. "We live upon good Terms with each other," Franklin wrote to Jay's secretary, Carmichael, "but he has never communicated anything of his Business to me, and I have made no Enquiries of him." Yet Adams was barely unpacked before he launched his diplomatic initiative. Adams asked Vergennes whether it would be "prudent" to inform the British ministry of his arrival and his status as a peace negotiator with additional authority to discuss a treaty of commerce. Adams pointedly noted that he was "the only Person in Europe" empowered to treat of peace. Any proposals that Vergennes received from their common enemy should be passed to him alone, not that other American in Paris. Vergennes's first response was to stall. Would it not be better, he replied, to await the arrival of the returning minister Gérard (Jay's shipmate), who would bring the formal instructions that Vergennes wrongly believed Adams had yet to receive?

Adams took the insult rather than the hint. Though he knew he would have to consult Vergennes should any peace initiative occur, he resented the insinuation that he was overstepping his duties. Vergennes had his own reasons for doubting Adams, derived from Gérard's reports that the Adamses and Lees formed a pro-English party who wanted a peace settlement realigning the United States with Britain. This was nonsense, but Adams's eagerness to open contacts with Britain made it plausible nonetheless. Congress had not imagined that Adams would launch a one-man diplomatic offensive. Nor had it sent him to offer free advice about the best deployment of the French navy, another matter Adams pressed upon Vergennes. Adams compounded his difficulties by falling into a dispute with Vergennes over the effects of the devaluation of continental currency upon the debts owed to French creditors. That was a matter that Vergennes should have referred to Franklin. But Adams took the bait and again acted without Franklin's knowledge. Perhaps he hoped the contrast between his vigor and Franklin's lassitude would persuade Congress that he was the right man for Paris after all.

If that was his plan, Adams was too clever by half. As their exchanges grew testier by the week, Vergennes decided that Adams should share the fate of Deane and Lee. To that end he gave Franklin copies of his correspondence with Adams, to be forwarded to Congress. Franklin took a few weeks to comply, but when he did, he let Congress know that Vergennes had cause to refuse further dealings with Adams. Franklin had spent four years translating his personal affection for France into a form of diplomacy that assumed that his nation indeed owed their ally a debt of gratitude. That assumption did not make Franklin a naif. Like Adams he believed that nations acted from interest, not sentiment. Yet he also thought that France and its "young and virtuous Prince" had shown "generous Benevolence" toward America. Adams disagreed. In his view, France "is more obliged to us than we to her," if not for the pleasure of having its coffers drained, then for the shift in the European balance of power that independence would produce. Franklin did not deny Adams his patriotism. But he did not want him undoing his own good work and deliberately avoiding his counsel, while writing the occasional letter addressing Franklin as "your Excellency," a note of false respect that Poor Richard, that shrewd judge of human nature, had no trouble detecting.

To his genuine relief, Adams and his sons left Paris in late July 1780 for Amsterdam, where he proposed to busy himself raising a loan from Dutch bankers. The "Air of Amsterdam," John complained to Abigail, "is not so clear and pure as that of France." But it seemed better in every other respect. Franklin was relieved when the first letter from his colleague proved a very Adams-like survey of the views of the American conflict he had already heard in Holland. Adams had calmed down, channeling his excess energy into a steady stream of dispatches to Congress. Though he had little to report and less to act on, Adams did not want to deprive Congress of the benefit of his reflections. "It is very difficult to discover, with Certainty the secret springs which actuate the Courts of Europe," he admitted, but that hardly deterred him from trying.

While Adams was souring his relations with Franklin and Vergennes, the Jays were settling into the forbidding atmosphere of Madrid. In July Sally gave birth to the infant daughter she had carried across the Atlantic and the mountains of Spain. Their Spanish maid ("the old Goody") proposed naming her for a saint, but the resolutely Protestant Jays thought it "most prudent to let her take the name of Susanna," honoring her grandmother and aunt as well as the biblical heroine who "nobly resisted the lascivious Attacks of two Inquisitors Generals." Three weeks later Susanna developed a fever that led to convulsions, a daylong "continued fit," and death. Rejecting their maid's belief that unbaptized children would languish in *"Limbo,"* the Jays sought consolation from a more benign form of Christianity. Their deeper strength lay in each other. For a while Sally had to grieve on her own while John followed the court to St. Ildefonso. "What can I fear, or how can I repine," she wrote to her mother, "when I behold him who is equally interested, composed in danger, resigned in affliction, and even possessing a cheerful disposition in every circumstance."

That bond grew all the more important as Jay met with other disappointments, less grievous but still frustrating. Jay enjoyed nothing like the resources that sustained Franklin in Paris. His official secretary, William Carmichael, proved "the most faithless and dangerous" man he had ever met. More surprising was the behavior of his private secretary, Sally's brother Henry Brockholst Livingston. Though the Jays tried to cure him of his sulking moods, her brother was unrepentant. The Jays were particularly upset after Brockholst regaled a French visitor with tales of congressional drunkenness. Here they were, still seeking Spanish recognition of American independence, and her brother was mocking his own employer.

Jay's two years in Madrid were an exercise in futility. Spain's entrance into the war as an ally of France further limited Britain's ability to reinforce its American navy. But with its own New World empire to protect, Spain had no incentive to support American independence. An optimistic Congress hoped that Spain could be persuaded to offer the same alliance with the United States that France had already made. Congress also wanted Spain to help alleviate its distressed finances and extend aid to the amount of five million dollars, preferably as a subsidy—but a loan would do nicely too. On the territorial side, the United States would recognize a Spanish reconquest of East and West Florida, yielded to Britain in 1763. In exchange it would insist that Americans be given the free navigation of the Mississippi and access to the Gulf of Mexico. On all these issues, Jay made no more headway with the Spanish foreign minister, Count Floridablanca, than he could with his surly brother-in-law. Audiences with Floridablanca were hard to obtain and invariably disappointing. Even a paltry loan of one hundred thousand pounds, which Jay needed to cover his expenses, would require the Americans to waive their rights on the Mississippi. For a while, in fact, keeping his mission solvent was Jay's most pressing concern, a nagging anxiety that lifted only after Franklin allowed Jay to draw upon a fresh loan he had received from the friendlier Bourbon court in France.

Behind that grant lay yet another story about the difficulty of coordinating diplomacy between Philadelphia and Paris. Even as Franklin was seeking this further support, Jack Laurens was en route on a special mission to France, charged by Congress with obtaining a whopping loan of twenty-five million livres. His appointment owed something to Franklin's protests over being overworked and something to the invective that Arthur Lee had directed against Franklin since joining the Virginia delegation to Congress. But by the time Jack arrived in March 1781, Franklin had already obtained *his* loan of six million livres—far less than Congress wanted, but the best any American could do, given how badly the war was straining French finances. Laurens still pressed for additional support. He spoke French as well as any American, but his brusque conduct resembled the bad manners of Arthur Lee more than the suave demeanor of Franklin. The one saving factor was that Vergennes discounted Jack's behavior as the product of immaturity, inexperience, and a soldier's impatience to rejoin the army. Laurens sailed home before summer, ignoring the opportunity to see his wife, child, or sisters. His indifference did not prevent Martha Manning Laurens from seeking a reunion with her husband. Somehow she crossed to France, where she contracted a fever and died at Lille in the fall, at age twenty-four.

The simple fact that Congress had sent Laurens to carry out a crucial task that so clearly lay within his own competence led Franklin to think that the time had come to resign his post. He couched his request in mild terms, pleading age, health, and overwork rather than insult to pride or reputation. But his letter arrived in May 1781, just as Gérard's successor, the chevalier de la Luzerne, was lobbying Congress to strip Adams of his status as sole peace negotiator. Luzerne used means both fair (letters from Vergennes and Franklin) and foul (outright bribes) to achieve two objectives. First, Congress enlarged its peace commission, giving Jay, Franklin, and the imprisoned Henry Laurens an additional portfolio. (Thomas Jefferson declined his appointment.) Luzerne's second objective was more insidious. At his urging Congress revised the negotiators' instructions so that they were now obliged to act as a virtual subsidiary of the French foreign ministry. While conceding that it was impossible "to tie you up by absolute and peremptory directions," Congress ordered its diplomats not only to keep the French fully informed of their work but also "to undertake nothing in the negotiations for peace or truce without their knowledge and concurrence, and ultimately to govern yourselves by their advice and opinion."

Before independence, John Adams had worried whether Americans sent overseas would resist the allures and temptations of Europe. He never conceived that Congress itself could be so easily suborned and seduced on its own terrain. Nor did he imagine that personality clashes among American diplomats could prove as disruptive as the wiles of their European counterparts. In his own mind he remained a monument of American rectitude, immune to the ambition and vanity to which others succumbed. Adams could hardly grasp that his sense of rectitude was its own form of vanity. He increasingly regarded Franklin as a nemesis. When Vergennes recalled him to Paris in July 1781 to consult on a new mediation offer from Russia and Austria, Adams came and left without seeing Franklin. He was back in Amsterdam in late August when a note from Franklin informed him of Congress's recent actions. The news may have triggered a grave illness that left Adams insensate for a week: a combination of an existing hyperthyroid condition, as one scholar plausibly speculates, and a nervous breakdown. As he recovered, letters from home revealed Franklin's complicity in the campaign to dilute Adams's authority as sole peacemaker. Here was further evidence that "the old Conjurer" was a character assassin, driven by "base Jealousy" and "Sordid Envy."

The two years after his return to Europe were thus a time of self-confessed anxiety for Adams, measured in the compulsive way in which he dashed off detailed but useless dispatches to Congress and the brevity of his letters home to Abigail, a far better correspondent than he ever pretended to be. But Luzerne's stunning success in carrying out Vergennes's orders could not be readily transported across the Atlantic. Congress could only instruct its diplomats, not supervise them. The three men it had in place in Europe to negotiate peace when the opportunity arose proved far more resilient and independent than either Vergennes or Congress could imagine.

While the diplomats' communications with Congress remained erratic, there was no shortage of news from America. When Jay and Adams first took up their posts in 1780, most of it was bad. The winter they missed was the harshest in decades, and it placed Washington's army in even worse straits than those it had endured at Valley Forge. Continental commissaries had to compete with their French allies, who could pay farmers in hard coin at a time when the Americans had essentially abandoned the very concept of money. After shutting down its printing presses in the fall of 1779, Congress devalued whatever currency was still circulating at a ratio of 40:1. It also adopted a new system of requisitioning the states for "specific supplies" of basic items such as beef, pork, blankets, and shoes. The results were predictable. After five years of war, military logistics remained an exercise in improvisation.

On the battlefield, the war had evolved into a prolonged struggle that neither side seemed capable of winning. Since the draw at Monmouth in June 1778, the northern front had stabilized, with the British safely fortified on Manhattan. The most sensational development was the treason of Benedict Arnold, exposed by the lucky capture of his British contact, the sensitive young major John André, who was carrying the plans for the key Hudson River fort at West Point that Arnold had given him. The most successful American campaign was General John Sullivan's brutal march of 1779 through the Iroquois heartland around the Finger Lakes of New York, which forced the hostile Seneca, Mohawk, Cayuga, and Onondaga nations to abandon their lands and regroup as refugees near Lake Erie. From victory in the field, Sullivan sought success in the cabinet. As a New Hampshire delegate to Congress, he became the chief beneficiary of Luzerne's generous bribes and a key player in the June 1781 revision of the peacemaking instructions.

It was in the south, however, that the war would be decided. For the Americans the low point came with the May 1780 surrender of Charleston and the humiliating defeat of Horatio Gates at Camden a month later. But under the new command of Nathanael Greene, Continental troops began to reorganize, while local militia rallied and fought effectively in both Carolinas. The new year of 1781 opened with two significant victories. At the Cowpens along the Broad River, Daniel Morgan and his famed riflemen broke the assault of Banastre Tarleton's brutal Legion of American loyalists and gave "a devil of a whipping" to a unit that even British officers reviled. Greene followed with a second triumph over Cornwallis at Guilford Courthouse on March 15.

By the spring of 1781 the British strategy of using loyalists to pacify the south had proved a delusion. In London the king and Lord Germain still daydreamed about reclaiming American affections and restoring royal rule. But in the lower south, where a genuine civil war was erupting, the reprisals of vengeful loyalists were ensuring that most of the population would support the American cause. The British army could march wherever it wished, but it was unable to restore imperial rule in the countryside. Once that political objective became unattainable, a military strategy that sent regulars rambling across southern pine barrens lost its rationale as well. Though the British troops were superior, their numbers were limited. The American regulars fought well enough, at times superbly. And unlike the British, their casualties were replaceable.

After the repulse at Guilford, Cornwallis could have turned south to help consolidate the British position in South Carolina and Georgia. Such a move could have enabled Britain to seek a peace settlement by which it might retain two valuable provinces. Instead Cornwallis invaded Virginia, joining a rampaging loyalist contingent under the command of the traitor Arnold. The depleted British forces farther south now had little choice but to withdraw to Charleston and Savannah. For a few weeks Cornwallis had free rein in Virginia. On June 4, Tarleton's Legion nearly caught Governor Jefferson at his mountaintop mansion above Charlottesville, the state's temporary capital. But by early July, with American strength gradually mounting, Cornwallis was fortifying Yorktown, a minor port on the north side of the peninsula between the York and James rivers. A few miles west lay the neglected site of Jamestown, England's first permanent settlement in North America. Cornwallis was about to give its neighbor its own chapter in British imperial history.

Washington was slow to perceive the opportunity. For three years he had been fixated on dislodging the British from New York—which gave the British ample time to make their positions there virtually impregnable. That was the view taken by the comte de Rochambeau, commanding the five-thousand-man army France sent to the mainland in 1780. Without naval support, an attack on Manhattan seemed impossible, but the bulk of French ships were committed to the West Indies. By August, however, Washington agreed with Rochambeau that the great chance lay along the Chesapeake, not the Hudson. Two developments were critical: British troops were now concentrated at Yorktown, and a French force of twenty-eight warships and additional soldiers, commanded by the comte de Grasse, was heading to the Chesapeake. Their timely arrival could seal off Cornwallis from reinforcement, resupply, or escape. Confirmation of both developments reached Washington's headquarters at Dobbs Ferry on August 14. He and Rochambeau quickly began the complex planning required to transfer their forces southward while masking their intentions, as long as possible, from British eyes. Six years of preparation had finally reached a decisive test.

The siege of Yorktown began in mid-September. Its outcome was never in doubt. De Grasse reached the Chesapeake first, and his ships soon repulsed a feeble British effort to enter the York River. Outnumbered and outgunned, the Royal Navy abandoned Cornwallis to his fate. On October 9 the hundred allied guns—primarily French artillery and naval cannon, but with effective fire coming from Henry Knox's veteran batteries—began bombarding the earthwork defenses the British had hastily erected. Much of the digging was done by escaped slaves whose loyalty Washington and the Laurenses had previously pondered. Within a week British strength fell almost by half, to barely three thousand effective soldiers against the allies' sixteen thousand. Surrender followed on October 18, 1781. For Washington the profound satisfaction of gaining the decisive triumph that had eluded him since Trenton and Princeton was soon marred by personal tragedy. On his way to Yorktown, he visited Mount Vernon for the first time in six years and invited his seventeen-year-old stepson, Jackie Custis, to accompany him to Yorktown. There Jackie contracted one of the virulent fevers that still flourished in most camps, defying his stepfather's strictures on sanitation. He died on November 5, a meaningless casualty to the great victory that secured the independence of the republic Jackie never had the opportunity to defend.

The news of Yorktown did not take long to reach Europe. Three French frigates carried it home in near record time. Franklin was at Passy on the evening of November 19, playing his famous glass harmonica for a young American guest before the talk turned to war and the grim prospect that the British might prevail in the Chesapeake. The visitor had just left when a note came from Vergennes, announcing Cornwallis's surrender. The next few days were passed in celebration. Only on the twenty-third did Franklin send a brief congratulatory note to Adams. He wrote again three days later, this time marveling at the American success in capturing two armies in a single war, and the "singular Circumstance" and "perfect Concord" of Franco-American cooperation—a modest reminder of his great diplomatic achievement. But neither Franklin, Jay, nor Adams thought that peace talks were nigh. Adams was particularly pessimistic. So what if the expansion of the peace mission had reduced him "to the Size of a Lilliputian, or of an Animalcule in Pepper Water," he wrote to his friend Francis Dana, the minister to Russia, whose private secretary was now John Quincy Adams. "There is no present Prospect of Peace" and Adams never expected to see one. The British attitude toward America would remain implacable, just as it had after Saratoga. Whatever protests for peace came from the people "will evaperate in a few frothy Speeches, and fruitless Remonstrances."

For peace to be made in the year ahead, significant developments would have to take place in each of the three capitals: on the Thames, the Seine, and the Delaware. First, and most important, the shift in attitudes in London that Adams found inconceivable would have to occur, and that would require a major political realignment in Westminster and Whitehall. Second, the American peace delegation would then have to convene in Paris and decide what its negotiating strategy and objectives should be—in particular, whether it should adhere to its congressional instructions to coordinate closely with Vergennes or open direct channels to the British emissaries. Third, by taking the latter course, the diplomats ultimately presented Congress, in its isolation at Philadelphia, with a delicate choice: whether to ratify a treaty that secured all its essential goals even as it alienated French sensibilities and impaired its own sense of importance by endorsing the independent-minded conduct of its emissaries.

When the Yorktown dispatches reached London on November 25, 1781, they went first to Lord Germain, the hawkish secretary of state for America. He carried them in person to North, who greeted the news with a heartfelt groan of despair: "Oh God! It is all over!" That was not Germain's opinion, and it was even less that of their sovereign, who rejected North's suggestion that he finally acknowledge American independence. Although Germain was forced to resign in late January, North's political and personal commitment to the king denied him that option. But the Commons that returned to London from its Christmas revels became the scene of repeated test votes, not only on American policy but effectively on North's twelve-year ministry as well. Parliamentary noses were being counted and re-counted, and the divisions grew narrower and narrower. On February 22, 1782, a motion to condemn the American war failed by a single vote. Five days later General Henry Seymour Conway, a "friend of America" with four decades in the Commons, again moved to end "the further prosecution of offensive warfare" in North America. Conway denied that his proposal unconstitutionally trenched upon the authority of the king and patiently explained how offensive and defensive operations differed. Debate went on until 2 A.M. on the twenty-eighth, Edmund Burke wrote to Franklin, and then the motion passed "in a very full house," 234–215. Strictly speaking, this was not a vote of no confidence, but North knew that his ministry was at an end.*

It took another three weeks to persuade George III of this fact. So firmly attached was the king to his American policy that he actually considered abdicating, until the weak-kneed but principled North convinced him that it was not "dishonourable" for a royal sovereign to yield to a parliamentary one. A grudging king asked the Marquess of Rockingham (Burke's patron, and the minister who repealed the Stamp

* This resolution, little known even to historians, should have deep constitutional significance for Americans today, particularly those concerned with the extravagant claims made in recent years for unilateral presidential authority over war making under the Commander-in-Chief Clause of Article II of the federal Constitution. It is difficult to imagine why its framers would have wanted to vest the executive with a command authority less subject to legislative restriction than their contemporaries in Britain deemed acceptable. Were they likely to have been greater defenders of this prerogative than parliamentarians such as General Conway?

Act) and the Earl of Shelburne to form a government. On March 20, with a late snow falling outside, the Commons heard North's resignation—but only after prolonged procedural wrangling prevented his opponents from gaining a potentially embarrassing vote demanding it. North prudently kept his carriage waiting just outside. While members milled about in the cold, waiting for theirs, he rode off with one last quip about the perquisites of office: "Goodnight, gentlemen, you see what it is to be in the secret."

The American diplomats initially doubted that the Commons vote of February 28 meant peace. Writing to Secretary for Foreign Affairs Robert Livingston, Franklin predicted the year would pass without negotiations. Recalling the commissions of 1776 and 1778, Adams deemed it more likely that Britain would "send Agents to America, to propose some mad Plan of American Vice-Roys and American Nobility." But the change in the ministry quickly led to the first feelers for a peace settlement, which Franklin encouraged by sending a personal note to Shelburne even before the new ministry was formed. The two were not only old friends but also mutual admirers of Madame Helvetius, the brazen, salon-keeping, and still beautiful philosopher's widow who had replaced the daughterly Madame Brillon as the chief object of Franklin's amorous affections. Shelburne immediately dispatched emissaries to meet Franklin and Adams. To Paris he sent Richard Oswald, a wealthy Scot (and a year older than Franklin) who was one of the slave traders Henry Laurens had represented in the 1750s. They had stayed friends, and Oswald had posted the bond for Laurens's release from the Tower. Laurens himself was the second emissary, sent to meet Adams in Haarlem and then return to England under his parole as a prisoner.

Oswald met Franklin in Paris on April 15; Laurens saw Adams the next day. The Paris meeting went more smoothly than the interview in Haarlem, where Adams was uncertain how much to tell Laurens, given his status as a parolee. Shelburne had instructed the messengers to ask whether the American diplomats were authorized to negotiate a separate peace between their nations and whether the Americans would accept a settlement "upon any Terms short of Independence." Franklin and Adams both made clear that they would not negotiate a treaty separate from France and that independence was the starting point of negotiations. Both intimated that Britain should consider ceding Canada to the United States. This had long been a pet project of Franklin's, but it also had potentially important implications for the future triangular relations among France, Britain, and America. Just as the removal of the French threat in 1763 made the colonies less dependent on British protection, so the accession of Canada could make the American alliance with France a superfluous security against a danger that no longer existed. In that case, the older affection that the British mistakenly thought they could restore in war might still revive in peace. At the same time, Franklin demonstrated that the French alliance was alive and well, escorting Oswald to Versailles for a joint audience with Vergennes.

Spring had come to Paris, and with it, thoughts of peace now vied in Franklin's breast with longings for marriage to Madame Helvetius. He was optimistic enough to invite Jay to abandon his quixotic Spanish mission and join him in Paris. "Spain has taken four Years to consider whether she should treat with us or not," Franklin wrote on April 22. "Give her Forty." Jay quickly agreed. Franklin wrote similarly to Adams, expressing the hope that both he and Laurens could be in Paris for Oswald's expected return, "for I shall much want your Advice, & cannot act without your Concurrence." This appeal proved unavailing. Laurens was still recuperating from his long confinement; if released from his parole, he intended to go to the south of France to reclaim the daughters he had not seen since 1775. In May he told Franklin that he meant to resign his post, though he might still assist Adams in concluding the Dutch treaties he had originally been dispatched to seek.

Adams was an even harder sell. Until April 1782 he met no more success in Holland than Jay had in Spain. Then, after Yorktown, the Dutch finally recognized American independence. With this action, the newly accredited minister wrote to his wife, "The American Cause has obtained a Tryumph in this Country more signal, than it ever obtained in Europe." It was, he wrote to Benjamin Rush, "the most Signal Epocha, in the History of a Century." So much for the French alliance of Doctor Franklin! His own "Tryumph" was greater. He would never have the "Leisure" nor "the Patience to describe the Dangers, the Mortifications, the Distresses he has undergone in accomplishing this great Work." So much for Washington's victory at Yorktown! Even greater success awaited, as Adams planned to seek the treaties and loan that now seemed possible. He had no time to spare for Paris. He still doubted that the new ministry in London would make peace. Nor could he bear facing that character assassin Franklin, who "has been actuated and is still by a low Jealousy and a meaner Envy of me." Adams meant to stay in Holland, pursuing a treaty that Yorktown made largely irrelevant.

There were two obstacles to peace, however, greater than Adams's aversion to seeing Franklin. One was the course of combat in theaters of war beyond America. In April the Royal Navy gained a decisive victory at Les Saintes, a group of islands south of the French island of Guadeloupe, ending the prospects for Franco-Spanish gains in the West Indies. But Spain still hoped to recapture Gibraltar, and until the fate of its siege there was decided, France would find it difficult to move for peace at the expense of its ally.

The second obstacle was the divided counsels in London. Not only did the king have to learn to get along with the unstable coalition of politicians who were now his ministers; the ministers had to get along with one another. As Secretary of State for Home, Colonial, and Irish Affairs, Shelburne claimed the authority to direct negotiations on the pretense that the American states were still colonies—and that Britain should seek a settlement short of independence. That doubtful notion had to contend with the competing claim of Charles James Fox ("the Eyebrow"), who held the newly created post of foreign minister and was an ally of the Marquess of Rockingham, Shelburne's chief rival within the cabinet. At first Fox gained the upper hand and soon sent his own man, Thomas Grenville (son of the Stamp Act's architect), to Paris. By indicating that he preferred to deal with Oswald (Shelburne's emissary), Franklin inadvertently helped swing the advantage back to Shelburne. On June 30 the cabinet endorsed Shelburne's position on delaying recognition of independence. The next day Rockingham died, a victim of the influenza epidemic sweeping Europe. The king now made Shelburne his chief minister, and Fox resigned.

Flu and other maladies also laid the Jays low after their arrival in Paris on June 23. Too ill to nurse, Sally Jay worried that her newborn, Maria, beset with the "hooping cough," was "on the point of leaving me." It was the beginning of an exceptionally wet and cold summer. With Jay bedridden well into July, Laurens unwilling to serve, and Adams doubting that Britain would make peace before 1784, Franklin was the sole American ready for active diplomacy. Whether the British were really prepared to talk remained a puzzle. Shelburne's reluctance to concede the basic point of independence implied that he was little less delusional than the king in assessing American attitudes. In one of several interviews with Laurens, Shelburne blithely wondered why Americans would want to cut all ties with Britain, when that would lose them the benefit of habeas corpus, the Great Writ of English liberty. Fresh from his stint in the Tower, Laurens was the wrong man to entertain a question that meshed imperial arrogance with stupefying ignorance of American law. Shelburne could succeed, Adams thought, only by adopting the views of the vanquished Fox. But the greater culprit remained Franklin. "Shelburne would not have opposed" recognizing American independence "if Franklin had not piddled" by agreeing to meet Oswald before Shelburne had authorized him to concede that essential point.

Back in Paris, Jay also deemed it essential to gain formal recognition of independence prior to actual negotiations. That fixation mattered as a rainy July gave way to a rainy August and a wet summer imperceptibly passed into a wet fall. The miserable weather remained unchanged, but not the health of the two Americans. As Jay regained his strength (despite a nagging pain in his chest), Franklin was felled by kidney stones. Jay was now the active negotiator, and Oswald soon learned that he had a different temperament than Franklin did. On August 7 Oswald rode out to Passy to present his newly arrived commission "for treating with the Commissioners of the Colonies." The formulaic reference to "the Colonies" did not bother Franklin so long as Oswald had the requisite authority. But an ensuing conversation with Jay went much less smoothly. Jay seemed "a man of good sense; of frank, easy and polite manners," Oswald reported. But that did not spare Oswald from a lecture treating the wording of his credentials as proof that Britain regarded independence as something America still had to bargain for, not a point to be conceded at once so that the real terms of peace—boundaries, reparations, fishing rights—could be pursued.

Oswald bore this harangue patiently. His amicable relations with Franklin made him hopeful that the two septuagenarians could reach agreement quickly. Jay was a new factor, and his measure had to be taken. The result proved disturbing. Like Franklin's rebuff of Hartley's and Wentworth's conciliatory inquiries in 1778, Jay began by recounting the "deep wounds" that a cruel war and Britain's knack for offering too little, too late had inflicted on American feelings. Oswald found "this Detail of particulars, as unnecessary" as it was "unpleasant." Jay doubtless intended this litany as a prelude to more compelling matters. But the sharp tone of his soliloquy was not mere rhetoric. In Jay's jaundiced view, Oswald's commission marked only the latest British trick to evade the absolute reality of American independence.

Although subsequent meetings revealed that Jay was less hostile to Britain than he at first seemed, he still insisted independence must be conceded before talks could begin. Franklin was unwilling to defy his colleague, only noting that "Mr. Jay was a Lawyer, and might think of things that did not occur to those who were not Lawyers." They were not playing a game of good diplomat, bad diplomat. Although Jay respected Franklin, he knew that his own stock at home stood at least as high as his colleague's—probably higher. Jay also knew Adams would support him. Occasional letters from Holland warned that Englishmen "hate us, universally from the Throne to the Footstool." America would gain nothing from "their Generosity, or Benevolence." A more pointed note followed. We may have differed in our political opinions, Adams wrote to Jay, but "there has never to my Knowledge been any Misunderstanding between Us." Only beware those "who will use all the Arts of the Devil to breed Misunderstandings between us." That could mean only Franklin and his co-conspirator, Vergennes.

Jay had his own reasons to accept this warning. He was irked that France had done little to aid his tortuous dealings with its Spanish ally. When it came to casting suspicion, Jay was generous. In early September he learned that Vergennes had sent his close aide, Gérard de Rayneval, to London. In Paris, Jay had also been holding talks with the Count de Aranda, the resident Spanish ambassador, over the future boundaries between Spain's colonies and the United States. Following his instructions, Jay argued that the United States was entitled to the same territorial claims that Britain had secured in 1763, which extended west to the Mississippi. The boundary Aranda proposed ran well to the east of that river. Jay now worried that Rayneval had gone to London to explore a peace that would leave the Americans in the lurch on independence while upholding the territorial demands of Spain.

Jay was wrong. Rayneval had not been dispatched with treachery in mind. His talks with Shelburne concerned European issues, though they did exchange some comments on America. But once formed, Jay's suspicions profoundly affected his diplomacy. He promptly broke off his talks with Aranda and then decided to approach Shelburne on his own.

He had a backdoor channel available in Benjamin Vaughan, yet another Franklin protégé, the son of a West Indian planter and a Boston-born mother—and, like Jack Laurens, a son-in-law to the London merchant William Manning. (It was Sarah Manning Vaughan who adopted Jack's daughter, Fanny, after her mother's death in Lille. Fanny was now fatherless as well. In a skirmish days earlier, Jack had rashly attacked a British foraging party at Chehaw Neck, South Carolina. At last he received the fatal wound he had long courted—a late and pointless victim to a long and bitter war.)

Fearing a French betrayal, Jay now prepared a stroke of his own. Without consulting Franklin or informing Vergennes, he sent Vaughan to London with an enticing message. It was in Britain's interest "to make friends of those she cannot subdue," Vaughan was to tell Shelburne, and the king's chief minister could best do this "by liberally yielding every point essential to the peace and happiness of America." It was also "the obvious interest of Britain immediately to cut the cords which tied us to France." Though the Americans were duty-bound to honor their alliance with France, *"yet it was a different thing to be guided by their or our construction of it."* Negotiators who heeded their own "construction" would not need to consult France about American terms of peace. Jay knew that Franklin did not share his suspicion of French motives, but he was willing to gamble. "Facts and future events must determine which of us is mistaken," he wrote to his old best friend, Robert Livingston, on September 18, in a remarkably curt letter announcing his démarche. Given that Livingston's view of the alliance more closely resembled Franklin's than Adams's, Jay knew he was declaring his independence of French guidance, his congressional instructions, and perhaps Franklin's support. Livingston's astonished response was to ask Jay what had become "of all that prudence and self-possession, for which you are happily distinguished."

Jay too had a temper, but it was calculation, not impulse, that set his course of action. Six months earlier he had put a barbed question to Livingston. Though it was too early to assess British intentions for peace, Jay noted, "can it be wise to instruct your commissioners to speak only as the French ministers shall give them utterance?" This was not a request for guidance. Jay knew or surmised too much about why Congress had issued its yoking instruction of June 15, 1781, to feel duty-bound to adhere to it absolutely. Accordingly, acting on his own counsel—or as his nation's independent counsel—he cast his vaunted "prudence" aside. Having stalled peace talks for two months over the formal recognition of independence, Jay now moderated his stance on that matter to see what other terms Britain was prepared to offer.

When Vaughan returned in late September, he carried a new commission for Oswald, which described the Americans as representatives "of the Thirteen United States of America." Finally satisfied on this key point, Jay dashed off a brief note to Adams, hoping they might meet very soon. Still wrapping up his Dutch dealings, Adams did not share Jay's sense of urgency. Amazingly he had only just learned (from Arthur Lee) of the congressional instructions to follow French guidance, having failed to decipher the relevant language when he received the dispatch a year earlier. Now that he knew what the instructions said, Adams refused to believe that a body as "enlightend" as Congress could act as supinely as Lee supposed—especially when it had such capable envoys in Europe: Jay, Laurens, and his personal friend in St. Petersburg, Francis Dana (modestly omitting his own name and pointedly excluding Franklin). "Those Chains I will never wear," Adams swore. He seemed in no rush to reach Paris. Incessant rains, bad roads, and a twice-broken axle slowed the trip. But so did the pleasures of tourism that Adams belatedly indulged. At Antwerp he admired "the famous Altar Piece of Reubens" and several private collections of paintings featuring more works by Rubens, Rembrandt, Vandyke, and Veronese. The best, in his opinion, was Rembrandt's painting of "an old Woman, his Mother, with a Bible on the Table before her." But it was a different vision of femininity that lingered with Adams when he reached Paris on Saturday, October 26. Earlier that day he was touring the Chantilly castle of the prince of Condé when twenty-four-year-old Louise Adelaide appeared, "dressed in beautifull White, her Hair uncombed hanging and flowing about her Showlders, with a Book in her Hand." Adams gaped at the vision until the future princess, abbess, and refugee from another revolution, sensing "that I viewed her more attentively than she fancied," rose, tossed her hair off her shoulders "in a manner that I could not comprehend," and retreated inside.

There were other dignitaries in Paris whom Adams was less eager to see. He promptly met with Jay but balked at visiting Franklin until another American, Matthew Ridley, insisted that as "last comer" Adams had to pay the first call. Chastened, Adams donned his coat and prepared to ride out to Passy. Even then his courage faltered until Ridley virtually pushed him out the door. One down, but there was still Vergennes to go. It took another ten days of dining regularly with Franklin and Lafayette's importuning to persuade Adams that he really must call on Vergennes, who was already miffed that word of his arrival came not from Adams but "from the Returns of the Police." Adams accordingly made the trip to Versailles, where Vergennes and his "remarkably attentive" countess poured on the charm, placing Adams at the lady's right hand at dinner while her husband "was constantly calling out to me, to know what I would eat and to offer me petits gateaux, Claret and Madeira." Others flattered Adams for his success in Holland, even calling him "le Washington de la Negotiation." None of this flattery erased his darker suspicions, but his mood improved enough to enable Adams to put aside his grudges and pursue the work at hand.

That work was already well advanced by the progress that Jay had made while Franklin was mending and Adams was making his way south. Once Britain recognized independence, Jay worked quickly, drafting a treaty that Oswald sent to London on October 8. Jay's first key objective was to secure American territorial rights reaching to the Mississippi and lying on both sides of the Ohio River. The second was to ensure access to the fisheries, including some rights for New England's mariners to dry fish on land. Jay was less familiar with the nuances of potential northeastern boundaries—that is, between the Maine district of Massachusetts and what is now New Brunswick, Canada—and he proposed a more modest line than Adams would have sought, had he arrived earlier. Similarly, Jay and Franklin had already decided not to seek British concessions in what is now western Ontario, above the Great Lakes. This was territory that Shelburne might have yielded in the summer, had Jay's insistence on the prior recognition of independence not stalled discussion. Shelburne was unlikely to tender it now, since the successful defense of Gibraltar had heartened British spirits. Looking southward, Jay made a stunning proposal by implying that the United States would accept, even abet, a British seizure of West Florida from Spain. This was a graceless way to treat a nation that was closely bound to America's great ally. Yet Jay's conversations with Aranda, and his two futile years in Madrid, had persuaded him that the United States could expect nothing from Spain.

Oswald was a merchant, not a diplomat, and in his zeal for peace, he did not press the Americans on two key points given in his instructions. One was to enable British merchants to recover pre-war debts owed by their American customers; the other, to gain compensation for loyalist refugees whose property had been confiscated by various American states. Both were politically sensitive issues that Shelburne, presiding over a tenuous ministry and a restive Parliament, could hardly overlook. Nor were Oswald's territorial concessions well received. Some members of the government wanted to adhere to the Quebec Act of 1774, which extended the government of Canada down the western line of Pennsylvania and along the Ohio River. Shelburne had no interest in preserving a claim to the American interior, but he thought these lands could be used to meet the needs of the loyalists. To stiffen Oswald's resolve, Shelburne also gave him a new adviser in the person of Henry Strachey, a ministerial undersecretary who had assisted the Howe commission of 1776.

Yet for all these reservations, Shelburne also made clear that Jay's draft remained the basic framework for negotiation. There was much yet to dispute, but good reason to think that agreement was within reach. The talks in Paris resumed on October 29. One imagines great negotiations being conducted in splendid palatial surroundings, but in this case the discussions rotated among the emissaries' lodgings. With Adams present and Franklin recovering from his bouts with kidney stones and gout, all three Americans were prepared to take an equal part in the discussions. Once he screwed up his nerve to pay Franklin a visit, Adams let him know that he supported Jay in his initiative to ignore their instructions and negotiate directly with Britain without allowing Vergennes to oversee their efforts. "The Dr. heard me patiently but said nothing," Adams recorded. But Franklin did not keep his own counsel for long. In their first joint conversation with Oswald, "Dr. Franklin turned to Mr. Jay and said, I am of your Opinion and will go on with these Gentlemen in the Business without consulting this Court." Franklin was good to his word, Adams noted. "He has accordingly met Us in most of our Conferences and has gone on with Us, in entire Harmony and Unanimity, throughout, and has been able and useful, both by his Sagacity and his Reputation in the whole Negotiation."

Two of the major issues in dispute, boundaries and debts, proved relatively easy to resolve. Two others, Tories and fish, did not. The Americans insisted on a western boundary at the Mississippi, and Oswald and Strachey acceded with only mild protest. There was more haggling over the Canadian boundary, but inadequate cartography made this a subject that would take decades of further surveying and negotiations to resolve. The question of pre-war debts owed to British merchants could have posed a greater obstacle—had Adams not yielded the point when it was first raised in conversation by flatly stating that he "had no Notion of cheating any Body." This brought a broad smile to Strachey's dour Scots visage, but without leading Adams, the lawyer who had tried more debt cases than he cared to count, to wonder whether he had yielded something for which the British should have had to bargain.

Adams took this course because he wanted to prevent the creditors "from making common cause with the Refugees"—the loyalists who had fled to Britain—in opposing peace until their demands were met. He was not reassured when Vergennes and Rayneval argued (just before the dinner where Adams was flattered mercilessly) that "all the Precedents" supported Britain's desire to restore "those who had adhered to the old Government in all their possessions." Adams begged to differ. If Britain really wanted to do right by the loyalists, it could pay them from its own coffers, with monies saved from the additional military expenses it would incur fighting on to bring Americans around to a point they would never accept. But his objection went deeper. There could be no justice in the loyalist claim, Adams insisted, because "Those People by their Misrepresentations had deceived the Nation" into pursuing a destructive and misguided policy that "had brought an indelible Stain on the British Name, and almost irretrievable Ruin on the Nation."

At some profound level, this was a reference to the late Thomas Hutchinson, whose death in 1780 Adams recorded in an insightful obituary for the *Boston Gazette.* To his mind the former governor remained a far more influential mover of independence than "the brace of Adamses"—John and Samuel—could ever have been. It was Hutchinson's treacherous accounts of the opposition he faced in Massachusetts, Adams believed, that had led Britain to its fateful errors of 1774–1775, persuading Lord North and other officials that a group of factious demagogues were whipping the colonists into a frenzy of political error. Franklin did not share Adams's personal animus against Hutchinson. But when he sent copies of Hutchinson's letters to Boston in 1772, he did so in the belief that the governor's political destruction was a small price to pay for maintaining the stability of the empire he hoped to preserve. Then too, Franklin had his own loyalist burdens to bear: his old ally Joseph Galloway, who had accompanied British troops to govern occupied Philadelphia, and far worse, his illegitimate son William, the last royal governor of New Jersey, who elevated obedience to king over duty to father. Franklin never forgave him. (William's own illegitimate son, William Temple Franklin, however, was now serving as secretary to the American delegation, much to the irritation of John Adams.)

To allow loyalists to return to America and regain property, Adams warned, would be to admit a discontented, subversive force into the republic, ripe for future manipulation by Britain or even, somehow, France. (Adams suspected that French agents, such as Rayneval, were somehow culpable for whipping "the Tories to set up their Demands" upon the king and his ministers.) But for all his anxieties, Adams ran second to Franklin in resisting the British claims for compensation. This became clear after Strachey, having returned to London for fresh instructions, came back to Paris with a revised treaty still seeking restitution of loyalist property. After it was read on November 25, an irritated Franklin spent the night drafting a response, which he shared with his colleagues over breakfast at Jay's rooms the next day. If Britain clung to this point, he argued, the Americans should respond by compiling accounts of the wanton damage wreaked by British arms during a war "brought on and encouraged" by the "Falshoods and Misrepresentations" of the very people Britain sought to protect. Franklin was "more decided" on this matter than Jay or Adams, and they agreed he should read his letter as a statement of "his private Sentiments." But then, having agreed on that strategy, the three men spent some time conversing about "the Conduct, Crimes and Demerits of those People." That they would do so suggests that their feelings on this point were genuine. Here was an emotional bond that united the American negotiators—a way of recalling the circumstances that had turned them from loyal subjects into active and committed revolutionaries.

With the Americans adamant, the most Oswald and Strachey could gain was a weak concession that Congress would merely *urge* the states to allow loyalists to seek restitution of their property. But in the final discussions the Americans did some yielding of their own on the other main issue in dispute: access to Canadian fisheries. Adams had spent hours explaining the nuances of fishing, down to quoting the well-known Boston saying, "when the Blossoms fall the Haddock begin to crawl, i.e., to move out into deep water." He vigorously defended the idea that full access to the fisheries was a right nature had bestowed on America, and one, if denied, "our Fishermen, the boldest Men alive," would ever assert, creating an unending basis for controversy. Yet in the final exchanges on this point, the Americans agreed to replace the bolder word *right* with the more tempered *liberty*—that is, a privilege Britain would grant rather than an entitlement Americans could naturally claim.

On November 29 the parties met again at Jay's lodgings at the Hôtel d'Orléans. Henry Laurens was now present, having come to Paris en route to see his daughters, and he supported his colleagues when Franklin proposed a new article calling upon the king to ask Parliament to compensate broad classes of the American population for property seized or destroyed during the war. This was a demand that neither king nor Parliament would ever accept, and faced with a united American front, the British team withdrew for their own consultations. After some time they returned to announce that they would accept the American peace terms—so long as the claim for compensation was abandoned.

Since that claim was made only as a threat, not a demand, the point of agreement had now been reached. The two sides "sat down and read over the whole Treaty and corrected it," and then asked their secretaries to prepare fresh copies to be signed the next day. When they arrived at Oswald's residence the next morning, the Americans were peeved to note that the British had silently omitted a provision limiting loyalist refugees to a year's sojourn while they sought "to recover their Estates if they could." The provision was restored. The loyalists had been Americans once, with a birthright to liberty; now they could return only as temporary residents. Laurens also took advantage of his belated presence to add "a Stipulation that the British should carry off no Negroes or other American property." Then the adversaries signed the preliminary treaty of peace and rode together to Franklin's lodgings at Passy for a celebratory repast.

A few days later, a set of dispatches arrived from Robert Livingston. From "private letters and common fame," Congress had some idea of what was transpiring in Europe, the secretary reported, but "Doctor Franklin has told us nothing." Then he registered a further complaint. "It is commonly said that Republics are better informed than monarchies of the state of their foreign Affairs" and require greater "vigilance and punctuality in their ministers." Americans "on the contrary seem to have adopted a new system, the ignorance in which we are kept of every interesting event renders it impossible for the Sovereign to instruct their servants, and of course forms them into an Independent privy Council for the direction of their Affairs without their advice or concurrence."

Knowing that Congress would soon learn that its diplomats had defied its instructions, Franklin admitted that this comment "may be in some Respects just." Congress did need to recall that "the extream Irregularity of Conveyances" might better explain its envoys' long silences than any "Desire of acting without the Knowledge or Orders of their Constituents." It should also realize that their situation differed from that of their European counterparts, who could write their court for fresh instructions and get a reply within a month. "Unless you leave more to the Discretion of your Ministers," Franklin concluded, "your Affairs may sometimes suffer extreamly from the Distance," especially in wartime, when Atlantic communications were even more prone to disruption.

Given the choice between seizing the main chance or deferring to French guidance, Jay and Adams never wavered in thinking that they understood the case for independent diplomacy far better than their superiors in Philadelphia. It helped that each was profoundly suspicious of France's ulterior ends. Adams never forgave Vergennes for promoting his recall, while Jay rightly sensed that the foreign minister fully supported Spain's desire to keep America's boundary well east of the Mississippi. In one respect Jay was the greater Francophobe, his efforts to imply otherwise to Conrad Gérard back in 1779 notwithstanding. "They are not a moral people"—Adams recorded Jay's words. "They know not what it is. He dont like any Frenchman." Even Lafayette was suspect: though "clever, he is a Frenchman." Jay's inner Huguenot, it seemed, still had a powerful voice. It was not prejudice, however, but calculation that set Jay and Adams on their independent course. A similarly prudent weighing of options carried Franklin in their wake, even when that meant laying aside the accommodating approach he had perfected over the past five years. When Vergennes asked him to delay sending the preliminary treaty to Congress until the parallel Anglo-French negotiations were further along, Franklin firmly replied that the diplomats had to place their duty to their constituents first.

As if to prove Franklin's strictures about "the extream Irregularity of Conveyances," it took nearly three months for the preliminary treaty and the accompanying dispatches to reach Philadelphia on the packet *Washington*. Their arrival on Wednesday, March 12, threw Congress into near uproar. As James Madison recorded, no one faulted the "extremely liberal" terms the diplomats had obtained. But objections were quickly sounded. "Many of the most judicious members thought" the Americans had been "ensnared by the dexterity of the British minister" in conducting separate negotiations. Jay earned special criticism for approaching Shelburne on his own, without consulting Franklin. The separate article of the treaty, dangling West Florida before the enemy, was deemed "most offensive" because it marked "a dishonorable departure from the candor rectitude & plain dealing professed by Congress" in its dealings with its French ally. The French envoy Luzerne stirred the pot further by complaining that the diplomats had knowingly deceived Vergennes about the progress of their talks. King Louis himself "had been surprized & displeased," exclaiming that "he did not think he had such allies to deal with."

Over the next fortnight, Congress agonized about its envoys' behavior. One hotheaded Virginia delegate called it "a tragedy to America & a comedy to all the world beside." Even Alexander Hamilton felt obliged to note that although his friend Jay "was a man of profound sagacity & pure integrity, yet he was of a suspicious temper, & that this trait might explain the extraordinary jealousies which he professed." On the other side, Arthur Lee, ever happy to take a slap at the French, thought that if any blame was to be laid it should "be done in such a way, as to fall ultimately on France, whose unfaithful conduct had produced & justified that of our Ministers." The controversy eased only when a French cutter brought word that "the preliminaries for a general peace" among all the belligerents had been signed on January 20. Though some delegates still wanted Congress to assuage French sensibilities, the fear that France might abandon America out of pique, should the war somehow continue, had evaporated. Congress moved on to more pressing business.

And soon it moved on in another sense. In late June Congress abandoned its original capital at Philadelphia for nearby Princeton after mutinous unpaid soldiers staged a threatening protest outside the State House and the Pennsylvania government refused to call out the militia in its defense. The ferrying of delegates over the Delaware was rather less heroic than Washington's famous crossings, which had saved the revolutionary cause just after Franklin's arrival in Paris in December 1776. Not only did Americans lack a permanent capital of any kind—much less a metropolis akin to London or Paris. Their national government was little more than a rotating pool of officeholders who served more for reasons of conscience and duty than from any deep ambition to wield power or make epochal decisions. With such a Congress to answer to, historians generally agree that the American peacemakers did right to bypass their formal instructions and negotiate separately. The results justify their decision. Perhaps Jay, Franklin, and Adams could have fared better on particular points. But a treaty that ended the war, recognized independence, secured a Mississippi boundary, and pledged Britain to abandon territory it still occupied was a peace worth having.

Assessing the conduct of the three main negotiators is more difficult. Jay did have a naturally suspicious cast of mind, and his excessive concern with the formal wording of Richard Oswald's credentials probably delayed the negotiations in a way that did not serve American interests. Yet Jay also demonstrated a lawyer's ability to remember that he served only one client, the United States, and that freed him to act against that client's express wishes when Jay concluded that separate negotiations with Britain offered the most promising course of action. When it came to assessing the alternatives, Jay was wholly unsentimental, immune both to Franklin's attachment to France and to Adams's brooding suspicions. In Jay's eyes, all Europeans were equally suspect, and Americans should simply seek the best bargain they could obtain.

Franklin too knew bargains when he saw them, and in his persona as Poor Richard, he had invented a character who could portray much of life as a set of calculations. Yet he had invested too much of himself in France to let his lawyer-colleagues provide a definitive representation of American character. On December 17, 1782, he sent Vergennes one of those inimitable missives that mark Franklin as the cleverest American of his (or perhaps any) age. Two days earlier he had written a brief note asking the foreign minister about the latest American request for a French loan. Vergennes answered promptly but with a flash of anger. When you have satisfied my doubts about your conduct in ignoring the instructions of Congress, "I will ask the King to place me in a situation [*état*] to answer your demands."

Franklin's letter of the seventeenth was gracious and generous to a fault. No one can "be more sensible than I am, of what I and every American owe to the King," he replied, speaking for himself and his entire country, if not his two colleagues. "All my Letters to America are Proofs of this; all tending to make the same Impressions on the Minds of my Countrymen, that I felt in my own." And what was that impression? "That no Prince was ever more belov'd and respected by his own Subjects, than the King is by the People of the United States." So Franklin again smoothed ruffled feelings, redeeming the American reputation while knowing that as guileful a diplomat as Vergennes could not fail to appreciate why the New World negotiators had acted as they did. Yet Franklin could easily have turned this sentiment around and applied it to his own situation. For he repeatedly wondered whether his countrymen would ever appreciate how much he too had done for them. That he still thought of them as his countrymen was itself remarkable. After a quarter-century lived largely in Europe, he retained few personal connections to his homeland. He had no real allies or patrons in Congress and was a stranger to the turbulent politics of Pennsylvania, the state whose public affairs he had once nearly dominated. He was still contemplating spending his declining years in Europe, and probably would have, had Madame Helvetius yielded to his marital entreaties.

Franklin was therefore troubled when a report from Boston implied that Adams had insinuated that Franklin had to be dragged along in his and Jay's wake to assert American rights to the fisheries and a Mississippi boundary. On September 10, 1783, he took the extraordinary step of writing identical letters to Adams, Jay, and Laurens, asking them to affirm that he had not been "behind any of them in Zeal and Faithfulness," lest he "suffer an Accusation, which falls little short of Treason to my Country." All three complied—though Adams did so by quoting the diary entry, incorporated in a letter to Congress, which noted that Franklin had said nothing when Adams first indicated he was siding with Jay, and only disclosed his intention to go along when they next met Oswald.

Yet when it came to gauging the insinuations of others, it was Adams, not Franklin, who felt the greater turmoil and who had fewer resources of temperament or philosophy to deploy against the nagging fear that he was not receiving his patriotic due. Perhaps that is why the reputation of John Adams remained dogged by the telling yet balanced comment that his nemesis attached to it in July 1783. "He means well for his country," Franklin wrote to Secretary Livingston, "is always an honest man, often a wise one, but sometimes, and in some things, is absolutely out of his senses." From the moment of his return to Paris in 1780, Adams regarded Franklin as a veritable persona non grata as far as his own diplomacy would be concerned. It is difficult to account for this attitude without recalling the pangs that Adams felt while watching Washington leave Philadelphia for his new command in June 1775, "to wear the Lawrells which I have sown." No one's horizons were more enlarged by the Revolution than those of the deacon's son from Braintree. Yet through some strange psychological equation, Adams conflated that cause with his career in a way that prodded his ambition and distorted his judgment. There was a vanity in this conflation he knew he should resist, but which he could never fully overcome.

PART III

LEGACIES

7

The Optimist Abroad

On the evening of November 20, 1783, two "courageous Philosophers" unexpectedly appeared at the lodgings of Benjamin Franklin in Passy. One was Étienne Montgolfier, the "very ingenious Inventor" of the hot-air balloons whose first flights were the talk of *tout Paris.* His companion was the marquis d'Arlandes, copilot with Pilâtre de Rozier of the latest test launch, which Franklin had observed just that day from the nearby Château de la Muette. Along with the "vast Concourse of Gentry in the Garden," Franklin caught his breath as the balloon brushed a tree, tilted to one side, and risked catching fire. But the philosopher-pilots quickly righted their craft, kept feeding the "Basket Grate in which Faggots and Sheaves of Straw were burnt," and soared above the roofs of Paris. Franklin sent an account of the flight to his friend Sir Joseph Banks, the botanist president of the Royal Society of Fellows who had sailed to the Antipodes with Captain Cook in 1768. Like everyone else, Franklin speculated about the uses of the new device. "Some fancied" they could be "anchored in the Air, to which by Pullies they may draw up Game to be preserved in the Cool." Another beneficial effect could be "Convincing Sovereigns of the Folly of wars." What king could afford "to guard his Dominions" against a strike force of "ten Thousand Men descending from the Clouds" to attack any object they (or the wind) chose?

Coming barely months after the signing of the definitive peace treaty, the balloon craze symbolized a hopeful transition from the needless destruction of war to the promise of peace. Ideas of air travel quickly seized everyone's imagination. Franklin, one of the great inventors of the age, doubted the balloon could "become a common Carriage in my time," much as he wished it would, now that gout and stones made it torturous to ride on land. John Adams was more optimistic. In October 1783 he and his eldest son, John Quincy, finally visited England, enduring eighteen wave-tossed hours and the worst seasickness these veteran sailors had known, to cross to Dover. After that wretched voyage a smooth sail through the air seemed enticing. Only come to Europe, Adams wrote to a still undecided Abigail, "and if the Balloon, Should be carried to such Perfection in the mean time as to give Mankind the safe navigation of the Air, I will fly in one of them at the Rate of thirty Knots an hour to meet you." Sally Jay had the same idea as she awaited her husband's return from a trip to the healing waters at Bath. "Don't you begin to think of taking your passage next Spring in a Ballon?" she asked. Jay had already sent "prints of the Rise and fall of the *Ballon*" to Robert Livingston and Robert Morris, who foresaw another use for the new wonder. "Pray cannot they contrive to send Passengers with a Man to steer the course, so as to make them the means of conveyance for Dispatches from one Country to another," Morris proposed, "or must they only be sent for intelligence to the Moon and Clouds?" For a merchant, information is always money, and at a moment when he could also report that he was "sending some Ships to China in order to encourage others in the adventurous pursuits of Commerce," Morris saw profits where Sally imagined other benefits.

Back in Annapolis, the new though temporary home of Congress, Thomas Jefferson also joined in the excitement, jesting over the commercial, military, and even political uses of balloons. "The French may now run over their laces, wines &c to England and duty free," he mused. Congress could use them "to move backwards and forwards" in its search for a permanent home. He also counted the study of natural "phaenomena of which the Atmosphere is the theatre" and "the discovery of the pole, which is but one day's journey, in a baloon," among potential applications. Three months later, Jefferson made this entry in his memorandum book: "Pd. for 2 tickets to see balon 15/." This exhibit of three balloons, which he and his eldest daughter, Patsy, caught at Philadelphia, featured a tethered ascent of barely three hundred feet—nothing compared to the trial flights that had taken place in France.

That was where Jefferson was himself now bound. He had just agreed to join Adams and Franklin in the new commission for negotiating commercial treaties with interested European countries. He hurried to Boston, hoping to join and escort Abigail Adams to her first reunion with her husband in nearly five years. Arriving, he learned that she was only hours from sailing, too little time for him to disassemble his phaeton carriage for shipment. He eventually booked passage on the London-bound brig *Ceres*, having been told "that I could with certainty get ashore on the coast of France somewhere" as they swung north toward the Thames estuary. Departure was set for July 4, the eighth anniversary of the Declaration of Independence, but regrettably for those who like their historical coincidences neat, *Ceres* did not weigh anchor until 4 A.M. the next day. The crossing took a brisk nineteen days. His wish to transfer at sea to some France-bound ship proved futile. Instead, Jefferson, Patsy, and his personal servant James Hemings disembarked at Portsmouth, where his daughter's illness forced them to spend a week. (Jefferson left his younger daughters, Mary and Lucy, in Virginia, with his late wife's sister Elizabeth Eppes.) The three crossed to Le Havre on July 31, reassembled the trusty phaeton, and proceeded to Paris.

Three previous opportunities to go abroad had come Jefferson's way. He declined the first in 1776, out of solicitude for his wife, Martha. A second came in June 1781, when Congress named him to the expanded peace commission through which France hoped to swaddle John Adams. This Jefferson refused on the grounds that he had to defend himself against pending charges that, as governor, he had failed to prepare Virginia against the British invasion. That appointment was renewed in November 1782, after Martha's death plunged him into a mourning as deep as that of George Mason a decade earlier. Hoping to rescue Jefferson from his gloom, his younger friend James Madison secured a fresh nomination from Congress. This time Jefferson promptly accepted. He hurried to Philadelphia and prepared to depart, only to have the Royal Navy and an ice-choked Chesapeake delay his sailing. Then word arrived of the signing of the definitive treaty, and Congress released him from his mission.

Thus it was not until midsummer 1784 that Jefferson finally reached Europe. For a while he wondered whether the trip was worthwhile. His spoken French proved wretched and his health little better. The cold, rainy Parisian fall and winter oppressed him. He missed his daily rides on horseback and the chance to shoot and hunt, until a burst of sunny weather in March 1785 enabled him to begin a program of daily walking, working up to six or eight miles a day. Many of those rambles were devoted to mulling over the sad news that Lafayette had brought on his return from a triumphal tour of America. Only midway through a letter from a Richmond physician, which first discussed (what else?) ballooning, did he learn that his youngest daughter, Lucy—his second child to bear that name, and the one whose birth led to Martha's final decline—had fallen "a Martyr to the Complicated evils of teething, Worms and Hooping Cough." Her first cousin and namesake, Lucy Eppes, had died as well. Jefferson's middle daughter, Polly, was also stricken "most violently" but recovered. Unlike the Jays, who treated the death of their infant daughter with the usual religious sentiments, Jefferson was too drained even to exchange consolations with his brother-in-law, Francis Eppes: "Doing so would pour balm neither into your wounds nor mine." In May a late-arriving letter from Eppes grimly reported that both Lucys "suffered as much pain, indeed more than ever I saw two of their ages experience." Jefferson immediately replied that Polly must join him in Europe.

Jefferson took his consolations in philosophy, not religion, but it was his aesthetic sense and sensibilities that rescued him from a lasting relapse into sadness. Among all his American contemporaries who journeyed abroad, none—not even Franklin—displayed anything like his interest in the attainments of European culture and civilization. For Franklin, Europe's charms lay chiefly in the company of intelligent friends and the doting attention of the cultured women he admired. As a tourist John Adams preferred sites that he could associate with historic events and personalities. John Jay seemed completely indifferent to Europe's attractions and mostly wanted to regain his health and bring his family home. "A Smile from you and the Caresses of our little ones are worth more to me Than all the Pleasures of this Town," he confessed to Sally from London. "I am really home Sick."

Jefferson, however, came to love Europe, even as he also believed that Americans had to be insulated, even quarantined against its allures, especially its feminine ones. In his dealings with Europe, he was like Odysseus resisting the Sirens. Under proper restraints, he could expose himself to its temptations and even fall in love again, as he plainly did in 1786 with the Anglo-Italian artist Maria Cosway, married, with curly, eighteenth-century-style big hair, and seventeen years younger than Jefferson. Though their romance surely went unconsummated, the surge of desire revealed that Martha's death had not withered his passions. He was smitten too by the buildings he saw on his travels. Most stunning was the Roman temple at Nîmes, the Maison Carrée. Jefferson insisted that it serve as a model for Virginia's new state capitol, which he was designing for Richmond. He studied countless other edifices on his travels through France, England, Holland, and northern Italy. Surprisingly, his Italian sojourn did not extend to the Veneto, where the work of Andrea Palladio, the architect he admired most, was best displayed. But it lasted long enough to convert him to the country's favorite *primo piatto*, and when he returned to Virginia he bore the design for a pasta machine "with holes of different shapes & sizes for the different sorts of Maccaroni." And who could savor a plate of pasta (or any other fine meal) without a good wine to accompany it? The serious *dégustation-degustazione* of French and Italian wines was also part of Jefferson's European project. Ways had to be found, he believed, for his countrymen to benefit from Europe's manifest superiority in architecture, painting, sculpture, music, cuisine. He fancied himself a reverse ambassador, an impresario of importation, deciding which aspects of European culture Americans could safely absorb and which were better limited to a gentry of his class and taste. But when he turned from aesthetic matters to consider the political and social factors that would give the new republic its character, his European sojourn pulled him in the opposite direction. The visionary ideas he had expressed in his great wartime project, supervising the revision of the Virginia legal code, found fresh confirmation in Europe. The best case for the notions of political and social equality he wished Americans to pursue lay in the evidence of all the evils that aristocratic dominance imposed on Europe.

Jefferson went abroad in no small part to assuage his grief over Martha's death. In that quest to recover a modicum of personal happiness, he largely succeeded. But his diplomatic career and his observation of European society had a deeper impact. They restored Jefferson to the public sphere he nearly abandoned in 1782 and revived political purposes he might otherwise have surrendered. Before 1789 few if any of his contemporaries would have imagined Jefferson assuming the roles that later fell to him, not merely as the leader of the nation's first opposition political party, but more important, as the symbol of egalitarian ideals and aspirations—most unlikely attributes for one who mysteriously sprang from the slave-owning class of wealthy planters. Today, when Americans seem uneasy with Jefferson's simultaneous embrace of the ideals of liberty and equality and the daily brutality of chattel slavery, it is easy to forget the greater puzzle of his life and legacy. How did a man of his class and stature, reared in the far reaches of a slave society, become an enlightened cosmopolitan thinker, with egalitarian commitments that seem all the more remarkable, given his origins? How did this dreamy idealist, who might have lapsed into a reclusive retirement devoted to family, house, food, music, and books, instead give his name to a political movement and the age it dominated?

We cannot resolve such questions by faulting Jefferson for his hypocrisy or dismissing him as the eighteenth-century version of a "limousine liberal"—or perhaps a "phaeton progressive." The term *hypocrisy* can only describe behavior but never explain it. And if used too casually, it can easily lead us to overlook or slight the real dilemmas with which individuals—or entire generations—wrestle. A good case can be made for viewing Jefferson as the strongest and perhaps the original embodiment of what Gunnar Myrdal called "the American dilemma" in his influential 1944 study of American racial attitudes and practices. "The Negro problem in America would be of a different nature," Myrdal wrote, "if the moral conflict raged only between valuations held by different persons and different groups of persons. *The essence of the moral situation is, however, that the conflicting valuations are also held by the same person.*" Jefferson epitomized that dilemma, and the years he spent in Europe in the 1780s, attracted to its culture and repelled by its social order, marked a critical moment in his own understanding of what made America different.

Jefferson would have done American diplomacy a great favor by joining the original commission to Paris in 1776 and thus preventing the tumult that flowed from the selection of Arthur Lee in his stead. His efficiency would have been a useful corrective to the inattention to detail that John Adams found so aggravating in Franklin. Jefferson and Franklin would also have made a splendid pair intellectually. America's two great polymaths shared broad interests in the natural and practical sciences and a passion for music and chess. Both were talented writers, with a knack for concise and vivid phrasing that few contemporaries matched. Franklin's *Autobiography* and Jefferson's *Notes on the State of*

Virginia (which is really the biography of a state) remain the two great literary monuments of the American Enlightenment.

But Jefferson had other priorities in the fall of 1776. He was back in Virginia when he learned that Congress wanted him in Paris. His former colleagues presumed he would heed their call. "That distinguished love for your country that has marked your life, will determine you here," Richard Henry Lee wrote. Jefferson kept the messenger waiting three days as he pondered his decision. Then he wrote regretfully to President Hancock. Though he would never let "cares for my own person, nor yet for my private affairs" interfere with public duty, "circumstances very peculiar in the situation of my family, such as neither permit me to leave nor to carry it," forced him "to decline a service so honorable" and vital "to the American cause."

The decisive factor was concern for his wife. Martha Wayles Skelton was a widow of twenty-three when they married on New Year's Day, 1772. She likely conceived that very evening, for their first child, Martha (Patsy), was born on September 27. "Every letter brings me such an account of the state of her health," he wrote of his wife in late July 1776, that he resolved to return to Virginia at the first possible moment. Even then he had to delay his departure until a replacement arrived in early September. After a few weeks at Monticello, the couple headed to Williamsburg, where he would attend the House of Delegates. That was where word of his nomination to Paris reached him. Unlike Abigail Adams, Martha could not face being left behind. Unlike Sally Jay, she could not imagine crossing the sea even with her husband as escort. Any misgivings Jefferson felt were offset by his belief that the real work of revolution remained to be done within the states. The new constitution Virginia had adopted in June marked only the start of a broader legal "reformation." Jefferson left Congress, he recalled in his autobiography, "in the persuasion that our whole code must be reviewed, adapted to our republican form of government," and purged of its monarchical residue.

He also left Philadelphia smarting over the editorial knife Congress had wielded on his Declaration. That knife deleted the final two words from his statement that "all men are created equal & independant." It similarly excised the first three words from his version of the trinity of great natural rights as "the preservation of life, & liberty, & the pursuit of happiness." But left untouched was that magical if elusive final phrase, whose use arguably owed more to Jefferson's reading of the Swiss jurist Jean-Jacques Burlamaqui than it did to his manifest debt to John Locke. Modern usage has stripped that last word, *happiness*, of the nuance it possessed in Jefferson's time. We think of happiness as a personal mood or state of mind. In the eighteenth century its connotations were broader. In his *Thoughts on Government*, for example, Adams used it as shorthand for "ease, comfort, security." Happiness was a condition that whole societies as well as individuals could enjoy. It implied a state of social contentment and not merely personal cheeriness and good humor. Happiness was one of those broad concepts that had both private and public meanings, a subject for philosophical inquiry rather than psychological babbling.

For Jefferson the concept of happiness was something to ponder as well as pursue. The word recurs repeatedly in his writings, often with intensely domestic associations. He planned to find—or rather, create—his own happiness in hearth and home, family and friends, music and books, food and wine. An early rhapsodic vision of this blessed state appears in a letter answering Robert Skipwith's humble request, as a "common reader who understands but little of the classicks," for advice on assembling a personal library. Ignoring Skipwith's disclaimer that he lacked "leisure for any intricate or tedious study," Jefferson volleyed back 150 titles bundled in nine categories, from "Fine Arts" to a final block of "Miscellaneous materials" that merely included "Voltaire's works" and "Locke on Education." But why build a library at all, Jefferson asked his future brother-in-law, when Skipwith and his bride, Martha's half-sister Tibby, could simply live nearby and borrow any work they fancied? The couples could meet nightly, by a "spring, centrically located," there to "talk over the lessons of the day, or lose them to Musick, Chess, or the merriments of our family companions," the two sisters. "In every scheme of happiness," Jefferson wrote of his own intended, "she is placed in the foreground of the picture, as the principal figure." That this vision of happiness was a *picture*, a tableau suitable for Gainsborough or Reynolds, seems evident from Jefferson's very description. He idealized family life as he idolized his wife. Their ten years of marriage were a time of "unchequered happiness." After her death he imagined his young friends James Monroe, James Madison, and William Short settling nearby in the same convivial fellowship he proposed to Skipwith. If they would do so, he wrote to Madison soon after his arrival in Paris, "I should still believe that life had some happiness in store for me. Agreeable society is the first essential in constituting the happiness and of course the value of our existence."

The site for this sociability would be the house Jefferson began designing and building in 1769, when he set his slaves to planing the mountaintop he inherited from his father, Peter, a dozen years in his grave. To a casual viewer, the idea that Monticello rests atop a mountain seems a stretch. Higher summits flank it, including the grassy Montalto directly south, which Jefferson acquired to enlarge his holdings. But Monticello does sit astride Virginia's Southwest Mountains, with the Blue Ridge a casual glance to the west, and modern geologists accord the name its due. Had Jefferson been more practical, he would have built along the Rivanna, which flowed through his estate—not one of the colony's great rivers, but a water source far more reliable than the sixty-nine-foot well his slaves dug in 1769 and the cisterns he added for those years when it ran low or failed. Like Franklin, he was a zealot for practicality and a maker of laborsaving gadgets. But placing his monument to domestic happiness on a mountaintop was a triumph of aesthetics over practicality.

Jefferson's vision of domestic happiness reflected sentiments common to his class. The view of the Potomac from his Mount Vernon veranda pleased Washington just as much as the Blue Ridge entranced the master of Monticello. George Mason took the same delight in the trompe l'oeil effect he obtained from four sharply pruned avenues of trees that radiated outward from Gunston Hall. Standing properly centered in the doorway, visitors would see only the first tree in each row; but if they shifted to one side, the rows suddenly stretched to an immeasurable distance while the proud owner chortled next to them. Neither Mason nor Jefferson, excellent horsemen both, could match Washington's equestrian skills. But when they mounted, they too could ramble for miles without leaving their estates.

Many Americans equated this vision with the words of the prophet Micah: "they shall sit every man under his vine and fig tree." Jefferson's ideas of happiness transcended the biblical image. Among his American contemporaries he was distinctive, perhaps unique, in the extent to which he studiously reflected on what would make him happy. Others doubtless shared his feelings and desires. None thought about how to fulfill them with as much care or imagination. That was already true before he made an undefined pursuit of happiness a natural right and a national motto. What seems remarkable about Jefferson after 1776 is how little inclined he was to let even the passing inconvenience of war interfere with his private pursuits.

The opportunity to serve as a revolutionary legislator offered Jefferson an optimal way to balance his desire for private happiness with the republican duty he was expected to honor. He was as enthusiastic a legislator as anyone who ever sat in the capitol at the eastern end of Duke of Gloucester Street. (The building Williamsburg tourists see today is a reconstruction of the original structure, which burned in 1747, not the second assembly hall of 1751–1779.) As Julian Boyd observes, Jefferson was "a veritable legislative drafting bureau." Though he had happily abandoned active legal practice to be a full-time planter, his legal skills and intellectual devotion to the science of legislation remained keen. On October 11, 1776—the same day he declined Paris—Jefferson was elected to the committee on religion. He soon began introducing the first of the scores of bills he drafted over the next few years. The first, proposed on October 14, was to abolish entail, a medieval mode of inheritance that kept estates intact by prohibiting successive eldest heirs from subdividing the entailed property among their descendants. That same day Edmund Pendleton introduced a bill "For the encouragement of foreign Protestants to settle in this Countrey." Jefferson insisted that "foreign Protestants" be changed simply to "foreigners." The alteration would allow Virginia to welcome both Jews, who would be an "advantageous" addition to the population, and Catholics drawn from Mediterranean countries, free immigrants who most likely Jefferson intended would replace the slaves he hoped first to emancipate and then require to emigrate.

One mark of Jefferson's legislative enthusiasm was the reading he did to aid his work on the religion committee. Here he first met James Madison, eight years his junior and his neighbor thirty miles north. But it was a shared commitment to religious liberty that formed their first bond. Madison's springtime success in amending the religion article of the state's declaration of rights to recognize that "all men are equally entitled to the free exercise of religion" had inspired Virginia's religious dissenters to flood the legislature with petitions urging it to take the even more radical step of disestablishing the Church of England, thereby depriving the province's official church of the public support on which it had always relied. The committee's task was to consider how to respond to these petitions.

Jefferson supported disestablishment with all the resources his legal training and reading could muster. Amid his other labors he found time to research the relevant acts of Virginia and Parliament, and to review leading authorities such as John Milton, Locke, and Locke's pupil, the third earl of Shaftesbury. His reading notes survive, and there, amid a summary of Locke's *Letter Concerning Toleration*, Jefferson inserted a remarkable aside at the point where Locke explained why "those who entertain op[inio]ns contrary to those moral rules necessary for the preservation of society" need not be tolerated. "It was a great thing to go so far (as he himself sais of the parl. who framed the act of toler[atio]n)," Jefferson noted. "But where he stopt short, we may go on." And so Jefferson did, soon drafting a resolution to disestablish the Church of England and an additional bill to exempt dissenters from having to contribute to its support.

Jefferson's aside on Locke echoes John Adams's joy at being "sent into life at a time when the greatest lawgivers of antiquity would have wished to live." It was also of a piece with his assertion that "the whole object of the present controversy" was not to recover the customary rights Britain had violated but rather to form new constitutions superior to the old colonial governments. In Jefferson's precocious view, reform was not only possible but necessary. It was not a project to defer until the war was won. Others might prudently object that there were more pressing objectives to secure first. What was truly necessary to the American cause in the fall of 1776 were the measures that Washington was desperately urging to keep his reeling army in the field. That was not Jefferson's view. On October 15 he proposed the measure that set his own duties for the next three years: the bill for the revision of the laws, which called for a thorough review of the entire body of Virginia law. It is a tribute to Jefferson's vision, or naïveté, that he could devote his prodigious energy to the high-minded project of legal reform at a moment when others were worrying that their lives and property—forget happiness—might soon be forfeit.

Well-timed or not, the bill proved uncontroversial. So did Jefferson's appointment to chair the committee elected to revise the state's laws. This was a substantial accomplishment. He was the youngest member of an eminent panel that included his intellectual and legal mentor,

George Wythe; Edmund Pendleton, also a distinguished lawyer and the living exemplar of political prudence; George Mason; and Thomas Ludwell Lee, the eldest of the five Lee brothers. The committee first met in January 1777, but its numbers soon shrank to three after Mason and Lee, the two non-lawyers, resigned. Their assigned tasks fell primarily to Jefferson as chair.

Jefferson performed his dominant share of its labors from his well-stocked library at Monticello. One-man drafting bureau that he was, he preferred to work from the privacy of his mountaintop. He attended the assembly irregularly: a dozen days in May 1777; another three weeks each in the spring and fall of 1778. The single session he attended at length was the fall 1777 meeting that ran into January 1778. Absence from Williamsburg made his heart grow fonder—for Monticello. Dispatching his important bill "for proportioning crimes and punishments in cases heretofore capital" to Wythe on November 1, 1778, he cavalierly noted that he had "heard little of the proceedings of the Assembly, and do not expect to be with you till about the close of the month." (Polly was born in August.) His absence from Congress was remarked by an even more famous Virginian. Lamenting the tendency of states "far distant from the scene of action" to think "that to regulate the government and police of their own State is all that remains to be done," Washington pointedly included Jefferson, Wythe, Pendleton, and Mason among the delinquents who should lend their talents to Congress. The commander in chief knew better than to think that the grand project of legal reform outweighed winning the war.

For Jefferson the revision of the laws offered a perfect way to pursue public and private happiness together. The contrast with his friend, John Adams, seems striking. Adams too felt repeated pangs over his separation from family and livelihood. If Abigail was less importuning than the younger Martha, she still let John know how sorely she missed him. Yet agonize as he did, Adams never permitted domestic cares to trump public duty. He knew the difficulty of balancing public and private affairs when he answered Jefferson's first letter to him in May 1777. "Your Country is not yet, quite Secure enough, to excuse your Retreat to the Delights of domestic Life," Adams noted in closing. "Yet, for the soul of me, when I attend to my own feelings, I cannot blame you." Replying three months later, Jefferson silently bypassed the hint.

A second contrast with Adams places Jefferson's concern with legal reform in its proper light. Adams too was an enthusiast for a republican revolution. Yet when correspondents pressed him to be a thorough reformer—whether it was Abigail urging him to "Remember the Ladies" or James Sullivan arguing for a broader suffrage—his reply was the same. Prudence dictated that a difficult war was not the moment to open "so Fruitfull a Source of Controversy and Altercation" as they proposed. For Jefferson, however, independence and reform were inseparable, and now was the time to act. Virginia was not Massachusetts. In Massachusetts, with its egalitarian distribution of property, Adams could credibly believe that hardworking ordinary farmers, like his deacon father, could acquire political respect and influence. Jefferson was born atop a genuine ruling class, and marriage only improved his standing within it. Yet by 1776, if not earlier, he was actively pondering the consequences of allowing a relatively small number of families to monopolize so much of the new state's land.

One fruit of this concern was the act abolishing entail, which Jefferson ushered through the assembly in the fall of 1776. Its passage was uncontroversial because it enabled the gentry to dispose of their property as they wished. More radical were Jefferson's plans for distributing land among the lower ranks of Virginia's free population. He incorporated his ideas into the fourth section (Rights Private and Public) of the state constitution he drafted in June 1776. Here he proposed endowing "every person of full age" with "an appropriation"of up to fifty acres of land, carved from the vast tracts that Virginia claimed under its colonial charter. The use of *person* rather than *freeman* meant that this provision covered women as well as men. The new commonwealth would become just that: a society premised on the idea that all its free members should possess enough land to enjoy the personal independence required to exercise the rights of citizenship. Nor would he restrict political rights to freeholders. By August 1776 he favored granting the vote "to all who had a permanent intention of living in this country." That intention could be proved in different ways: "either the having resided a certain time, or having a family, or having property, any or all of them" would do. Jefferson thus aligned himself with Mason's original proposal to enfranchise tenant farmers with seven years to run on their leases or "housekeepers" with three children.

Rather than lower the standard of political citizenship, the provincial convention retained the traditional property requirement for voters. But Mason and Jefferson soon collaborated on another measure directed to the same end. In December 1777 both served on a special committee charged with drafting a plan to open a land office "for granting waste and unappropriated lands" in the state's western reaches. The bill they introduced in early January included a novel provision, likely drafted by Jefferson. Upon marriage and a year's further residence in the state, every native "free-Born" Virginian—male or female—would receive seventy-five acres of land. A Virginia couple would therefore start out with a 150-acre freehold, enough land to provide the decent "competence" that Americans defined as the standard of self-sufficiency and personal independence—and thus of happiness too.

This proposal fared no better than Mason's expanded suffrage of 1776. The land office bill was tabled, not to be enacted for another year, and then in much revised form. The use of Virginia's extensive western land claims "to encourage Marriage and Population" was another provision rejected. Its inclusion in the original bill indicates how far Jefferson's notion of public happiness and legal reform ran beyond what was politically attainable. The task of the original committee, after all, was to devise ways and means of paying the mounting costs of war, not to adopt some visionary social policy better pondered in peace.

By the spring of 1779 the revisal was nearly complete. Its two most celebrated measures were bill 79, "for the More General Diffusion of Knowledge," and bill 82, "for Establishing Religious Freedom." Whole volumes of commentary could be written about each, and Jefferson himself intended to guide their proper interpretation with the preambles he provided for both. (Indeed, in the religion bill the preamble dwarfs the body of the statute. Even in the 1770s, general-purpose statutes affecting the entire commonwealth were still regarded as exceptional legal acts, and preambles of this kind offered a way of explaining the reasons for their enactment.)

Bill 79 presented a scheme for establishing public schools to provide a minimum of three years of education to all of Virginia's free children, girls as well as boys. Such schools had existed in New England since the 1630s, initially justified by the Puritans' commitment to biblical literacy. Jefferson's scheme had a different, even opposite purpose, which he tied directly to the "publick happiness" of a republican society. "That people will be happiest whose laws are best, and are best administered," the preamble explained. This end would be obtained in two complementary ways. As citizens, a people had to be given "knowledge of those facts, which history exhibiteth," the better to detect the ways in which ambitious rulers in all ages had artfully "perverted" their legitimate powers into tyranny. But bill 79 also sought to create a new and merit-based pool of political leaders, by progressively winnowing the advancing classes to identify those students—now boys only—deserving further education. The goal was to produce a cadre of leaders qualified for office by a "liberal education" and "without regard to wealth, birth, or other accidental condition or circumstance."

As he later explained in *Notes on the State of Virginia*, Jefferson had further reasons for making historical literacy the foundation of a common education. Rather than "putting the Bible and Testament into the hands of the children, at an age when their judgments are not sufficiently matured for religious enquiries, their memories may here be stored with the most useful facts from Grecian, Roman, European and American history." Substitute *stuffed* for *stored* and this sounds like the formula that regularly deadens students to the delights of historical study. But the kinds of historical facts that Jefferson deemed useful and his ideas of education carry over into the subject of bill 82, the bill for religious freedom. Though its formal purpose was to disestablish the Anglican Church, its deeper animus was to free individuals from any obligation to adopt religious views they found unpersuasive. In Jefferson's view, all religious belief was finally a matter of individual opinion. The history of religious establishments was an unrelenting story of corrupting alliances between churchmen and rulers, abusing their power to impose their opinions and modes of thinking on others. This too was a form of tyranny, as inimical to liberty as anything else the Stuarts and other execrable autocrats had attempted. For Jefferson as for Locke, religion was not a matter of children inheriting the faith of parents. It was instead a subject of *inquiry*, and no one could simply adopt another's convictions. The point of reading history first, scripture later, was to empower individuals to judge the claims of all religions by teaching them that much of what passed for orthodoxy in other times and places depended on the impure alliance of church and state.

Here was another twist on Jefferson's recurring concern with issues of inheritance, whether from parent to child or from one generation to the next. For Jefferson the concept of posterity was more than a vague reference to those who would come later. His schemes of education and religious freedom, like the abolition of primogeniture and entail and his plans for using public lands to sustain newly married couples, all rested on a concern with the legacy one generation bequeathed to another. That he would think of this in wartime, when the liberty his contemporaries were seeking remained at risk, was again a tribute to a view of the American future that was either wholly optimistic or terribly naive.

The revisal was nearly complete when Jefferson succeeded Patrick Henry as governor in June 1779. He was not the assembly's consensus pick. It took two rounds of voting for Jefferson to edge out his boyhood and college friend John Page, the better-prepared candidate. While Page had spent the past three years on the state's executive council, Jefferson's preoccupation with the revisal insulated him from the real work of wartime governance. Like other republicans he viewed executive power with suspicion. His draft constitution replaced the reprobated title *governor* with the subservient name *administrator*, and pointedly listed all the traditional royal prerogatives the executive would not possess. Now that position had come to him unsolicited.

His election coincided with the two great developments that shaped the last major phase of the American war: the collapse of continental paper currency and the new demands upon the states to which it led; and the shift in British strategy that made the south the main theater of combat. Jefferson's second term as governor expired while the British invasion of Virginia was in full cry. Given his complete lack of military experience, he had resolved that someone better qualified should take his place—most likely Thomas Nelson Jr., the militia commander who was also a contender in 1779. But the enemy's approach made any orderly transition impossible. In May 1781 the assembly fled Richmond, which, thanks to another of Jefferson's pet reforms, had just become the state capital. It reconvened at Charlottesville on May 24, while Jefferson and his family returned to nearby Monticello. A week later Cornwallis dispatched Tarleton's loyalist cavalry to Charlottesville. Only a timely alert from a watchful militia officer enabled the lawmakers to flee west to Staunton. By Jefferson's account "the enemy were ascending the hill of Monticello" when he mounted his horse and fled "thro' the woods along the mountains" after his family. Rather than rejoin the assembly as his term was expiring, Jefferson stayed with his family until they were resettled at his Poplar Forest retreat, eighty miles southwest. In the confusion Virginia lacked a governor until the reconvened lawmakers elected Nelson on June 12.

From his knowledge of Locke's *Second Treatise,* Jefferson could have chosen another course of action. In a key passage, Locke defined the *prerogative* of the executive not in terms of the time-encrusted powers of a king, but rather as a discretionary power to act "for the public good, without the prescription of the law, and sometimes even against it." To this definition he added a shrewd observation: "Whilst employed for the benefit of the community, this true use of prerogative never is questioned: for the people are very seldom or never scrupulous or nice in the point." Had Jefferson rallied to Staunton and acted until a successor was named, no one would have blamed him for stretching constitutional propriety. By seeming to favor private happiness over public safety, he could be faulted for abandoning his post in a moment of dire need.

The lawmakers had barely reconvened when George Nicholas, a new member, moved for an inquiry into Jefferson's conduct. The assembly approved the request. When Jefferson pressed to know "the unfortunate passages in my conduct" for which he was faulted, Nicholas supplied a short list stressing the state's feeble preparation against invasion. In early August, Jefferson cited the inquiry as one reason to decline the peace commission, even though a European post offered the only opportunity he might ever have "of combining public service with private gratification, of seeing countries whose improvements in science, in arts, and in civilization it has been my fortune to admire at a distance but never see." Privately the looming inquiry elicited an even stronger conclusion. A trip to Paris would be most agreeable, he admitted to Edmund Randolph. "But I have taken my final leave of every thing of that nature, have retired to my farm, my family and books from which I think nothing will ever separate me." His appearance at the assembly in December 1781 only confirmed his resolve. Nicholas did not even attend, and Jefferson was left both to recite and refute a lengthier bill of particulars that his absent accuser had provided. The lawmakers quickly closed the inquiry and unanimously adopted a vote of thanks in its place.

Jefferson returned to Monticello intent on pursuing his private happiness. When Martha first fell dangerously ill, he had just declined his election to the assembly. Though he welcomed this expression of confidence from his Albemarle County neighbors, he was still bitter over the charges the assembly had leveled against him, which "inflicted a wound on my spirit which will only be cured by the all-healing grave." No state, not even a republic, "has *a perpetual* right to the services of all it's members." To think otherwise would be "to contradict the giver of life who gave it for happiness and not for wretchedness." Then Martha lapsed into the final decline of her frail health and his world turned upside down. "All my plans of comfort and happiness reversed by a single event," he wrote to his sister-in-law, Elizabeth Wayles Eppes, "and nothing answering in prospect before me but a gloom unbrightened with one chearful expectation"—save for "the temporary abstractions from wretchedness" afforded by the care of his three daughters.

Yet as he dragged himself out of grief in November 1782, Jefferson turned the page on his longing for retirement. Before the catastrophe of Martha's death, "my scheme of life had been determined," he wrote to the marquis de Chastellux, a recent visitor to Monticello with whom he had spent a memorable night reciting poetry. "I had folded myself in the arms of retirement, and rested all prospects of future happiness on domestic and literary objects." Martha's death "wiped away all my plans." Now Madison's intervention in securing his reappointment to the peace mission, and "the change of scene proposed," rescued Jefferson by restoring the claims of public duty. His daughters could remain with their aunt Elizabeth. Jefferson could have reverted to his original plan of retirement once his diplomatic appointment proved abortive. Instead, the time he spent preparing for his mission and awaiting passage to Europe revived his interest in politics. His growing friendship with Madison was another factor. At Philadelphia they lodged at the Fifth and Market boardinghouse kept by the aptly named Mrs. Mary House. There Jefferson happily observed Madison's courtship of Kitty Floyd, daughter of another housemate, the New York delegate William Floyd. The two seemed close to betrothal, exchanging miniature portraits, and Jefferson imagined them settling down as his neighbors. Then Kitty, half her suitor's age, dropped Madison for a medical student. The two men also shared their concerns over the erratic dispatches of John Adams. "His vanity is a lineament in his character which had entirely escaped me," Jefferson confessed. "His want of taste I had observed." Still, he thought that Adams had the "integrity" and "honesty" to serve usefully. Madison was less sure.

The two men corresponded regularly after Jefferson returned home. In June 1783 he sent Madison a new draft of the state constitution, written in the belief that the one Virginia had framed in his absence in 1776 was deficient. Another letter to Edmund Randolph reveals a nationalistic Jefferson worrying that the states might seek European patrons to pursue their mutual quarrels. "Can any man be so puffed up with his little portion of sovereignty," he asked, "as to prefer this calamitous accompaniment" to recognizing the importance of Congress? This was the voice of political engagement, not private retirement. When the Virginia assembly elected him to Congress, Jefferson promptly accepted. He arrived in Philadelphia in November, hired "the best tutors in French, dancing, music and drawing" for Patsy, and then rode back with Madison to the new if temporary capital of Annapolis.

One topic the two men likely discussed on their four days' journey was the completion of the Virginia land cession to Congress, the last step in the formation of a national domain above the Ohio River. From 1778 until the winter of 1781, Maryland had crabbily delayed ratification of the Articles of Confederation—the nation's original federal constitution—until Virginia finally agreed to cede to Congress its vast claims to the territory west of Pennsylvania and north of the Ohio. Once created, this national domain posed two broad issues for the vagabond delegates who were struggling to maintain a congressional quorum. One was how quickly its orderly sale and settlement could alleviate the Union's chronic financial woes. That public purpose was already being threatened by an unregulated and virtually unstoppable flow of squatter-settlers across the Ohio. The other issue, less urgent but not less profound, involved the political status of the new territory. Would it be governed as an internal colony of the existing Union? Or would it be formed into future states that would enter the Union with the same rights of republican self-government that its original members enjoyed?

Jefferson had actively considered these questions since his first stint at Congress in 1775–1776. It is fitting that the additional months of service he spent at Annapolis are best recalled for the land ordinance drafted by the three-man committee he chaired. This was exactly the kind of legislative project that Jefferson loved to take on. Not that the ensuing report of March 1, 1784, and the initial ordinance that Congress approved on April 2—forerunner of the Northwest Ordinance of 1787—were uniquely a stroke of Jeffersonian genius. They were not. The evolution of congressional land policy in the 1780s reflected a broad consensus that the west would be the nursery of future republican commonwealths, not a domain to be controlled by the original seaboard states.

Yet there was a sense in which Jefferson's vision of the west did reflect his distinctive social thinking. In the land office bill of 1779, he and Mason had sought to commit Virginia to providing all its citizens with land. Privately he held an even more radical view. Back in 1774 he inserted in his *Summary View of the Rights of British America* an intriguing digression protesting the Crown's recent decision to consolidate future land grants under its sole authority. Such a claim could never be proper for Virginia, Jefferson asserted, a country that had only been settled under English authority but never conquered. The power to grant land rightfully belonged either to the people of a society collectively or their elected lawmakers. But if neither acted, another option existed: "each individual of the society may appropriate to himself such lands as he may find vacant, and occupancy will give him title." In the potential conflict between squatters and speculators, Jefferson, a substantial landowner himself, cast his philosophical lot with the squatters.

There was one other philosophical question that Jefferson contemplated as he imagined the role frontier settlement would play in sustaining a republican society. One of the stock topics of the age was whether the stern, ascetic values of republicanism were compatible with the acquisitive vices and refined tastes of a commercial society. In the abstract it was a great subject for belles-lettres essayists. But as a matter of policy, the debate was over. The reality, as Jefferson wrote to Washington in March 1784, is that "all the world is becoming commercial. Was it practicable to keep our new empire separated from them we might indulge ourselves in speculating whether commerce contributes to the happiness of mankind." But such a separation was wholly impracticable. "Our citizens have had too full a taste of the comforts furnished by the arts and manufactures to be debarred the use of them." By *citizens* he meant *Virginians* just as much as *Americans*. For Jefferson was writing to Washington in the hope of persuading the retired hero to put "the sweets of retirement and repose" far enough aside to lobby the state legislature to devote funds to making the Potomac navigable far inland. If Virginians could overcome their aversion to taxation, the state could challenge New York, with its Hudson, Mohawk, and Lake Erie channels, for "the whole commerce of the Western world."

The canals of France and Britain were one of the improvements that Jefferson looked forward to studying as he prepared for his European trip two months later. At the same time, he hoped to go abroad stocked with the best information he could carry on the state of American commerce. One reason to sail from Boston was to pass "thro' the Eastern states in hopes of deriving some knolege of them from actual inspection and enquiry," the better to represent them overseas. Jefferson accordingly prepared a questionnaire, with twelve distinct headings and nearly sixty subheadings, which he used to gather information about each of the states, from New York through New Hampshire.

It was, however, the final completion of another questionnaire that preoccupied him during his first gloomy months in Paris, where he was largely confined by poor health and worse weather to the house he had taken on the cul-de-sac Taitbout, north of the Louvre. Back in 1780, he and other governors had received a set of queries from the French legation, asking each to describe his state. The twenty-two topics that François Marbois listed began with "The Charters of your State" and "the present Constitution" and ended with "extraordinary Stones found in the state's mines" and "A description of the Indians established in the State before the European Settlements and of those who are still remaining." The list reflected the impulse to compile and classify knowledge that gave the French Enlightenment its distinctly *encyclopédiste* character. It was also just the kind of query that Jefferson, a born researcher, would naturally want to answer. Unlike other governors, he treated Marbois's request as an opportunity, not an imposition.

Whether he would have completed or published the project had he remained at home is an intriguing but unanswerable question. Friends like Madison and Wythe knew he was at work on some kind of manuscript. Charles Thomson proposed that he present a report to the American Philosophical Society, which Franklin had founded in 1743, the year of Jefferson's birth. But Jefferson was still mulling it over when he sailed for France. In those first housebound months in Paris, distance from home sharpened his sense of the uses to which his researches might be put. The result was his *Notes on the State of Virginia.* Of the reading (and purchase) of many books, Jefferson (like Madison and Adams) knew no end. But of their making, this prolific author of thousands of letters attempted only one, and that was the *Notes.*

The initial printing numbered two hundred closely guarded copies. Several he distributed in France. Madison got two more, with a request that he "peruse it carefully because I ask your advice on it and ask nobody else's." Outside France, its chief intended audience was "the young men at the college" (his alma mater, William and Mary), but Jefferson first wanted Madison to show it to a few others "whose judgments and information you would pay respect to." If they approved, "I will send a copy to each of the students at W.M.C. and other friends." If they did not, Jefferson would keep a few copies "and burn the rest." The work was not meant for "the public at large." Madison took the request seriously and "consulted several judicious friends in confidence." They agreed that Jefferson's observations were "too valuable not to be made known, at least to those for whom you destine them." Madison offered a prudent addendum to Wythe's idea of keeping the limited stock of copies in the college library, with access left to "the discretion of the professors" because he worried that "an indiscriminate *gift* might offend some narrow minded Parents." Jefferson was, of course, an ardent advocate of freedom of expression and the combat of ideas. But when it came to launching his own work on the sea of Virginia opinion, he took a decidedly cautious tack.

Why such caution? "I fear the terms in which I speak of slavery and of our constitution may produce an irritation which will revolt the minds of our countrymen against reformation," Jefferson explained to Monroe, "and thus do more harm than good." He was right to worry that his remarks might give offense. Against the constitution, "formed when we were new and inexperienced in the science of government," he raised two main objections. One was that by concentrating all effective power in the legislature, it met "precisely the definition of despotic government" laid down by Montesquieu. "173 despots would surely be as oppressive as one," he noted, in one of those overstatements that often made Madison blanch. "An *elective despotism* was not the government we fought for." His second leading objection challenged the authority of the constitution itself. Because the convention that adopted it had done other legislative business, it was not the "perpetual" charter "unalterable by other legislatures" that it ought to be, but a mere legal enactment that subsequent assemblies were free to amend, ignore, or even violate. Merely calling this ordinance a constitution would not make it one.

These were trenchant objections, and raising them placed Jefferson in the vanguard of American thinking. Still, they were not so grave as to justify squirreling his *Notes* away in the recesses of a library. Jefferson could worry that his remarks on legislative despotism might be read as a vengeful slap at the body that once considered his impeachment. But Americans had been freely debating their new constitutions since 1776, and his reputation was secure enough to allow him a voice in that discussion. Privately he also despaired of constitutional reform while Patrick Henry was the dominant force in Virginia politics. "What we have to do I think is devoutly to pray for his death," he wrote to Madison, only half-jokingly, while we work "to prepare the minds of the young men" for the moment when Henry passed to his just reward.

It must therefore have been the other momentous question of slavery that better explains Jefferson's concern about disseminating the *Notes.* Here a disturbing irony (or worse) refracts our reading of the two famous passages in which he first explained (in Query XIV) his scheme for slave emancipation, to be followed by colonization of the freedmen to some other country; and then (in Query XVIII) trembled for his country in conjuring the wrath that a just God might wreak upon Virginians for fastening the evils of slavery upon their bondsmen and themselves. Jefferson expected later readers to applaud the two queries. In his mind he was endorsing emancipation, proposing to commit Americans to the creation of a new nation of former slaves, and insisting that the dictates of religion and republicanism alike militated against slavery. If he meant the *Notes* primarily for a younger generation, as his letters suggest, then it was posterity he was most intent on addressing.

Yet many modern readers plausibly draw a completely different moral from these famous passages. They see Jefferson's muddled comparison of the physical, mental, and moral characteristics of the two races in Query XIV as a harbinger of the virulent racism that gripped nineteenth-century America and continued to plague the nation through the era of Jim Crow segregation and beyond. Read this way, the religious and political appeal of Query XVIII attests less to his desire to stir the conscience of his contemporaries than to his own hypocritical failure to free his own slaves. Indeed, given the skepticism about the authority of divine revelation that informs the preceding Query XVI, "The different religions received into that State," Jefferson's appeal to a recognizably biblical, judgment-wielding God could itself be queried as a resort to religious beliefs the author did not wholly share.

The chief problem lies with Query XIV. Midway through a discussion of the state's legal system and the revisal of 1777–1779, Jefferson veers off to describe his plan of emancipation. He begins somewhat cryptically, alluding to a bill "To emancipate all slaves born after passing the act," then notes that the revisal "does not itself contain this proposition, but that an amendment containing it was prepared," to be offered "whenever the bill was taken up." The scheme was first to train the youthful slaves for freedom, teaching them "tillage, arts or sciences," and then emancipate them as adults (women at age eighteen, men at twenty-one). But rather than enjoy their freedom in Virginia, they would be sent away to colonize an unspecified land, after first being provided with "arms, implements of household and of the handicraft arts, feeds"—and in perhaps an unconscious play on theories linking Africans' dark skin to the curse against Noah's son Ham—"pairs of the domestic animals." But there was nothing unconscious about Jefferson's next point: we shall "declare them a free and independant people, and extend to them our alliance and protection." This was an explicit echo of the Declaration of Independence, with one notable inversion. Independence would be declared *for* the freedmen, not *by* them. As such, they would gain the same "separate and equal station" to which all self-governing nations were entitled.

Jefferson grounded that equality *among* nations, however, on a troubling inequality *between* them. If asked, "Why not retain and incorporate the blacks into the state" and thus avoid recruiting European peasants to replace them, Jefferson defended colonization on two sets of reasons. The first could be labeled political and psychological: "Deep rooted prejudices entertained by the whites; ten thousand recollections, by the blacks, of the injuries they have sustained; new provocations; the real distinctions which nature has made; and many other circumstances, will divide us into parties, and produce convulsions which will probably never end but in the extermination of the one or the other race." Relations between black and white were already so poisoned, Jefferson assumed, that even the most enlightened policy could not repair them. Had Jefferson stopped there, we might fault him for a failure of moral nerve, yet still concede that he was wrestling with a genuinely difficult and wholly novel problem. Nowhere in his world could one encounter a biracial society organized as a republic, which presupposed the existence of a cohesive citizenry bound together by shared interests and attitudes. Jefferson's pessimism on this point sounds a counterpoint to his upbeat pronouncements on so many other matters. Even so, his grim realism also marked an attempt to confront a problem, not evade it.

But Jefferson did not stop there. Instead he plunged on, treating "the real distinctions which nature has made" as evidence of innate racial differences that raised "physical and moral" objections against retaining the freedmen within the society where their parents had labored. In many ways the resulting analysis is a mess. Jefferson conflated aesthetic ideals of beauty with other physical distinctions. He speculated that blacks need less sleep, then cited "their disposition to sleep when abstracted from their diversions" as evidence of inferior mental capacities. The entire passage is replete with casual impressions that ill accord with the statistical data that Jefferson, an adept mathematician, presented in other queries. Then he concluded this rambling discussion by confessing that "after a century and half of having under our eyes the races of black and of red men, they have never yet been viewed by us as objects of natural history." All of his conclusions, in other words, rested not on a well-grounded, systematic anthropology, but on impressionistic speculation.

These musings placed Jefferson well ahead of contemporary thinking about race and are disturbing precisely because they anticipated and even helped legitimate the virulent racism of the next century. Yet there is a crucial difference between his fumbling hypotheses in Query XIV and the invidious racial science to come. Unlike his successors, Jefferson invoked these differences not to defend slavery but to make the case for emancipation—albeit followed by colonization. Had he been offering racial or psychological justifications for enslavement, he would not have worried about the criticism of fellow Virginians. His fears rested on the idea that he was proposing an end to slavery—a heresy his own generation would hardly accept, but one in which the rising generation might be educated.

To argue the case for emancipation, however, Jefferson left the realm of natural science to appeal to old-time religion. This was the note he sounded when he returned to slavery in Query XVIII, "the *particular* customs and manners of the state." Under a heading that broad, he could have taken the easy path of describing the Virginia gentry's fondness for good horses, strong grog, and gracious hospitality—daily customs and manners all. Instead, after noting how hard it was to "determine the standard by which the manners of a nation may be tried"—that is, judged and not simply described—Jefferson made the daily experience of slavery the defining feature of Virginia society. Without further ado, in perhaps the most powerful language he ever used, he denounced the corrupting effect that slavery had on the morals of both master and slave. Sounding much like John Dickinson in London thirty years earlier, and George Mason (elder brother of the Farmer's study-mate) in 1773, the master of Monticello adopted the best Lockean psychology to explain how the "unremitting despotism" that parents exercised over slaves taught the first and deepest lessons their children would ever absorb. Here Jefferson portrayed slaves rather differently from the inferior part of creation he had just alleged they formed. Now they were a trampled part of the citizenry, persons who might feel "amor patriae" [love of country] if only their masters had not destroyed any attachment they could have by expropriating their labor and "entail[ing their] own miserable condition on the endless generations to come." Jefferson ended this denunciation by invoking the justice of a God "who has no attribute which can take side with us should a revolution of the wheel of fortune, an exchange of situation between white and black ever occur—as well it might, even by supernatural interference!"

Here was an appeal to a God of judgment instantly familiar to believers more orthodox than Jefferson ever professed to be. It was this passage that Abraham Lincoln assuredly drew upon in his Second Inaugural Address four score years later. "If we suppose that the sufferings of this terrible war were only the woe due to those by whom the offense [of slavery] came," Lincoln asked, "shall we discern therein any departure from those divine attributes which the believers in a living God always ascribe to Him?" Like Jefferson, whom he deeply admired, Lincoln had abandoned the tenets of orthodox Christianity. Yet he too was the product of a profoundly Protestant culture whose influence and idiom he inevitably absorbed.

Taken as a whole, *Notes on the State of Virginia* could profitably be read by any enlightened member of the international republic of letters that counted Jefferson, Franklin, and Adams among its citizens. But on the sensitive topics of constitutions and especially the nexus of slavery and race, Jefferson really did intend his thoughts for a generation younger than his own or even the cohort to which Madison, Monroe, and the martyred John Laurens (children of the 1750s) belonged. In "the interesting spectacle of justice in conflict with avarice and oppression," he wrote to the Welsh radical Richard Price, he took heart "from the influx into office of young men grown and growing up. These have sucked in the principles of liberty with their mother's milk, and it is to them I look with anxiety to turn the fate of this question." Here again was an expression of Jefferson's deep, even pioneering interest in the concept of a generation, the motif that recurs in his ideas of entail, inheritance, the use of public land, education, religion, and now emancipation. In effect he imagined the future of Virginia as a tale of two rising generations: one to be taught to let their bondsmen go, the other reared for freedom in some unknown Canaan.

From distant Paris the idea of directing his *Notes* at a younger generation of Virginians gained an idealized allure. Beyond recruiting acolytes like Madison and Monroe, Jefferson could imagine the students of William and Mary as the immediate heirs of his agenda of legal reform. But he lost control of the *Notes* when one of its original recipients, Charles Williamos, died. A rascal of a Parisian bookseller got hold of his copy before it could be retrieved and arranged a French translation so wretched—"a blotch of errors from beginning to end"—that Jefferson felt he had no choice but to see the job done right. The idea that his book would have a targeted, private audience dissolved. Instead it became a public text and a tribute to its author's extraordinary range of intellectual interests. Yet its most durable legacy may be the damaging implication that Jefferson's belief in the equal rights of self-governing peoples masked a theory of racial difference that, perverted, could allow one people to permanently dominate another.

Just as he was preparing the *Notes* for publication in May 1785, Jefferson learned that Congress had finally granted Franklin's long-pending wish for retirement. Jefferson would take his place as minister to France, and Adams would hold the same post in Britain. Franklin might have stayed in Europe had the widow Helvetius ever agreed to marriage. But she did not, and his thoughts turned homeward too—so much so that he even imagined visiting his Boston birthplace, perhaps "to lay my bones there." For his part, Adams had been orbiting around his clashing moods since war's end, pining to go home to his family while petitioning Abigail to join him in Europe, if Congress gave him duties commensurate with his talents. Then there was his health. He was still suffering the effects of his "Amsterdam Fever," with "swelled Ankles, Weakness in my Limbs, a Sharp humour in my Blood, lowness of Spirits, Anxieties &c." London was the post he desired most, "the only Place where I could go with Honour." But if he went there, Adams feared, "The Vanity, Pride, Revenge of that People, irritated by French and Franklinian politics"—for Adams knew that his vengeful senior colleague had many English friends—"would make it Purgatory to me." So the complaints poured out, making their erratic way to Braintree, to compound the anxieties of his long-suffering wife. "The whole tenor of your Letters rather added to my melancholy, than mitigated it," she wrote to him in October 1783. John wanted Abigail to join him, even if her stay lasted bare months. Then they could "return to our Cottage, with contented minds." But Abigail would hazard the crossing only if her European sojourn would be of some duration. She could not understand why John was still risking health and reputation when the greatest services he could do for his country were now concluded. In the decade since his departure for the First Continental Congress of 1774, they had spent roughly twenty-one months together—and only three since his first voyage in 1778. Abigail remained, as ever, his strongest supporter. But when she wrote to him now, she did so with a new independence, confessing her weaknesses but asserting her own judgment. She agreed to come only when she learned that Congress did indeed want her husband in London.

John was in Amsterdam when Abigail's first letter of July 23, 1784, announced her arrival in London. He sent John Quincy on ahead, then followed a few days later, reaching the British capital on August 7. "Met my Wife and Daughter after a Separation of four Years and an half," he noted in his diary. "Indeed after a Seperation of ten Years, excepting a few Visits." Eager to catch up with Jefferson, who had just arrived in Paris after a voyage far smoother and quicker than Abigail's, Adams allowed his family to spend only one more night in London before they hustled off for Paris. Abigail's presence had a soothing effect on her husband, but her companionship proved important to his Virginia colleague as well during those months when Jefferson was often housebound. When the Adamses went to London in May 1785, Abigail was barely settled in their Grosvenor Square house before she opened her own affectionate correspondence with Jefferson. She enclosed a squib from the *Publick Advertiser* ridiculing John's appointment as "such a phenomenon in the Corps Diplomatique that it is hard to say which can excite indignation most, the insolence of those who appoint the Character, or the meanness of those who receive it."

When Jefferson replied, he confirmed a news item he knew Abigail had already heard. Back in January 1785, Jean-Pierre Blanchard and his copilot, Dr. John Jeffries—an American but a loyalist—made the first successful aerial crossing of the English Channel to France. But a second attempt in mid-June, this time starting from France, ended near Boulogne in a crash that killed Pilâtre de Rozier instantly, while "Romain his companion" survived insensate for ten minutes before expiring. "The arts instead of advancing have lately received a check," Jefferson wrote to Charles Thomson, "which will render stationary for a while that branch of them which had promised to elevate us to the skies."

Jefferson included this same news as a postscript to a letter to Monroe. He had closed the letter with the standing invitation he also extended to Madison: visit me, please! But the terms in which he couched this request were unusual. "The pleasure of the trip will be less than you expect but the utility greater," Jefferson warned.

> It will make you adore your own country, it's soil, it's climate, it's equality, liberty, laws, people and manners. My god! How little do my countrymen know what precious blessings they are in possession of, and which no other people on earth enjoy. I confess I had no idea of it myself. While we shall see multiplied instances of Europeans going to live in America, I will venture to say no man now living will ever see an instance of an American removing to Europe and continuing there. Come then and see the proofs of this, and on your return add your testimony to that of every thinking American, in order to satisfy our countrymen how much it is their interest to preserve uninfected by contagion those peculiarities in their government and manners to which they are indebted for these blessings.

Like many of Jefferson's most cited passages, this one revealed his characteristic propensity for overstatement. No American would ever settle in Europe? How about Dr. Jeffries the balloonist? Or Bancroft the naturalist turned spy? Or even Franklin? But Jefferson was right to predict that European emigration to America would encounter no corresponding immigration in return. It is equally noteworthy that in writing to his Virginia protégé, he described their "countrymen" as Americans. It was America as a whole he wished to "preserve uninfected by contagion" from European institutions and habits.

The restoration of health and good weather in the summer of 1785 gave Jefferson a new beginning in Europe. Unlike many American tourists to come, he found the French to be a "polite, self-denying, feeling, goodhumoured people," as he wrote to Abigail Adams in a good-natured comparison of the merits of Paris and London. English and American manners might seem to be alike, he wrote to her future son-in-law, William Stephens Smith. But "I doubt whether in benignity of disposition we do not find a greater resemblance here" in Paris. This conviction deepened with time. The French are "a people of the most benevolent, the most gay, and amiable character of which the human form is susceptible," he wrote to Wythe a year later.

Yet the limits of this comparison became apparent when it reached his pet subject. French manners might be "the best calculated for happiness to a people in their situation," he conceded. But the French "fall far short of effecting a happiness so temperate, so uniform and so lasting as is generally enjoyed with us." There were two distinct explanations for this disparity. One drew on his observations of the elite classes who set the tone for a gawking society. "The domestic bonds are here absolutely done away" with, he complained. "Conjugal love having no existence among them," he warned, "domestic happiness, of which that is the basis, is utterly unknown." In its place, the French enjoyed occasional "moments of transport above the level of the ordinary tranquil joy we experience, but they are separated by long intervals" when "all the passions are at sea without rudder or compass." They knew only "moments of extasy amidst days and months of restlessness and torment."

The image of "moments of transport" and "extasy" had avowedly sexual connotations. Jefferson's aversion to the rampant sexuality of French high society could have elicited the kind of prim moral condemnation that Abigail Adams pronounced after dining with Franklin and Madame Helvetius, whose "careless jaunty" demeanor and brazenly provocative speech "highly disgusted" her. Instead, he explained the dalliances of the French, "fallacious as these pursuits of happiness" were, as the best distraction they could find "from a contemplation of the hardness of their own government."

Here was the second, public source of the disparity between America and France. At first glance the French seemed destined for natural happiness. They had "the finest soil upon earth, the finest climate under heaven," he wrote to Washington, and a "compact state" that should be easy to administer. "Surrounded by so many blessings from nature," their people were "yet loaded with misery by kings, nobles, and priests." All of these natural advantages, he observed at another time, were rendered "ineffectual for producing human happiness by one single factor, that of a bad form of government." Out of an estimated population of twenty million in France, he believed "nineteen millions more wretched, more accursed in every circumstance of human existence, than the most conspicuously wretched individual of the whole United States"—a judgment that presumably included chattel slaves. "If any body thinks that kings, nobles, or priests are good conservators of the public happiness, send them here," he wrote Wythe. "They will see here with their own eyes that these descriptions of men are an abandoned confederacy against the happiness of the mass of the people."

These impressions became more palpable when Jefferson made his first foray beyond Paris in October 1785. It was only to nearby Fontainebleau, where the king liked to adjourn for the fall hunt, carrying the court and diplomatic corps along with him. Already fretting over his expenses, which he rarely controlled, Jefferson planned "to come occasionally" from Paris "to attend the king's levées." With thoughts of another distant mountaintop in mind, he used his first visit to hike "towards the highest of the mountains in sight," the better to survey the tableau. En route, he wrote to Madison, "I fell in with a poor woman" who proved to be "a day labourer" and mother of two. Like a good social statistician, Jefferson asked about "her vocation, condition and circumstance." After a mile they parted, and Jefferson pressed twenty-four sous into her hands, leaving her wordless and in tears. This "little attendrissement"—meaning her "unfeigned" tears—"led me into a train of reflections on that unequal division of property which occasions the numberless instances of wretchedness which I had observed in this country and is to be observed all over Europe." The sources of the problem seemed obvious. The bulk of landed property in France was "absolutely concentered in a very few hands." This landed elite kept small legions of servants and "a great number of manufacturers, and tradesmen" at work. But the mass of landless peasants lacked steady subsistence, all because so much land was "kept idle mostly for the sake of game" by owners who did not need the added income they could gain by its cultivation.

There were, of course, aspects of French society that Jefferson admired. "Here it seems a man might pass a life without encountering a single rudeness," he wrote to Charles Bellini, the Italian-born professor of Romance languages at William and Mary. The French knew how to combine "the pleasures of the table" with a degree of "temperance" that hard-drinking Virginians lacked. "They do not terminate the most sociable meals by transforming themselves into brutes," he pointedly noted. The one decided European advantage that Jefferson freely conceded lay in the arts: "architecture, painting, sculpture, music." But as much as he confessed to "covet" Europe's superiority in this aesthetic version of a medieval quadrivium, he denied that these were proper subjects for the next American generation. "The objects of an useful American education" were "Classical knowledge, modern languages . . . Mathematics; Natural Philosophy; Natural History; Civil History; Ethics." All of these could be learned just as well at home, with the obvious exception of the value of studying foreign languages overseas. Any American doubting the adequacy of a native education need only ask, "Who are the men of most learning, of most eloquence, most beloved by their country and most trusted and promoted by them"—those trained at home, or those educated abroad? Every balance sheet he drew proved that whatever Americans coming to Europe might gain from learning European "manners" would be "poor compensation" for what they "will lose in science, in virtue, in health and in happiness."

From these observations, Jefferson drew obvious morals that confirmed and reinforced the social vision of the revisal of the laws. The "train of reflections" that followed his walk with the poor woman of Fontainebleau drew him back to his plans for using the landed patrimony of Virginia to secure the material independence of the next generation. "Legislators cannot invent too many devices for subdividing property," he wrote to Madison, so long as "their subdivisions" did not clash "with the natural affections" that led parents to give some children larger legacies than others. But equal inheritance among all children, of both sexes, remained his preference. Jefferson's concerns ran beyond the private rights of inheritors, however. A commonwealth, like a patriarch, had to regard its lands as a source of potential wealth for all its members. "Whenever there is in any country, uncultivated lands and unemployed labor, it is clear that the laws of property have been so far extended as to violate natural rights," Jefferson stipulated. "The earth is given as a common stock for man to labour and live on." In the abstract he was willing to ask whether "the unemployed" might reclaim "the fundamental right to labour the earth" whenever a society endowed with unused land failed to provide for their sustenance. In America it was not yet time "to say that every man who cannot find employment" could simply settle some piece of "uncultivated land," he concluded. "But it is not too soon to provide by every possible means that as few as possible shall be without a little portion of land"—just as he had proposed to do with his draft constitution and land office bill.

So too Jefferson's contrasting observations on the common intelligence of the Americans and the French reminded him of his plan for public education—"by far the most important bill in our whole code" of revised Virginia laws, he wrote to Wythe. Why? Because the misery he saw in Europe owed less to the outright domination of "kings, nobles and priests" than to the collective ignorance in which they kept their subjects. Writing to Bellini in 1785, he calculated that "in science, the mass of people [in Europe] is two centuries behind ours, their literati half a dozen years before us"—the time it took for useful ideas to get published, disseminated, and "acquire just reputation" on either side of the Atlantic. A year later he advanced America's lead time in intelligence to a whole millennium. Americans would not have attained their own "common sense," he explained, "had they not been separated from their parent stock and been kept from contamination, either from them, or the other people of the old world, by the intervention of so wide an ocean."

Most Americans were in no risk of "contamination" because they would never return to the lands of their ancestors. But there was a group of Americans whose potential hankering for European knowledge still worried Jefferson: the young men, candidates for William and Mary and other colleges, whom he saw as the republic's future leadership. Once he had planned to have his nephew Peter Carr, son of his late brother-in-law Dabney Carr, pursue his studies in Europe. "But I am thoroughly cured of that Idea," he wrote to Peter's schoolmaster back in Virginia. "No American should come to Europe under 30 years of age." His objections were less pedagogical than moral, less concerned with the ideas Americans might absorb than the bad behavior they would encounter. A young American taking his cue from the upper classes would become ensnared in the web of sexuality that Jefferson found so unsettling. As passions went, sexual lust was not merely "bad." It was in fact "the strongest of all the human passions." The American student would be drawn "into a spirit for female intrigue destructive of his own and others happiness, or a passion for whores destructive of his health, and in both cases learns to consider fidelity to the marriage bed as an ungentlemanly practice and inconsistent with happiness."

There was that word again. Dangerous liaisons with European women would forever unfit American men for the affectionate bonds they should form with their countrywomen, and thus for the domestic happiness Jefferson had known with Martha. Yet he might not have found the threat to the next generation so alarming had he not found the women he met in Paris so alluring. For like Franklin, Jefferson loved the company of women. Like many a widower of a happy marriage, he found that the need for female companionship outweighed his duty to Martha's memory. On her deathbed, their daughter Patsy recalled, he promised Martha never to remarry, lest a stepmother rear her children. But that pledge left him free to rediscover the charms of female company amid the refined conversation of many a dinner and the salon society where bright and witty women presided over the best minds of the ancien régime. Some of his most expressive letters in Europe went to female correspondents. The letters were flirtatious to the point of being risqué, mixing flattering compliments to their beauty with frank praise of their intelligence, taste, and judgment. If the particular customs and manners of the French offended his moral sense, they yet reawakened him to life's possibilities. One was that agreeable female society could still bring pleasure even when his full vision of domestic happiness was beyond reach.

Before Maria Cosway taught him this lesson in fuller measure, a summons from John Adams called Jefferson to England in March 1786. The object was to conclude a treaty with an ambassador from Tripoli, one of the "pyratical states" of North Africa that regularly preyed upon American shipping while enslaving the seamen they captured. The tribute Congress was prepared to offer was "but as a drop to a bucket" of what was demanded, and the talks proved fruitless. The two diplomats also made a last effort to negotiate a commercial treaty with Britain before their commission from Congress for this task formally lapsed. It was treated "with derision," Jefferson noted, and he knew why. "That nation hates us, their minister hates us, and their king more than all other men." The two Americans had personal evidence of this when Adams escorted Jefferson to a royal levee, where George III turned his back on them in a public snub. (Perhaps he had been reading Jefferson's *Summary View* of 1774, or the Declaration of Independence.) Even if British thinking was not warped by prejudice, what incentive did the old empire have to seek a mutually beneficial treaty with thirteen barely United States? Both diplomats knew that their countrymen were all too happy to consume British imports and content to let their own exports be carried by British ships even while American merchantmen were barred from imperial ports. Both knew too that their diplomatic frustrations were a tribute to the weakness of Congress—often described as "imbecile" in these years. Until it received adequate powers to regulate commerce, they routinely wrote home, and used that power to retaliate against Britain, there was little hope of securing the lasting commercial ties that John Adams's model treaty of 1776 envisioned as the framework of American foreign relations.

With diplomacy a venture in futility, Jefferson joined Adams in an April tour of famous English gardens. He carried Thomas Whately's *Observations on Modern Gardening*, a celebrated work that laid out the principles of the natural style of landscaping that English taste decidedly preferred to the mathematical symmetry beloved of the French. Whately had been a major drafter and defender of the Stamp Act, but Jefferson did not hold that against him. Instead he marveled at how accurate Whately's descriptions remained. His own purpose, Jefferson noted, was to "enable me to estimate the expence of making and maintaining a garden in that style." Adams carried a copy of Whately too, but saw little practical use for it. "It will be long, I hope, before Ridings, Parks, Pleasure grounds, Gardens and ornamented Farms grow so much in fashion in America." Why bother to improve upon the "nobler Materials" that raw nature had already provided America? Adams admired the "residences of Greatness and Luxury" they saw but drew another moral as well. "A national debt of 274 millions sterling accumulated by Jobs, Contracts, Salaries and Pensions in the course of a Century might easily produce all this Magnificence." He was more taken by the historical associations to be made at Stratford-upon-Avon, where he mused on the great bard, and at Edgehill and Worcester, battlefields of the English Civil War. At Worcester the locals seemed so "ignorant and careless" about the site that Adams launched into an oration on English patriotism. This was "holy Ground," he told the bemused if friendly natives, where "All England should come in Pilgrimage" each year.

By early May 1786 Jefferson was back in Paris. His overall impressions of England were not favorable. He loved its gardens, "the article in which it surpasses all the earth," but despised its architecture, "the most wretched stile I ever saw." He made one other concession: "The mechanical arts in London are carried to the highest perfection." He and Adams saw impressive evidence of this at the newly opened Albion Mills, at the south end of Blackfriars Bridge in London. There James Watt and his partner, Matthew Boulton, were using Watt's steam engine, a milestone in the industrial revolution, to grind wheat into flour. The significance of this innovation was not lost on Jefferson. He relished talking shop with "the famous Boulton," both at London and again at Paris in December, where Boulton showed the calculations proving that "a peck and a half of coal perform exactly as much as a horse in one day can perform." Jefferson dutifully conveyed the results to Charles Thomson in Philadelphia.

This was one of the first letters he had been able to compose in his natural hand for three months. The problem, he wrote to Madison, was "an unlucky dislocation of my right wrist" that even now allowed him to write "a little, but with great pain" and then only for short stretches. He was planning a therapeutic trip to mineral baths in the south of France. He did not explain the circumstances of his injury, which were less "unlucky" than foolish. Perhaps (though it is not certain) while trying to impress Maria Cosway with his athleticism, Jefferson took a nasty tumble in the Champs Élysées on September 18. Badly set, the wrist was so painful that it sometimes prevented him from sleeping. By the time of the mishap he had been squiring Cosway around Paris and its environs for a month, occasionally with her husband, Richard, a man his own age, but usually not. Smitten from the moment that John Trumbull, the American painter, introduced them, Jefferson sent "lying Messengers" across Paris to beg off a slew of prior engagements so he could spend that first day and evening with his new acquaintances.

Of English descent, Maria Hadfield Cosway was Florence-born and -reared and more fluent in Italian than English. Trained to art and music, she married Richard Cosway, an acclaimed miniaturist painter, in 1781, more for convenience than love. Their London household had a libertine reputation, and rumors of extramarital liaisons swirled around Maria herself. To a less infatuated observer she might have embodied all the vices of fast European women against which Jefferson wanted to protect young Americans. But she was also devoutly Catholic, and their whirlwind travels around Paris suggest that art, music, and mutual pleasure in each other's company formed the basis of their romance. Its most lasting product was the much studied "Dialogue between my Head and my Heart" that Jefferson dragooned his left hand into writing soon after the Cosways departed Paris in early October. In this charming and witty missive, he gives the Head its rational due while the Heart gets to deliver nearly all the good lines. The Head is the straight man, the Heart the performer the audience really wants to hear.

As one would also expect of Jefferson, he makes the definition and pursuit of happiness the decisive theme in the exchange twixt Head and Heart. The Head sets up the colloquy by warning the Heart (and perhaps the Wrist) that "the most effectual means of being secure against pain is to retire within ourselves, and to suffice for our own happiness." The Head treats "Everything in this world" as "a matter of calculation," and turns "The art of life" into "the art of avoiding pain." Friendship, the Head warns, only multiplies our woes by forcing us to share the misfortunes and griefs of others. The Heart's reply is so vivid as to amount to a refutation. It begins with an allusion to the death of Martha and the consolation that comes from sharing grief with another. "When Heaven has taken from us some object of our love, how sweet it is to have a bosom whereon to recline our heads, and into which we may pour the torrent of our tears." But friendship exists for purposes greater than the relief of misery, the Heart adds. "The greater part of life is sunshine," and for this "I will recur for proof to the days we have lately passed." To this token of affection Jefferson adds more philosophical reflections. We cannot equate happiness with "the mere absence of pain," the Heart declares. Nor is it even a proper subject for the Head to ponder, much less decide. By "denying to you the feelings of sympathy, of benevolence, of gratitude, of justice, of love, of friendship, she has excluded you from their controul." These were moral matters, not amenable to precise calculation. "Morals were too essential to the happiness of man to be risked on the incertain combinations of the head."

Here the dialogue ended but not the letter in which it was inserted. Jefferson's sorely taxed left hand still had to balance an apology for "such a tedious sermon" with a promise of future letters. And correspond they continued to do, with open avowals of affection punctuated by pangs of disappointment that arose from gaps in writing and inevitable misunderstandings. (English was her second language and her syntax showed it.) The most flagrant came in April 1788, when Maria missed the sexual allusions in his musings on "the promontory of noses"—a play on a bawdy passage in Laurence Sterne's *Tristram Shandy*, the novel he carried regularly on his person and she knew, if at all, rather less well. In an epistolary pout that she could easily have sung as an aria, Maria took her admirer to task for his failures as a correspondent, asking him to tender the same "proof of esteem" that he expected from her. Then without pause she expressed her own wish to see him again at Paris while confessing that "I am afraid to question my Lord and Master on this subject." Perhaps Jefferson could visit London instead? They could go to the opera, where he could hear the great Marchesi and discover "what Italian singing is."

Jefferson wrote his "nosy" letter after completing the second of his major European trips, to Amsterdam and Strasbourg, in the spring of 1788. He began the first a year earlier, spending three months traveling through France and into northern Italy before returning to Paris in June. One purpose of the trip was to try the mineral waters at Aix-en-Provence as a cure for the ailing wrist. That proved "useless." The wrist improved enough, though, for Jefferson to compile forty-four pages of notes on his travels. "Architecture, painting, sculpture, antiquities, agriculture, the condition of the laboring poor fill all my moments," he wrote to his secretary, William Short, from Lyon. The notes indicate that agriculture in all its aspects—from the quality of the soil and choice of crops to the design of hoes and the production of Parmesan and "Mascarponi, a kind of curd"—was his chief concern. Viticulture too received its due, and perhaps more, as Jefferson visited vineyards whose appellations remain familiar today. Though he left few comments on the condition of the peasantry, he was attentive to their dress, diet, and the division of labor between the sexes, particularly in Languedoc, where the social order seemed upside down. Men were filling all the women's proper trades, turning the women into "porters, carters, reapers, wood cutters, sailors, lock keepers, smiters on the anvil, cultivators of the earth &c." No wonder those who retained "a little beauty" were "driven to whoredom."

From Nîmes and Marseille, Jefferson found time to write flirtatious letters to his two great female friends in Paris, Madame de Tesse and Madame du Tott. The first opened with his oft-quoted description of himself "gazing whole hours at the Maison quarree, like a lover at his Mistress," while nearby artisans wondered whether he was "an hypochondriac Englishman, about to write with a pistol the last chapter of his history." It was history writ large, though, into which Jefferson cast his imagination, fancying himself an ancient Roman "filled with alarms" over the invasions of Goths and Vandals, "lest they should reconquer us to our original barbarism." Marseille dragged him back into his own time, with "four thousand three hundred and fifty market-women (I have counted them one by one) brawling, squabbling, and jabbering Patois" beneath his window, along with "three hundred asses braying and bewailing" and "four files of mule carts." But his noisy rooms were not to be confused with the kind of rural auberge where he had spent most nights. There a different fancy took hold. Now he is a solitary traveler, anonymous, with a few possessions, and all the time he needs to write, read, and think. He discovers "how cheap a thing is happiness, how expensive a one pride," and comes to "pity the wretched rich" who pass their days in the ostentatious, pointless display of wealth.

This was less fancy than fantasy, a literary conceit suitable for a letter to a genteel woman friend in Paris, but hardly consistent with his own pursuit of happiness. Jefferson could never view a French chateau or an Italian villa without thinking of Monticello and the improvements he might make there. An idea was taking shape in these years abroad to transform the house he had built for Martha into something grander. It would not be an isolated retreat where a retired statesman could live content with his memories, his books, and his daughters. It would be the site of the sociability that still defined his ideal of happiness, open, in the great Virginia tradition, to friends like Madison and visitors like the marquis de Chastellux. It would be furnished too with the numerous *objets* that Jefferson spent considerable time pursuing and greater sums purchasing. As the journalist cum historian Garry Wills archly notes, "Jefferson went on a buying spree in France that was staggering in its intensity. At times it must have looked as if he meant to take much of Paris back with him to his mountain 'chateau.' The eighty-six large crates of goods he shipped to the United States included sixty-three oil paintings, seven busts by Houdon, forty-eight formal chairs, Sèvres table sculptures of biscuit, damask hangings, four full-length mirrors in gilt frames, four marble-topped tables, 120 porcelain plates, and numberless items of personal luxury." These were the purchases of an absentee planter who faced an uphill struggle to master a substantial personal debt, but who never allowed obligations to creditors—especially when they were British merchants—to mar his vision of domestic happiness.

Was Maria Cosway ever part of that vision? Since he spoke of Monticello often to her, he must have fantasized that one day she might not only visit it but inhabit it as well. He thought of her often on his two tours. But unlike *les madames* de Tesse and du Tott, he prudently wrote to Maria only after he was back in Paris. He could not trust messages this intimate to the post. Returning to Paris in June 1787, he waited to write until his servant Adrien Petit went to London to fetch Polly, finally arrived from America and now in the doting care of Abigail Adams. When he did write, he followed an apology for not venturing further into her native Italy with the remark that "I am born to lose everything that I love. Why were you not with me?" And then, "When are you coming here?" In the reply she quickly dashed off, Maria predictably chastised Jefferson for his silence, then closed by regretting that he had denied her an opportunity to meet Polly.

Here was one of the two fatal obstacles that Jefferson also contemplated when time and distance allowed his Head and Heart to pursue an interior colloquy: the deathbed pledge to Martha. Even had that promise not been given or extracted, Maria had a "Lord and Master" of a husband already. If her use of that term implied a private bond with her admirer, it no more altered her status as a married woman of Catholic convictions than it licensed Jefferson to snatch her from Richard's control and sail away to his mountaintop aerie. The ways in which Jefferson would liberalize the legal status of women did not extend to divorce.

When Maria, alone, did return to Paris in the late summer of 1787, the reunion that he promised did not occur. They did not "breakfast every day a l'Angloise" or "dine under the bowers of Marly, or forget that we are ever to part again." They saw each other less often than they did the year before, and conversational intimacy eluded them. By joining the household of Princess Lubomirska of Poland and surrounding herself with a "domestic cortege," Maria made it impossible for Jefferson to "enjoy [her] company in the only way it yeilds enjoiment, that is, en petite comité." They planned to breakfast on December 8, the day Maria set for her departure for London, but she impulsively left town at the early hour of 5 A.M. Had he actually appeared for a farewell? she asked from London. Indeed he had, Jefferson replied. Then he faulted *her* for not making herself more available.

Their correspondence continued in fits and starts, with expressions both of affection and disappointment unlike the mock emotions that merely playful letter writers easily feign. But Jefferson never saw Maria again. Perhaps, as many biographers conclude, the Head finally triumphed, or perhaps his passions cooled. Perhaps, with Polly now in Europe, his role as father overtook any other fancies he entertained. Polly came to Europe resenting her father, first for having left her behind, then for making her abandon her new home with the Eppeses. Once deposited with the Adams family in London, she bawled at being parted from the ship's captain, who had squired her passage more effectively than Sally Hemings, the young slave girl (and her half-aunt by Jefferson's father-in-law) whom the Eppeses pressed into service to escort her. Nor did she part easily from Abigail Adams, who wanted to keep her in London. Polly resented Jefferson again for sending his servant to fetch her when any truly caring, long-absent father would have come himself. But nine-year-old Polly could not carry a grudge too long. Once settled in Paris, the charmer replaced the sulker.

Call it what one will—flirtation, romance, unconsummated affair, or a "verbal jousting match" that Cosway could never win—Jefferson's attachment to Maria should not be dismissed as a Parisian lark. Other than his daughters, she remains the one important woman in his personal life for whom a documentary record survives. He did not ignore her memory, as he did his mother's, nor destroy her letters, as he had Martha's. The evidence of his liaison with Sally Hemings, which likely began after their return to America, is inscribed in the oral reminiscences and Y-chromosomes of the Hemings descendants, which confirm that Sally mothered one, and probably more, of his children. The emotional content of their relationship remains open to speculation and surmise but is not subject to the textual tests that historians prefer. Jefferson's letters to Cosway reveal a man whose grief has subsided though not wholly vanished, who wrestles with his own passions even as he seemingly subdues them.

But the Jefferson who grew enamored of the married Maria Cosway was the same moral authority who counseled younger Americans against the sexual wiles of European women. Here lies yet another facet of the Jefferson personality that so disturbs Americans now. This is the Jefferson who existed comfortably, even happily, with contradictions, evasions, and hypocrisies that someone so committed to living by the light of reason and philosophy should have found intolerable and even morally indefensible. This is the Jefferson who locked the principle of equality into the nation's political creed, but whose vision of domestic happiness depended on the exploited slaves whose freedom he imagined only as an abstraction. It was the same Jefferson who railed against the evils of debt, but whose own profligacy, measured in all those Parisian purchases and his endless projects of home improvement, led not to the *post nati* emancipation of his slaves* but their *post mortem* sale (that is, after his own death), along with his beloved Monticello, the unique architectural jewel he began thoroughly rebuilding in 1794. That sale broke up many a slave family whose happiness he never weighed against his own.

For those who regard history as an exercise in moral judgment, where a wiser present summons a flawed past to the bar of its superior rectitude, Jefferson makes an easy mark. But moral condemnation, deserved or not, is not historical explanation. So too hypocrisy is merely a label for how we act, not an explanation of why we profess one thing and

* Under *post nati* rules, only slaves born after the enactment of an emancipation statute would receive their freedom, typically upon reaching adulthood. Incredible as it may seem to modern readers, the belief that a master deprived of his legal property right in a slave would require compensation for his loss constituted a major objection to general schemes of emancipation.

do another, such as lamenting abuses against human rights committed around the world—sometimes in our name, or under conditions we have the power to alter—while pursuing our lives just as we privately wish to do. Thus it was with Jefferson. He was a willing, obedient servant to an exquisite and refined vision of happiness. Many members of his class doubtless shared his vision, but few or none ever contemplated it with his philosophical rigor or self-knowledge. Recalled to an active life by his return to public service and the stimulus of Paris, he buried his grief and freshly conceived of Monticello as the architectural fulfillment of his domestic vision. Martha's death forever altered the form that happiness would take. But it could not dam or drain the refined impulses that prolonged exposure to the finer things of Europe had so powerfully revived.

Jefferson's two walking trips in 1787–1788 overlapped the great debates that led to the adoption of a new federal Constitution. Had he been home, he certainly would have joined the eminent delegation that Virginia sent to the Philadelphia convention. From a distance he seemed detached from the process and lukewarm about its result. "I confess there are things in it which stagger all my dispositions to subscribe to what such an Assembly has proposed," he wrote to Adams in November 1787. A few "new articles" added to "the good, old and venerable fabric" of the Articles of Confederation should have sufficed. He was not sure whether its "very good" parts "preponderate" over its "very bad" ones, principally the novel office of the presidency. He suspected that Americans had overreacted to the accounts of their "anarchy" that European papers had been propagating since 1783. Though the Adamses had already taken him to task for his casual dismissal of Shays's Rebellion, the upheaval that wracked Massachusetts in the winter of 1787, Jefferson clung to his belief that these protests were a sign of the health, not the decay, of the body politic. Compare the well-ordered Massachusetts uprising with the political violence that regularly erupted in France, he wrote to Madison, and you will see how exaggerated your fears have been. Jefferson moderated his views once Madison's accounts of the making of the Constitution began to reach him in December. Soon their correspondence, which suffered from unusual transatlantic delays, focused on Jefferson's other major objection: the omission of a bill of rights from the Constitution. Madison was not amused when another letter from Jefferson was published, expressing a hope that four states would hold out until suitable amendments were adopted. Here was Madison, trying to reconstitute the Union, and there was Jefferson, indifferent to how hard that work had been, nonchalantly endorsing a scheme of temporary separation. Still, the two men were well along in the process of converting their intellectual friendship into a lasting political alliance. Their letters of 1787–1789 on the specific issue of constitutional rights are arguably the single most fascinating exchange of views between any two leaders in all of the nation's history.

Had Jefferson been at home, he would have been less cavalier. Abroad he could allow his dominant impressions of Europe to shape his judgment and confirm his opinions. Ultimately the strength of the republic did not depend on constitutional reforms, he wrote in answer to Madison's first reports. Keep our people agricultural and prevent their being "piled upon one another in large cities, as in Europe," and they will avoid the corruption and servility he observed in the lower classes of France. "Above all things I hope the education of the common people will be attended to"—as in bill 79 of his revised legal code—for "their good sense" was the best "security for the preservation of a due degree of liberty." As proof, Jefferson cited the French monarchy's recent refusal to relax the civil disabilities imposed on the nation's Protestants, solely on the grounds of their religious beliefs. "Compared with the authors of this law," Jefferson thought, "the most illiterate peasant" in America was "a Solon." The real problem was that Americans "do not know their own superiority."

For Jefferson, then, the moral constitution of a people had become more important than their constitution of government. Or so it seemed at this moment, when Americans were engaged in an intense debate that he could view only from afar, more as a freethinking critic than an avowed partisan. His distaste for serious political discussion extended to the ferment he found in Paris upon returning from his Amsterdam-Strasbourg tour in April 1788. "Men, women, children talk nothing else," he complained to Anne Willing Bingham (daughter of Thomas Willing, longtime business partner of Robert Morris). "Society is spoilt by it." This was one interest he did not want his female friends to pursue, and another instance in which American women—possessing "the good sense to value domestic happiness above all other"—enjoyed the advantage over their European sisters. He did not share with ravishing young Anne (later engraved as the goddess Liberty on American coins) the darker view he conveyed to Washington in December. One dangerous source of political influence "will elude" France's reformers, he predicted: "the omnipotence" of women who could readily "visit, alone, all persons in office, to sollicit the affairs of the husband, family, or friends, and their sollicitations bid defiance to laws and regulations." Here again, American women had the advantage in probity.

Jefferson was not yet sure whether "a thorough reformation of abuse" could occur in France. Events after the new year soon convinced him that a "constitutional reformation" of the ancien régime he so detested was at hand. Though anxious to receive leave for a visit home, he was still in Paris as the Estates-General met for the first time in generations. Nor did he remain the mere onlooker he claimed to be the year before. He regularly attended the debates at Versailles and actively counseled Lafayette on matters of strategy as well as the ways in which this genuine aristocrat with republican sympathies could retain the confidence of the people. Their friendship allowed Jefferson to join the discussions of a potential declaration of rights to guide the new compact they hoped to form. Many of the meetings of Lafayette and his circle of liberal reformers were held at the Hôtel de Langeac, the house Jefferson had leased since 1785. These conversations prompted Jefferson to write "the most famous of his thousands of letters," the September 6, 1789, missive to Madison addressing "The question Whether one generation of men has a right to bind another." He took as his starting point the "self-evident" (meaning axiomatic) principle "'that the earth belongs in usufruct to the living.'" His immediate concern, as the heart of the letter makes clear, was narrower than the breadth of his question implied. Did one generation have the moral right to bequeath its public debts to its successors? This was a topic of some concern to Lafayette and his circle. The moment for "constitutional reformation" they hoped to seize had been brought about by the financial crisis of the monarchy, which was itself a product of the war it had fought on behalf of America. And debt, private as well as public, was a matter of great moment to Jefferson. Not only was he struggling, and failing, to rein in the expenses caused by his Parisian profligacy. He was also wrestling with the financial obligations he and his brothers-in-law had assumed in 1773 by dividing the assets of John Wayles, Martha's father, before his estate's debts were cleared. That disastrous decision made them liable for the debts of the previous generation, imposing a burden Jefferson never surmounted but always resented. Legally he grasped his plight all too well. Morally he felt there must be some limit to an obligation that interfered with his pursuit of happiness.

Yet if debt was the central theme of this letter, Jefferson's query did not stop there. "This principle that the earth belongs to the living, and not to the dead, is of very extensive application and consequences," he told Madison, "in every country, and most especially in France." It could also apply to the duration of political constitutions and a host of other legal practices, such as entail or the vested rights of corporations, that protected great concentrations of property against fresh regulation or beneficial use. In this remarkable letter, Jefferson converted his longstanding concern with the upbringing of the next generation into a philosophical statement of every generation's self-evident right to self-government—or even more, to exercise that right "to alter or abolish" governments that the Declaration of 1776 had proclaimed. That sense of futurity, which was always so pronounced in Jefferson, attained a new force and coherence in Europe. That did not make his principle persuasive, either to a skeptical Madison, who politely but firmly rebutted his conclusions the next year, or to most commentators since. But it may explain his lasting hold on the American imagination, and perhaps even the political preeminence he soon attained.

Whether Jefferson would have formulated this more advanced notion of the rights of the living generation had he not gone to Europe is one of those counterfactual questions that scholars like to raise but can rarely answer. The stimulus for the letter did not emerge full-blown from his own brain. A note from Richard Gem, the eccentric and elderly physician-philosophe who had just been treating him, perhaps for one of his migraines, stated the same basic questions Jefferson posed to Madison. Yet critical, indeed essential elements of this basic idea had long been present in Jefferson's thought: in his precocious belief that the goal of the Revolution was not to restore lost rights but to write improved constitutions of government; in his plans for public education, inheritance, and the distribution of public lands; in his desire to train children to question, not simply inherit, the religious beliefs of their parents. They were also present in his plan for emancipation, so unsatisfactory to us but nonetheless an attempt to address the fact that American slavery had become a hereditary condition, a debt of lost freedom imposed on "endless generations" to come.

Three weeks after writing the letter, which he still had with him, Jefferson, Patsy, Polly, and the Hemings siblings James and Sally set out for America. In late July, before his permission to take a leave of absence ("my Congé") had arrived, he wrote to Maria in London. The Bastille had fallen ten days earlier, and Jefferson briefly relaxed his ban against talking politics with women to marvel that "my fortune has been singular, to see in the course of fourteen years two such revolutions as were never before seen." Then he turned affectionate and chided Maria that it had been "a long time since you told me" that he retained a place "in your affection." For his part, "I feed on your friendship" and "therefore have need of it in all times and places." Pressed for time, Maria sent the briefest note in reply. But she wrote again ten days letter, coupling a sudden wish, "Pray take me to that Country" of America, with another: that he would stop in England on his expected return the next year, "since you will not in your way to America. 'Tis very cruel of you."

Late in August, the awaited permission for a temporary leave from his diplomatic post finally arrived. Jefferson's party went to Le Havre in the phaeton, planning to cross to Cowes, on the Isle of Wight, where they would catch an America-bound ship. Fierce winds kept them stranded in France for a fortnight. On a stormy day, Jefferson hiked ten miles on a "fruitless" quest for a pair of shepherd dogs to take home. Returning to Le Havre he stumbled upon a suicide, the man's face completely blown off by the pistol lying at his feet. At last they crossed to Cowes. From London, a bedridden Maria sent a note through the painter John Trumbull, intimating that Jefferson might still visit her. That was impracticable, but on October 4 Jefferson replied that perhaps they could meet at Paris in the spring, "with the first swallow." Then he bid "adieu under the hope which springs naturally out of what we wish. Once and again then farewell, remember me and love me."

The day before, the most eminent Virginian of the age was writing to Jefferson with a different proposal: to serve as secretary of state in the new government over which he was presiding. For reasons of his own, Madison concurred with Washington's plan. With some misgivings, Jefferson acceded to their requests. He never visited Europe again. For Maria, Monticello and its splendid views remained an imagined tableau. For Jefferson, the complete and never-ending rebuilding of his mountaintop home became the chief avenue for his private pursuit of happiness. There Madison and Monroe were his frequent guests, Sally Hemings his most intimate companion, and their enslaved children (freed at his death) a legacy of the American dilemma that Jefferson embodied, lived, and still represents.

8

The Greatest Lawgiver of Modernity

THROUGHOUT HIS EUROPEAN SOJOURN, Jefferson relied on Madison and Monroe for American news. His friendship with his two successors as president lasted until that day in 1826 when Jefferson and John Adams, perhaps with providential aid, managed to time their deaths for the fiftieth anniversary of the signing of the Declaration of Independence. (Monroe picked the same date in 1831. Madison fell six days short in 1836.) It was Madison with whom Jefferson enjoyed the closer friendship and more profound correspondence. As friends and allies they agreed on most principles and policies, yet their minds worked very differently. Jefferson reacted more quickly, even impulsively; his interests ranged more widely; he wrote more vividly, concisely, and directly. Madison by contrast had a deserved reputation for prudence and thoughtfulness. As a thinker and writer, he loved his studied distinctions and qualifications. One could never say of him what he later said of Jefferson: that he had the habit "as in others of great genius of expressing in strong and round terms, impressions of the moment."

Like intellectuals in all ages, Jefferson loved rummaging in bookshops. From Paris he supplied Madison with a "literary cargo" of history, political philosophy, and jurisprudence. Many of these books discussed the workings of other confederacies, some dating to antiquity, such as the Amphyctionic and Achaean leagues of Greek city-states, others of more recent vintage, such as the Dutch confederation, which had been the object of John Adams's diplomatic attention. Madison turned to these works in the winter of 1786, when he was home at

Montpelier, the family plantation near Orange, Virginia. His notes on "ancient and modern confederacies" survive, in his incredibly fine hand. They marked a vital step in his preparations for the great Constitutional Convention of 1787.

Madison's project echoed another that Niccolò Machiavelli undertook after the restoration of Medici rule in Florence in 1512 abruptly ended his eighteen years of able service to his city. In involuntary retirement, hoping to curry favor with the new rulers, he composed his masterpiece of statecraft, *The Prince.* Early in his labors, Machiavelli wrote a celebrated letter describing his daily routine on his farm south of Florence. He visits his wood lot to "kill some time with the cutters, who have always some bad-luck story ready." He hears more gossip at an inn, dines at home with his family, then returns to the inn to play cards and "sink into vulgarity for the whole day" as the players squabble and kibitz, as card players always have. In the evening, a metamorphosis occurs as Niccolò retires to his study, removes his dusty clothing, and dons "garments regal and courtly." Only then does he deem himself ready, through his reading, to "enter the ancient courts of ancient men, where, received by them with affection, I feed on that food which only is mine and which I was born for, where I am not ashamed to speak with them and to ask them the reason for their actions; and they in their kindness answer me." The questions Machiavelli posed to the ancients were the ones he answered in his next project, *Discourses on the First Ten Books of Titus Livy.* How had the Roman republic survived for so many centuries, and done so while expanding its empire? Could lessons from antiquity be applied to the even more daunting task of founding a modern republic?

At home in the mid-1780s, Madison pondered similar questions. Just as Machiavelli gazed on the mountains surrounding Florence and wondered what threats to his city and Italy lay beyond, so Madison could step out the door at Montpelier and know that the destiny of the American Union manifestly lay somewhere beyond the Blue Ridge crisply visible twenty miles west. A union of American republics had been established, its independence recognized. But could these republics preserve the "perpetual union" that the nation's original constitution, the Articles of Confederation, proclaimed? Could they secure the generous Mississippi River boundary that Jay, Franklin, and Adams had gained at Paris? Would an American empire one day extend even farther? If it did, could it also remain a republic—as Rome had finally failed to do?

Like Machiavelli, Madison knew the tedium of rural life. With no ambition to become a planter and little taste for idle amusements, he spent his days immersed in reading. He also knew what it meant to be forced into political retirement. In 1783 forty-two months of constant attendance at Congress ended when the Articles of Confederation made him one of the first casualties of term limits in American history. Once home he quickly gained a seat in the Virginia assembly. But attendance there was not the full-time vocation that membership in Congress had been. Why not undertake a sustained course of historical reading, akin to Machiavelli's?

Years later Madison recalled how "the curiosity" and the frustration "I had felt during my researches" into other confederacies led him to compile the daily notes of debates at the Constitutional Convention of 1787, which students of the Constitution have consulted since their publication in 1840. Like Machiavelli, Madison viewed history as a useful source not only of examples and anecdotes but also of lessons to learn and apply. Unlike the Florentine, he was less certain that the ancients offered the answers he sought. But there was a greater difference. Where Machiavelli aspired only to teach others how to act, Madison seized the opportunity to apply his experience and learning to the transformation of a republic. When the convention met at Philadelphia in May 1787 to draft the new federal Constitution, he was supremely prepared to set its agenda and direct its discussions. Though he lost many key points, the Constitution still bore the imprint of his probing intellect. All our attempts to recover its original meaning begin with his plans for Philadelphia and the lessons he learned there.

That Madison was an active learner of political lessons offers another critical clue to his success. Today Americans often encounter him most directly as the co-author of *The Federalist*, the eighty-five newspaper essays that Madison, Alexander Hamilton (the project's organizer and main author), and John Jay wrote to support the adoption of the Constitution. Deciphering those essays, with their eighteenth-century syntax, can be a demanding task, and students can never quite forget that they are reading Great Thoughts—the founding ideas of American Constitutional Theory. Many scholars treat Madison the same way.

A cottage industry of scholarship is devoted to explaining his key propositions in *Federalist* Nos. 10 and 51: that an extended national republic will cure "the mischiefs of faction"; that "ambition must be made to counteract ambition" for the separation of powers to work.

Yet these essays were not the work of an "ingenious theorist" toiling away in splendid isolation "in his closet or in his imagination" (to use Madison's language in *Federalist* No. 37). They were far more the product of experience and reflections on it. In this sense the comparison to Machiavelli a quarter-millennium earlier or to Locke an exact century previous seems both apt but also potentially misleading. If we read these men only as political philosophers, writing timelessly for the ages, we forget how much their ideas grew out of an impassioned involvement in the controversies of their day. So it was with Madison. He was an actor as well as—indeed, more than—a thinker or writer, and his ideas grew out of his action. Capture him in the act of *thinking*, trying to master the flow of events around him, and his famous texts become more than a justification for a Constitution already written. For the opportunity Madison seized in 1787 had no parallel in the lives of Machiavelli, Hobbes, Locke, Montesquieu, and Rousseau. His publications supporting the Constitution were not his chief political legacy, as *Il principe, I discorsi, Leviathan, Two Treatises of Government, De l'esprit des loix,* and *Le contrat social* were theirs. That greater legacy was the Constitution, whose "father" Madison is said to be. But it is not so much the text itself as the deliberations that made it possible that were the real product of the great project of creative political thinking that Madison conducted in the months preceding the momentous gathering at Philadelphia in the spring of 1787.

If Jefferson, Adams, and Washington were members of the charter generation who made the Revolution, Madison was part of the cohort who came of age with it. Like Hamilton, his ally of the 1780s and opponent of the 1790s, he was one of its "young men," still entering a delayed adulthood in the year of independence. These were leaders the Revolution made, and their extraordinary political creativity owes much to that fact.

Madison was in Philadelphia when copies of the vengeful Boston Port Act reached America in May 1774. After taking his younger brother, William, north to school, he tarried there to visit William

Bradford, a friend from their days at the College of New Jersey. Jemmy (as family and friends called him) first made that trip in 1769, when he was eighteen. By colonial standards he was late to start college. Madison made up for it by completing his degree in two years. Then he stayed on for some months to study Greek and Hebrew. Modern Princeton University claims Madison as its first graduate student and most distinguished alumnus. More likely he was fretting about his health and delayed his return to recover from whatever ailed him.

But in a later age Madison *would* have been a natural candidate for graduate school. His letters to Billey Bradford portray a slightly priggish youth lacking any clear sense of purpose. Madison moped about his rustic isolation and his health, confessing that his "sensations for many months past have intimated to me not to expect a long or healthy life." When Billey sought advice on whether he should practice law, Madison answered thoughtfully but revealed nothing of his own plans. Nor did he feel any need to choose a vocation. As the eldest son of the largest landowner in Orange County, he had no immediate worries about making a living. From his youth he had acquired a moral distaste for slavery and an aversion to the life of a planter. But with no desire to practice law or enter the ministry—obvious choices for someone of his temperament—Madison in his early twenties was aimless.

He was thus the son of privilege, the fifth-generation beneficiary of a family saga that illustrated how upward mobility worked in Virginia. The family had modest origins. John Maddison was a ship's carpenter who invested his wages in the indentured servant trade to the Chesapeake, earning a fifty-acre headright for every servant transported. Over time his holdings grew from six hundred acres in 1653 to thirteen hundred three decades later. Every successive son added to the estate. James Madison's grandfather Ambrose carried the family into the Virginia Piedmont in 1721. His father (James Sr.), Orange County's largest landowner, married Nelly Conway, the seventeen-year-old daughter of a merchant planter, in 1749. James Jr. arrived on March 16, 1751, the first of eleven children whom Nelly bore over the course of a quarter-century.

At eleven Madison was sent to a school kept by a University of Edinburgh graduate in neighboring King and Queen County. After five years there and two more tutored at home by an Anglican minister, Madison went off to Princeton. There he fell under the tutelage of John

Witherspoon, newly arrived from Scotland to ensure that America's leading Presbyterian college upheld orthodox Calvinism. But Witherspoon's real contribution proved to be intellectual, not doctrinal. Under his leadership the College of New Jersey became a major outpost of the eighteenth-century Scottish Enlightenment. Its students read widely in the works of David Hume, Adam Smith, Adam Ferguson, Francis Hutcheson, Thomas Reid, and Lord Kames. It was the best learning available in the English-speaking world, and Madison absorbed much of it at an early age.

The college in the late 1760s was a hotbed of patriot sentiment, which Madison imbibed. There, as at other schools, students gave earnest declamations on patriotic themes such as the nobility of resisting tyranny. Once home, however, Madison's surviving letters offer few hints of any interest in politics. When Bradford relayed news of the Boston Tea Party, Madison primly replied that he hoped "Boston may conduct matters with as much discretion as they seem to do with boldness." A few equally bland sentences exhausted his concern. "But away with Politicks!" he abruptly declared. "Let me address you as a Student & Philosopher & not as a Patriot now." So much for political passion.

Watching Philadelphia mobilize to defy Parliament in 1774 galvanized Madison's sense of political engagement. At home in late June he relapsed into "my customary enjoyments Solitude and Contemplation." The new militant note in his letters took time to find an outlet. In December he joined the local committee of safety formed to implement the Association of the First Continental Congress. The next fall he received a commission as colonel in the county militia. Both appointments owed much to his father's positions as committee chair and militia commander. And it is difficult to imagine Madison, bookish, slight, and soft-spoken, as a candidate for military adventure.

In the spring of 1776, however, he was elected to Virginia's fifth provincial convention, the body that adopted the new constitution and declaration of rights that George Mason largely drafted. Madison was only an apprentice legislator whose ability remained to be judged by his elders. But he did attain one noteworthy victory. Fittingly, it involved freedom of conscience, the one issue that genuinely engaged the young graduate. His best-known letter from these early years denounced the local magistrates ("Imps") who had jailed a covey of itinerant Baptists for preaching without a license. "I have squabbled and scolded abused and ridiculed so long about it," he wrote to Bradford, "that I am without common patience." He envied Bradford his good fortune to live in Pennsylvania, where freedom of conscience had reigned ever since William Penn founded his commonwealth on radical Quaker principles.

As proposed by Mason's constitution committee, the religion article of the declaration of rights affirmed that "all men shou'd enjoy the fullest Toleration in the Exercise of Religion, according to the Dictates of Conscience." That promise did not go far enough, Madison thought. Toleration implied that the state was only permitting its subjects to worship freely, not recognizing their inherent right to believe as they wished, free from public control. Madison accordingly offered an amendment stating that "all men are equally entitled to the free exercise of religion." His change was quickly approved, and Madison notched the first of many legislative victories.

Madison had at last found his vocation, and the next three years marked his apprenticeship. He was elected to the fall 1776 session of the new House of Delegates; he first met Jefferson while serving on the committee on religion. Only one obstacle blocked the pursuit of a political career: the voters of Orange County. Standing for reelection in April 1777, Madison learned that fidelity to republican principles did not ensure political success. Virginia elections had well-established rituals. Candidates not only courted individual voters but treated all present to refreshments, some quite potent. But Madison deemed the "personal solicitation" of voters and "the corrupting influence of spiritous liquors" to be "equally inconsistent with the purity of moral and of republican principles." The local planter and tavern keeper opposing him lacked such scruples. So did the thirsty voters. Once again Madison had time for contemplative reading at Montpelier.

But not for long. Madison's legislative service had left a favorable impression, and the assembly soon appointed him to the council of state, the executive board that formally advised Governor Patrick Henry. Six days a week, Madison faithfully attended its meetings, missing only those "sickly" summer months when Williamsburg's mosquitoes feasted on every mammal in town. The council took the mundane but vital decisions needed to maintain Virginia's contribution to the war effort.

Two years on the council led to Madison's election to the Continental Congress in December 1779. He was just turning twenty-nine when he took his seat in March 1780, his journey delayed by that winter's heavy snows. He found agreeable lodgings in a rooming house at Fifth and Market kept by the widow Mary House. But his real home for the next three and a half years lay a block over, at Fifth on Chestnut, in the State House where Congress had met since 1775.

Madison's arrival at Congress marked the true beginning of the advanced political education that ultimately made him the premier constitutionalist of the era. That education had two major phases: three and a half years as a member of Congress (1780–1783), followed by three terms (1784–1786) representing his Orange County neighbors in the Virginia House of Delegates. Madison's training was essentially complete by January 1786, when he rode the hundred miles home from the fall session at Richmond and began preparing for the interstate convention that he and his fellow Virginia commissioners scheduled for Annapolis in September. This was when he began reading the history of confederacies. But Madison did not undertake this project as a diversion from an unwanted retirement or because he planned to write a treatise on politics. He had more urgent motives for action, and these involved synthesizing his reading with the lessons of rich personal experience, well balanced between national and state governments. That experience was not unique, for many other leaders shuttled between Congress and the states. What did prove distinctive and consequential were the care and insight with which Madison translated his experiences and reflections into a comprehensive critique of "the vices of the political system of the United States."

While serving in Congress, Madison dealt with the usual array of issues and demands that this overworked institution confronted daily. Madison now became a master of legislative deliberation. He served on countless committees, drafted numerous reports and motions, gave carefully prepared speeches, and fashioned key compromises. He began keeping notes of debates, less to preserve an archive for history than to prepare for future discussions by knowing what had been said in past ones. Deliberation for Madison was not a matter of elocutionary flash and rhetorical flourish, the oratorical arts that his era prized and studied, which made reputations for Burke and Fox in the House of Commons and for Patrick Henry and Richard Henry Lee in the Virginia assembly. Madison never tried to perfect those skills. His ideal of deliberation instead meant canvassing alternatives, weighing exigencies, asking what was realistically attainable, forgoing what was not, and being ready to rebut objections and refute fallacies.

The greatest political fallacy that Madison detected during his congressional years was the underlying premise on which the American federal system originally rested. When he first entered Congress, the Articles of Confederation—which Congress had sent to the states in November 1777—still lay a year away from ratification. Madison soon became involved in the final maneuvers needed to secure ratification, encouraging Virginia to cede its western land claims to Congress and thereby persuade "landless" Maryland, the last holdout, to approve the Confederation. But its formal adoption would only confirm, not alter, the basic method of federal governance. Congress ruled (if it ruled at all) by proposing measures to the states, which then determined how best to implement its recommendations and adjust national policies to provincial circumstances. In theory the states were obliged to execute those recommendations. In practice, for reasons good and bad, the states often fell short.

Madison identified the flaw in this system in his first letter to Governor Jefferson, written barely a week after his arrival in Congress. "From a defect of adequate Statesmen," Congress struck him as a body "more likely to fall into wrong measures and of less weight to enforce right ones, recommending plans to the several states for execution and the states separately rejudging the expediency of such plans, whereby the same distrust of concurrent exertions that has dampened the ardor of patriotic individuals must produce the same effect among the States themselves." Given his newcomer status, this was both a presumptuous remark and a precocious insight. Years had passed since Congress was last hailed as "the collected wisdom of America." But this harsh judgment was now the common view. The deeper insight lay in Madison's recognition of the problem of "rejudging." Agreement within Congress was often only a prelude to a second cycle of decision making in the states. Congress had little influence over the state legislatures, and it could hardly prevent them from wondering whether or why they should do more than their backsliding neighbors.

The greatest test of this dilemma during Madison's congressional term came when Congress debated the ambitious revenue plan that Superintendent of Finance Robert Morris presented in July 1782. Morris urged Congress to ask the states to grant it permanent revenues in the form of land, poll, and excise taxes to supplement the duty on foreign imports it had requested in February 1781—and which now seemed doomed to defeat after its rejection by the Rhode Island legislature. Madison supported the Morris program on its merits. But like other delegates, he bridled at the superintendent's heavy-handed efforts to pressure Congress by mobilizing public creditors throughout the states and encouraging unrest among Continental army officers cantoned at Newburgh on the Hudson. Even if Morris prevailed in Congress, there was little chance that similar tactics would be effective with the state legislatures. Following a critical caucus in late February 1783, Madison led a group of delegates who broke with the superintendent and fashioned a compromise set of revenue measures, which Morris, in a sulk, promptly denounced. On April 18, Congress adopted the resulting compromise as a package of amendments to the Confederation that would require the consent of all thirteen state legislatures.

The thirty-two-year-old statesman who rode home to Virginia that fall resembled the returning college graduate of 1771. Now as then, no ordinary occupation beckoned him. But at the spring 1784 elections, the voters of Orange County returned Madison to the House of Delegates. Madison easily settled into a new routine, attending the assembly, spending his spare time at Montpelier, and heading north every summer to avoid the Virginia heat. These annual trips enabled him to supplement James Monroe's regular reports on congressional doings with personal intelligence. They also preserved his deep engagement with national affairs.

One of Madison's main projects during his three terms in Richmond was to persuade the legislature to follow "pro-federal" policies. Initially that meant ratifying the revenue amendments he had guided through Congress in April 1783. A year later, Congress proposed two further amendments to the Articles of Confederation, seeking limited authority to regulate foreign commerce. These were meant to buttress the bargaining position of Adams and Jefferson as they struggled to negotiate commercial treaties in Europe, while enabling Congress to restrict the flow of European goods into American harbors and thus rescue artisans and merchants from the trade depression sweeping the country. Madison also wanted to see Virginia comply with the controversial provisions of the peace treaty asserting the rights of British creditors and loyalists to sue for recovery of pre-war debts and confiscated property. The obstacles the states erected against these two detested provisions gave Britain a pretext for retaining control of its forts at Niagara, Oswego, and Detroit. That in turn threatened the great national objective of developing the northwest territory to which Congress now held title.

Madison's second great project as a legislator was to prod his colleagues to take up the revised legal code that Jefferson and his committee had completed in 1779. At first he did make substantial progress on getting the code enacted. One admirer noted that Madison had "by means perfectly constitutional become almost a dictator upon all subjects that the House have not so far prejudged." Then a reaction set in as the lawmakers balked at his leadership. The entire revision could have been dispatched "with great ease," Madison grumbled, "if the time spent on motions to put it off and other dilatory artifices, had been employed on its merits." Ordinary legislators were simply not up to his high standards of deliberation.

Madison's greatest success came on the issue that he personally valued most: religious liberty. At its fall 1784 session the legislature debated but then delayed action on a bill to levy a tax to support all "teachers [ministers] of the Christian religion," a measure that would allow the state to support all denominations equally, under what we now call "non-preferential" principles. Patrick Henry, the state's most popular politician, strongly supported the assessment. Madison opposed it just as ardently, viewing it as an improper establishment of religion that violated the spirit of the declaration of rights. Once the assembly adjourned, Madison and his allies among Virginia's religious dissenters circulated petitions, eventually gathering over ten thousand signatures—a remarkable show of public opinion by eighteenth-century standards. Madison's personal contribution to this debate was the anonymous *Memorial and Remonstrance Against Religious Assessments*, a petition that demonstrated his intellectual debt to Locke's *Letter Concerning Toleration.* A chastened assembly rejected the general assessment at its fall 1785 meeting. It went on to adopt Jefferson's bill for religious freedom, which Madison then called up. With its passage, Madison wrote to the bill's author, "I flatter myself" that we have "in this Country extinguished for ever the ambitious project of making laws for the human mind." No longer would the private religious convictions of ordinary men and women be subject to the oversight of the state.

Even with that signal success, Madison was increasingly disturbed over the tenor of legislative proceedings and the character of his fellow lawmakers. He first recorded his concerns in a revealing letter to Caleb Wallace, a college friend now living in Kentucky who sought his advice on the constitution its settlers might adopt when their territory finally separated from Virginia. Madison made a number of suggestions, but his opening comments best capture the direction of his thinking. A lack of "*wisdom* and steadiness to legislation," he observed, was "the grievance complained of in all our republics." The best way to correct that deficiency would be to establish a true senate, with the political confidence to check the impulsive acts of the people's immediate representatives in the lower house. "As a further security against fluid & indegested [*sic*] laws," Madison endorsed the example of the New York council of revision, a joint executive-judicial body with a veto over legislation. Or perhaps it would be a good idea to appoint "a standing committee composed of a few select & skilful individuals" to take care that all bills were properly drafted and not written with a haste that led to error and confusion.

This dual concern with the *quality* of legislation and the *character* of legislators identifies another major motif in Madison's intellectual progress during the mid-1780s. The more he saw of legislative politics, the better he understood why a federal system requiring steady cooperation between Congress and the states was prone to failure. Repeatedly Madison had to defend Congress against petty attacks while trying to convey to his parochial colleagues why it was in Virginia's long-term interest to support the Union. But the ills of Virginia politics were not limited to federal matters. "A considerable itch for paper money discovered itself" in the previous session, Madison wrote to Jefferson in January 1786, and he feared that itch could well turn into an inflammation when the assembly next met. Paper currency served a recurring need in the American economy, where hard coin flowed steadily across the Atlantic to pay for imports. But the proposals for paper money and debtor relief that were now appearing had a more sinister intent, Madison thought: to deprive creditors of their rightful property by forcing them to accept payment in the depreciating currency the states were likely to emit. Lawmakers who would approve such measures might earn a different kind of credit with voters, but only at the expense of violating fundamental rights of property.

This concern with property illustrates a final aspect of Madison's thinking in the mid-1780s. There is no question that the populist tenor of politics in the years just after the Revolution left him fearful that demagogic legislators might prove all too willing to enact measures inimical to wealthy families like his own. This is the Madison whom the great Progressive historian Charles A. Beard made the pivotal figure in his *Economic Interpretation of the Constitution*, the Madison who recognized that "the regulation of these varying and interfering interests forms the principal task of modern legislation, and involves the spirit of party and faction in the necessary and ordinary operations of the government." Yet this reactionary fear of the threat to property also converged with his youthful commitment to freedom of conscience to produce one powerful insight about the protection of rights in republican America. These two concerns enabled Madison to perceive a truth that the political theory of the age did not yet properly recognize. In a republic, unlike a monarchy, the problem of rights would not be to guard the people as a whole against the arbitrary power of government, but rather to secure individuals and minorities against the legal authority of popular majorities.

In this and other critical respects, then, Madison was already extremely well prepared for the unprecedented constitutional enterprise he was about to help launch. His experience since 1780 provided him with a deep grasp of the inherent problems of maintaining a federal system that required the decisions of one deliberative institution (Congress) to be "rejudged" and administered by others (the state legislatures). Three terms in the Virginia assembly gave Madison a much richer—but also far more critical—understanding of the defects of the legislative process, including the haste and carelessness with which too many bills were drafted and enacted. Here were potent sources of inspiration for a fresh approach to the great project of constitution making that John Adams and others had celebrated a decade earlier, and which Madison had joined as a young delegate to Virginia's fifth provincial convention. All that was needed was an occasion to turn his hard-earned reflections and musings into an exercise in applied political thinking.

How and when would such occasions arise? Writing new constitutions had become a priority during the crisis of independence. But a decade later there was no equally obvious imperative for action, at least at the *state* level of governance. Nor did the pending efforts to strengthen the Confederation yet require a deep rethinking of the premises of republican government. In 1785 the agenda for *federal* reform was limited to ratification of the revenue and trade amendments Congress had proposed in 1783 and 1784. If these could obtain unanimous state approval, Americans might become receptive to further modifications. Gradually, the existing Confederation might evolve into an effective national government.

That calculus began to shift in 1786. The first key development took place within the Virginia legislature. At the fall 1785 session Madison worked hard to pass a resolution granting Congress general authority to regulate commerce. But the proposal was so weakened in debate that its supporters "chose rather, to do nothing than to adopt it in that form." The substitute measure they salvaged "in this extremity," as the session was ending in January, was to call "a general meeting of Commissrs. from the States to consider and recommend a foederal plan for regulating commerce." John Tyler was its author, and Madison initially doubted it would do any good either, even though he was one of the first three commissioners appointed of a delegation of seven. "The expedient is no doubt liable to objections and will probably miscarry," he wrote to Monroe, his confidant at Congress. Still, "it is better than nothing, and it may possibly lead to better consequences than at first occur."

This proved one of the great understatements of American history, but it took Madison some months to appreciate the opportunity that the assembly's last-minute action had created. Back at Montpelier, he renewed his reading on the history of other confederacies. His notes illustrate how he merged the curiosity of the historian with the analytical perspective of the social scientist, for they combine basic facts about the origins of each confederation with summaries of their key characteristics and a concluding list of the "Vices of the Constitution." In politics as in morality, the point of studying vice is to point the way to virtue, and Madison understood his research project as an effort to distill lessons about the proper design of confederations.

In March the Virginia commissioners set mid-September and Annapolis as the time and place for the projected meeting. Events beyond Virginia soon led Madison to reconsider his qualms about the conference. In March, the New Jersey legislature flatly refused to pay its quota of the federal expenses until its neighbor, New York, ratified the revenue amendments of 1783. The next month New York rejected those amendments, seemingly dooming the compromise Madison had helped to produce three years earlier. Late in May, Secretary of Foreign Affairs Jay asked Congress to revise the instructions under which he was negotiating a commercial treaty with Spain. Rather than insist that Spain open the Mississippi to American navigation, as he had been instructed, Jay suggested that the United States yield this claim for twenty years. Gaining that concession might help northern merchants get the treaty they desired—but only at the grave risk of curbing the expansion of southern-style agriculture into the fertile lands south of the Ohio. Jay's request sparked a sharp sectional split in Congress, and Monroe kept Madison fully informed of the details. He grew even more alarmed when the eight northernmost states approved Jay's request, even though they lacked the ninth vote necessary to ratify any treaty he might conclude. For the first time, Madison feared that the revolutionary Union of thirteen states might actually devolve into two or three regional confederacies.

The strategy of gradually strengthening the Confederation now seemed risky, and that made the Annapolis Convention a more attractive gambit than Madison had first judged it to be. The idea of holding a special convention to consider problems of federal governance was not itself new. There had been some discussion of it among Continental army officers toward the end of the war, and Alexander Hamilton had prepared a resolution to that effect while serving in Congress in 1783. But "pro-federal" politicians like Madison were wary of endorsing an extra-constitutional procedure whose very use might impugn the authority of an already weakened Congress.

By the late spring of 1786, Madison had changed his mind on this point. The Annapolis Convention had been summoned with one specific objective: to find a way to enlarge federal authority over commerce. But now, with a host of new political facts to consider, Madison recalculated the strategy of reform. In principle he still opposed "temporizing or partial remedies." But fearing that Virginia would reject any change originating in Congress, the idea of an extra-constitutional convention became a plausible alternative. "If the present paroxysm of our affairs be totally neglected, our case may become desperate," he wrote to Monroe. When Monroe informed him that Congress was discussing yet another set of amendments, Madison replied that it would be better to await the results from Annapolis. If that "partial experiment" went well, then Congress could build on its success to propose additional measures. But for Congress to take the lead, at a time when its reputation was so low, would be a strategic error.

Madison headed north in late June 1786, well before he needed to be in Annapolis. Fording the Potomac at Harpers Ferry, he observed fifty wraithlike slaves at work in a dense fog, moving boulders in an effort to make the river navigable without the expense of building canal locks. He spent nearly five weeks in Philadelphia, and another three in New York, where he and Monroe discussed their plan to buy land in the Mohawk Valley. No less an expert than Washington thought this was the best spot the two young gentlemen could pick. All they lacked were funds to make the purchase. Perhaps Jefferson could contribute, they hoped, or perhaps he could help them secure a loan from private French investors.

Madison arrived in Annapolis on September 4 and took a room at Mann's Tavern to await the eight delegations that were supposed to attend. From conversations on his travels, he could report to Jefferson that "Many Gentlemen both within & without Congs. wish to make this Meeting subservient to a Plenipotentiary Convention for amending the Confederation." So did he, he confessed, "yet I despair so much of its accomplishment at the present crisis that I do not extend my views beyond a Commercial Reform. To speak the truth *I almost despair even of this.*" He probably despaired even more on September 11, when a dozen commissioners from Virginia, Delaware, New Jersey, Pennsylvania, and New York called the meeting to order. A few others were thought to be en route. Yet even if they appeared, a meeting so thinly attended would look more like a club of concerned citizens than a dignified interstate conference.

Yet a club can cabal whereas a conference might not—and this was not just any club. Those present included John Dickinson; Abraham Clark, a veteran New Jersey leader; Madison's close friend Edmund Randolph; and Alexander Hamilton. Rather than do nothing, Hamilton suggested, the conference should exploit a clause in the New Jersey delegates' credentials, implying that their agenda could go beyond commerce. With editorial help from Madison, Hamilton drafted an address that called upon their states to issue a general invitation to a second convention, "to meet at Philadelphia on the second Monday in May next, to take into consideration the situation of the United States, to devise such further provisions as shall appear to them necessary to render the constitution of the Foederal Government adequate to the exigencies of the Union."

This call for a second convention was either a brilliant gambit or a desperate gamble. Once played, however, it made little difference whether audacity outweighed hope in this stroke of political entrepreneurship. By the fall of 1786, no one who favored national constitutional reform still believed that the prescribed rules for amending the Confederation could ever be made to work. For one thing, Rhode Island was under the control of an "anti-federal" party that was predisposed to veto any change in the Confederation. For another, Congress itself seemed so discredited in the public eye that any new amendment it proposed risked dismissal.

Exactly how to pursue this opportunity was the great question that Madison began considering once the abortive conference adjourned. From Annapolis he returned to Philadelphia. A month later he rode south, joining Monroe and his bride, Elizabeth Kortright, as they traveled to their new home at Fredricksburg. (The only good news Americans heard about Congress in 1786 was that its members were busy wedding the daughters of New York's merchant elite.) Elizabeth was a dark-haired, eighteen-year-old, very pregnant beauty, and the journey must have been full of banter on subjects other than politics. (Much later James and Dolley Madison came to dislike her company and the airs she took on during her husband's diplomatic tours in Europe.)

The banter must have ceased when the party spent two nights visiting Washington at Mount Vernon. Madison was intent on enlisting the general in this new campaign. There was no question about his sympathies. Washington was the nation's most fervent nationalist. His June 1783 circular address to the state governments remained the strongest statement of national principles any American had produced. But when he resigned his commission six months later, he also declared his intention to return forever to Mount Vernon and the private pursuits that eight years of command had forced him to neglect. To go back on that promise, even for a cause so noble, risked sullying the public character the American Cincinnatus had worked so hard to earn.

From Mount Vernon, Madison rode straight to Richmond for the fall session of the legislature. He quickly moved a bill to appoint delegates to the proposed convention and invite the other states to do so as well. It passed unanimously. The next step was to appoint delegates. "It has been thought advisable to give this subject a very solemn dress, and all the weight which could be derived from a single state," Madison wrote to Washington. "You will infer our earnestness on this point from the liberty which will be used of placing your name at the head of them." Madison had not raised that point at Mount Vernon, and Washington could not possibly reply in time to ward off election. His only consolation was that he could later decide whether to attend, as circumstances dictated. The rest of the delegation was nearly as distinguished, including George Wythe, George Mason, Governor Edmund Randolph, and (of course) James Madison.

Washington's initial reply was discouraging. He had already told the Society of the Cincinnati—the fraternity of Continental army officers founded in 1784, amid great controversy over its "aristocratic" pretensions—that he would not attend their annual May meeting in Philadelphia. How could he attend this other convention, Washington fretted, "without giving offence" to his fellow officers? That was not a fatal objection. The society would hardly protest if he now chose to honor them with his presence. Still, the general needed to be courted as well as consulted. Madison had occasion to visit Washington again when the Virginia legislature sent him back to Congress (his three years of ineligibility having expired). Passing Princeton, he ran straight into a nor'easter blizzard and was soon crossing rivers "clogged with Ice, and a half congealed mixture of snow & water which was more in the way than the Ice itself."

Madison was in no hurry to get to Congress, which could barely muster a quorum and had little business to transact—with one exception. On February 21, 1787, nine days after his arrival, it voted its approval of the coming convention. Obtaining this endorsement removed a critical obstacle to the convention. By now seven states had agreed to attend. But among the six yet to act there were mixed opinions as to whether Congress could properly approve a meeting unknown to the Confederation from which its own authority derived. Once Congress acted, the other states fell into line, save Rhode Island. But its lone absence would prove a blessing. A state that would not even send a delegation to Philadelphia was unlikely to approve anything done there, and that provided a further incentive to abandon the rule of unanimous state ratification that had thwarted the previous efforts to amend the Articles of Confederation.

Promising as these developments were, they hardly guaranteed that the convention would be able to agree on a plan of reform. Perhaps the most to be expected "from the Measure," John Jay wrote to John Adams, was "that it will tend to approximate the public Mind to the Changes which ought to take place." Madison hovered between cautious optimism and prudent skepticism. The prospects for the convention remained "among the other arcana of futurity," he wrote in late February, "and almost as inscrutable as any of them." As late as April 15 he still worried whether Washington should "postpone his actual attendance, until some judgment can be made of the result of the meeting." Yet undeniably an opportunity now existed to rethink the basic premises of the Union, to imagine it as something more than a loose federation of sovereign states. But what form might its reconstitution take? That was the question Madison set out to answer in the late winter and early spring of 1787, in an exercise in applied political thinking for which the history he had studied offered neither precedents nor guidance.

Madison recorded the progress of his thinking in three letters to trusted Virginia correspondents (Jefferson, Randolph, and Washington) and a fourteen-page memorandum, "The Vices of the Political System of the United States," which he originally wrote solely for his own use. The letters laid out his agenda for Philadelphia, but the real intellectual breakthroughs appear in the memorandum, which reveals its provisional nature in the ample space for further jotting that Madison left between its twelve subheadings. Scholars rightly read "The Vices" as a first draft of the famous *Federalist* No. 10. But it is also much more: the visible evidence of how Madison thought his way through a complicated set of problems to fashion a diagnosis of the political evils besetting the republic. No other document in American history is quite like it.

To grasp the significance of these texts, it is essential to recall that the agenda of federal constitutional reform *before* 1787 had been defined solely in terms of adding a limited number of additional powers—primarily over revenue and trade—to those that the Confederation already vested in Congress. There was little discussion of changing the structure of national government or empowering Congress to enact laws binding the American people as individuals. The existing unicameral Congress would remain "a diplomatic assembly"—as John Adams described it in his *Defence of the Constitutions of the United States*, published in London just as the delegates were preparing their journeys to Philadelphia—in which each state cast a single vote (even though its members voted as individuals, not unified delegations).

Madison's analysis of the vices of the political system radically transformed this agenda in several ways:

First, rather than dwell on the familiar weaknesses of Congress and the formal limitations of the Confederation, Madison directed the brunt of his criticism against the failings of the states for their lapses in political judgment and pursuit of short-sighted policies. Asking what was wrong with Congress, he implied, was the wrong question. The real issue should be what was the matter with the states.

Second, in a single brilliant passage (item 7 in the memorandum), Madison devastated the foundational principle of the Confederation: the idea that the states could be trusted to implement national decisions in good faith while pragmatically adjusting congressional directives to local circumstances. Any federal system that relied on the states' voluntary compliance with national measures was bound to fail, he concluded. National government had to operate not by recommendations or requisitions addressed to the states but by *laws* binding individuals.

Third, the convention should not limit its agenda to the manifest problems of national government under the Confederation. It also needed to deal with "those which are found within the states individually," in particular with the "evils" arising from the "multiplicity," the "mutability," and worst, the "injustice of the laws of [the] States."

From this diagnosis of the vices afflicting the *states*, Madison then derived the broad principles that guided his constitutional project for the *nation* and ultimately shaped the agenda of the convention.

First, and perhaps most idealistically, ways had to be found to improve the quality of deliberation and decision making at the national level of government, so that its quest to identify and pursue the public good would reduce the play of interest, opinion, and passion that seemed so prevalent within the states. The deep goal of constitution making was not simply to assign powers and duties to institutions. It was also to foster the best deliberation possible. That required insulating the people's elected representatives from the erratic currents of popular feeling that seemed to surge and swirl all too forcefully within the states. It also meant protecting the weaker branches of the executive and judiciary against the "impetuous vortex" of legislative domination.

Second, and arguably most important for the agenda of the convention, if the federal Union was to operate by real laws and not by mere recommendations, it had to be reconstituted as a government in the normal sense of the term, with the same bicameral legislature and independent executive and judicial departments that sound republicans expected any well-balanced regime to possess.

Third, and potentially most radically, if the convention also presented an opportunity to correct the injustice of state legislation, then the national government had to enjoy some power to oversee the state legislatures. The most provocative proposal that Madison carried with him to Philadelphia thus became the idea that it was "absolutely necessary" to arm the national government with "a negative *in all cases whatsoever* on the legislative acts of the States, as heretofore exercised by the Kingly prerogative" to veto provincial laws. The italicized phrase (emphasis Madison's) exactly echoed the detested parliamentary Declaratory Act of 1766, and thus marked the radical—or reactionary—turn of Madison's mind.

The reasoning supporting this drastic measure rested on the same argument that he would publicize eight months later in *Federalist* No. 10. "Contrary to the prevailing Theory," Madison observed, in a phrase that illustrated his sense of discovery, smaller republics were more likely to enact unjust laws than larger ones were. Why? Because the constitution writers of 1776, and advocates of republican government more generally, had erred in thinking that the best security for a republic lay in the civic virtue of its citizens. That expectation was wholly naive, for most citizens acted on opinions and passions dictated by their private interests and fallible judgments. "Whenever therefore an apparent interest or common passion unites a majority," Madison asked, "what is to restrain them from unjust violations of the rights and interests of the minority, or of individuals?" The short answer was nothing—not a "prudent regard" for the public good, not "respect for character," not even religion. A better solution, he hypothesized, was "an enlargement of the sphere"—a larger republic that would embrace a greater array of interests, in which the "requisite combinations" would prove far more difficult to form than in the smaller compass of the states. This extended national republic would be entrusted with the authority to veto the unjust laws that state-based majorities would still enact.

Accounting for the public vices of the political system, then, required taking a healthy but realistic measure of the private vices of Americans. Unlike the constitution makers of 1776, Madison no longer presumed that his countrymen possessed heroic stocks of civic virtue. Yet his hypothesis about the advantages of large republics also reflected his particular commitments and experiences. The sweeping critique of state legislatures and legislation clearly drew upon his disillusionment with the petty ambitions of his fellow Virginia assemblymen. His belief that the real problem of rights was not to shield the people as a whole against arbitrary acts of government, but instead to protect minorities and even individuals from the democratic acts of popular majorities, was rooted in his early impassioned commitment to freedom of conscience: the one right that placed the greatest value on an individual's sovereign capacity to think and act for herself. That commitment was further strengthened by Madison's abhorrence of the paper money and debtor relief legislation that so many states seemed poised to enact in the mid-1780s. If popular majorities of small farmers, tenants, and artisans thought they would gain by such measures, no moral concern or political obligation could restrain their unjust conduct. What was to stop such self-interested majorities from eventually redistributing property in other ways?

These are the famous concerns and propositions that students in American history and government courses are expected to master. Madison certainly thought they marked a major shift in the received political wisdom of his age. He was obviously proud of the challenge he was posing to the orthodox views of leading writers such as "the celebrated Montesquieu," the author most associated with the idea that a stable republic could safely operate only in small homogeneous societies, where citizens' personal knowledge of their common interests would promote the virtue that Madison now deemed unattainable. And the proposed national veto on state laws was a critical feature of his larger plan of reform: "the least possible encroachment on the State jurisdictions" the convention could make.

Yet for all the attention these elements of Madison's program have attracted, the greater transformation in the agenda he was preparing for Philadelphia stemmed from his critique of the dynamics of federalism. Perhaps no other passage in his six decades of political observations better illustrates his distinctive qualities of mind than the single-paragraph seventh item of "The Vices," which he subtitled "Want of sanction to the laws, and of coercion in the Government of the Confederacy." Madison packed more analytical force into this paragraph than most writers capture in whole treatises.

Madison opened his analysis with two simple historical observations. First, "the compilers" of the Articles of Confederation were not to be faulted for failing to give Congress the power to compel or coerce the states to do their federal duty. The dominant republican assumptions of the mid-1770s imbued them with "a mistaken confidence that the justice, the good faith, the honor, the sound policy" of the state legislatures would be enough. They shared "the enthusiastic virtue" that was the reigning value of the mid-1770s and that "the inexperience of the crisis" only reinforced.

Thus far Madison was reasoning like an academic historian, asking, Why did people in the past act as they did? His second observation was less academic but still historical: What have we learned since the Articles were written? A simple lesson: "Even during the war, when external danger supplied in some degree the defect of legal & coercive sanctions, how imperfectly did the States fulfil their obligations to the Union?" Since the peace, compliance had naturally grown even feebler.

From this point his thinking shifted in a very different direction, away from lessons of the past and toward a line of reasoning that looks very much like modern game theory, which studies how strategic choices affect a wide array of human actions. Now Madison considered the Confederation not as a historian asking what went wrong, but rather as a social scientist probing its basic structure. This analysis viewed American federalism as a continuing cooperative game in which different players—the state legislatures—would repeatedly ask whether it was in their interest to enforce or shirk their federal duties.

The starting point for this analysis was the deceptively simple question Madison posed after noting how badly the states had performed since 1783: "How indeed could it be otherwise?" In rapid succession he delivered three answers to this question:

> In the first place, Every general act of the Union must necessarily bear unequally hard on some particular member or members of it. Secondly the partiality of the members to their own interests and rights, a partiality which will be fostered by the Courtiers of popularity, will naturally exaggerate the inequality where it exists, and even suspect it where it has no existence. Thirdly a distrust of the voluntary compliance of each other may prevent the compliance of any, although it should be the latent disposition of all. Here are causes & pretexts which will never fail to render federal measures abortive.

Translated into the language of game theory, Madison's three answers could be restated as follows. A system of federalism requiring the voluntary compliance of the states with national measures will never work,

> first, because states have different interests, and thus different incentives to execute, support, or shirk any particular measure; second, because some set of politicians in every state will always have personal incentives to oppose national measures in order to increase their own influence; and third, because even where the states do recognize common interests, doubts about each other's likelihood of complying will discourage every one from stepping forward to do the right thing (why should any state go first if it doubts whether others will act at all?).

Perhaps these conditions would not operate with equal force on every occasion. But they would do so often enough to foster a crippling attitude of "distrust" that could only worsen over time. The federal game Americans were currently playing would *always* expose the Union to the whims and interests of the member states.

It was this capacity to think like a historian and predict like a social scientist that led Madison to the radical conclusion that the Articles of Confederation had to be abandoned, not amended. The Union would have to be reconstituted on a completely different principle: as a government that would act by law, and not on states but on citizens. To do that it would have to be empowered to enact, execute, and adjudicate its own laws, and that meant replacing the single constitutional institution of a unicameral Continental Congress with the three independent departments that any well-constructed republican government should possess. To create such a government, the convention would have to consider all the political and constitutional lessons Americans had been compiling since 1776—the same lessons that Madison had elaborated on in his letter to Caleb Wallace.

So Madison spent the early spring of 1787 reviewing his notes on ancient and modern confederacies, reflecting on his experience in state and national government, and doing political theory in the most creative sense of the term—all with the calculated intention of setting the agenda he believed the convention would need to follow. That agenda included one other critical calculation, which Madison described to his three Virginia correspondents as "the ground work" for everything else: "that a change be made in the principle of representation." Here too the Articles had to be abandoned, not modified. The equal state vote, which had been the rule since 1774, would give way to a rule of proportional representation. It would certainly apply to the lower house of the new legislature, and probably to the upper house as well—though Madison was toying with the notion that this future senate might not really be a representative body at all, but just a deliberative chamber where the nation's political wisdom could be concentrated.

Madison saw this change in the principle of representation as a matter of simple justice. Representation was about citizens, not governments, and thus it was important to design a national government that would act directly on the people, not through the states. As Madison calculated the chances of getting the convention to take this radical step, he thought he saw a way to prevail over the predicted objections of the smaller states. The change should appeal to the northern states because of "their present populousness; to the Southern by their expected advantage" in the future. As for the less populous states, they "must in every event yield to the predominant will." But the most important calculation was that the larger states would never make "the necessary concessions of power" unless this fundamental change took effect.

So Madison spent the final weeks before the convention simultaneously doing political theory as an applied art and making political calculations grounded in his theory. Just turned thirty-six, he embraced politics as his true vocation, though he and Monroe still had plans for becoming New York landlords. Madison did not share Jefferson's yearnings for private happiness nor the goad of vanity that gnawed at John Adams. Perhaps the convention would have taken a different course had either or both of those luminaries been available to attend it. In their European absence, Madison was free to take the lead. A decade of public service had taught him the advantages of seizing the initiative that less industrious colleagues would be happy to yield, and distinctive faculties of mind left him uniquely qualified to frame an unprecedented debate.

As at Annapolis, Madison was first on the scene. He left New York on May 2, 1787, and reached Philadelphia on the fifth. Nine days later he was happy to hear the city bells chime in honor of Washington's arrival. Washington had kept busy at Mount Vernon until the last moment, supervising spring planting and instructing his trusted nephew George Augustine Washington what to do in his absence. The ride north was marred by "a violent hd. ach & sick stomach" south of Baltimore and a choppy crossing of the Susquehanna. At Grays Ferry on the Schuylkill he traversed the same bridge his men had used en route to Yorktown in 1781. Washington meant to take rooms at the boardinghouse where Madison was staying. But as soon as he "alighted" there, he was "warmly and kindly pressed by Mr. & Mrs. Rob. Morris to lodge with them I did so and had my baggage removed thither"—three doors down.

The general arrived just as the convention was scheduled to begin—except that Virginia and Pennsylvania were the only states whose delegates were present on the appointed day, May 14. If America's greatest man could be punctual, Madison must have wondered, could not others be so as well? The general was not amused. "These delays greatly impede public measures," he grumbled on May 20, while they were still awaiting a quorum, "and serve to sour the temper of the punctual members, who do not like to idle away their time." In one respect, though, delay proved a boon to Madison. It freed the Virginians to meet "two or three hours a day, in order to form a proper correspondence of sentiments." Their meetings produced the eleven-article Virginia Plan that Governor Randolph, as titular head of the delegation, introduced on Tuesday, May 29, four days after a quorum of seven states finally appeared. Had the other states been punctual, the convention might well have meandered into discussing general goals and principles or naming a committee to shape an agenda before finally getting down to deliberation.

Instead, Madison's spring labors provided the basis for the Virginia Plan, and that plan in turn enabled the delegates to get to work right away. Once Randolph spoke, they realized that their task was not limited to amending the Articles of Confederation. True, the first resolution benignly stated that the Articles "ought to be so corrected & enlarged as to accomplish the objects proposed by their institution." But as Randolph went on, the delegates sensed that "corrected & enlarged" really meant "altered and transformed." Succeeding articles proposed creating a bicameral legislature, with a lower house elected by the people and an upper house elected by the lower. Beyond the powers the Continental Congress already enjoyed, this legislature could "legislate in all cases to which the separate States are incompetent, or in which the harmony of the United States may be interrupted by the exercise of individual Legislation," as well as "negative all laws passed by the several States, contravening in the opinion of the National Legislature the articles of Union." There would be a national executive and judiciary, and an executive-judicial council of revision with a limited negative over acts of the legislature.

For the next fortnight the Virginia Plan provided the framework for discussion. The convention first sat as a committee of the whole, sparing Washington from presiding over the initial debates. (Nathaniel Gorham took the chair instead, honoring Massachusetts as the second-oldest province.) The convention met in the same main-floor assembly room of the State House where Congress had sat from 1775 to 1783. Like Congress, its deliberations were secret. Windows and doors were shut, and the air grew stale as the hours passed. But the weather that summer was moderate, not exceptionally hot, and the delegates were used to the humidity. Most of the fifty-five who attended sat silently, day after day, showing no ambition to be memorialized in the notes they saw Madison compiling from his front-and-center seat. The active speakers numbered no more than fifteen, but the engaged listeners embraced the entire company. There were moments when attention wandered, eyelids drooped, and perhaps a few doodles scratched their way across parchments lost to the archives. Then there was Luther Martin's numbing defense of states' rights on June 27–28, "delivered with much diffuseness & considerable vehemence," Madison noted, a polite way of implying that his fellow Princeton alumnus had tippled a bit too much to settle his nerves. Over the course of three and a half months, however, the delegates sustained a seriousness that amply justified the "profound & solemn conviction" Madison voiced years later: "that there never was an assembly of men, charged with a great & arduous trust, who were more pure in their motives, or more exclusively or anxiously devoted to the object committed to them."

That was the judgment of decades, as it remains that of posterity. But at the outset the delegates had to feel their way into their "trust." The first time the executive branch of government came up, for example, opening comments by Charles Pinckney and James Wilson gave way to prolonged silence until Benjamin Franklin and John Rutledge gently chided others into speaking. On one critical point, however, the basic choice confronting them was quickly evident, and it drove the politics of the first seven weeks of deliberation. Madison's strategy rested on the belief that the powers to be given the new government depended on a satisfactory solution to the issue of representation, and that meant that *both* houses of the national legislature should abandon the one-state, one-vote rule of the Confederation. Delegates from the smaller states immediately countered that if the new government was to be so powerful, their constituents would be mad to yield the point Madison insisted they must surrender. Without an equal state vote in at least one house, they argued, the vital interests of the small states would never be safe.

How could that concern be allayed? Madison first laid out his strategy while Virginia and Pennsylvania were waiting for the other delegations to appear. In joint discussions, Gouverneur Morris suggested "that the large states should unite in firmly refusing to the small States an equal vote" even in the convention. Worried "that such an attempt might beget fatal altercations," the Virginians countered that it would be better to persuade the small states to relinquish their rights "for the sake of an effective Government." But rather than insist that the small states simply had to sacrifice for the greater good, Madison believed he could convince them that they did not truly need the equal state vote.

The argument Madison intended to use flowed directly from his general ideas about faction. This application of the broader argument remains one of the least understood aspects of his theory. When he first jotted down his observations about the sources of faction in his memorandum on "The Vices," his target was the settled belief that the stability of a republic depended on the virtue of its citizens. But virtue could be trusted only so far. The more homogeneous a society, the easier it should be for its citizens to perceive their common interests and thus to act virtuously. But the experience of the past decade proved that republics could not rely on virtue alone. Men had to be taken as they were, as creatures of interest, passion, and fallible opinion. Replacing this naive image of virtue with the superior sociology of faction would enable republican realists to wed their commitment to popular rule to a more accurate account of what modern societies were really like.

This improved sociology, however, also affected how Madison thought about the interests of the member states of the Union. Simply put, those interests were not a function of the *size* of a state in which a citizen happened to live. Neither a voter nor his representative would ever again ask, What's good for the citizens of the small or large states? Instead they would ask, What's good for the merchants or farmers of our community, for fishermen or artisans, creditors or debtors, mainstream Protestants or the dissenting Baptists a county over, industrious German farmers or hard-drinking Scots-Irish backwoodsmen?—all the diverse forms of economic interest or social identity that would account for the *individual* sources of political faction. The size of a state would never determine the real interests of its citizens—with one ironic exception: when voting on rules of voting. Small states had a stake in retaining the equal state rule of the Confederation, large states in adopting the different notion of equality we call "one person, one vote." But once such a rule was adopted, representatives would never again act on the basis of the size of their state. At most size could be a crude marker of diversity, but even then, the social sources of that diversity were what truly mattered.

Madison's great goal, then, was to convince the small states that they neither deserved nor needed an equal vote in either house of the legislature. He laid down his challenge on May 30 and held to it until the decision of July 16. "Whatever reason might have existed for the equality of suffrage when the Union was a federal one among sovereign States must cease," he flatly declared, when the government was to operate directly on the people. That was a simple matter of justice, for otherwise the basic equality of citizens would be denied. But there was a political argument to be made as well. The small states need not fear that populous Pennsylvania, Virginia, and Massachusetts would gang up against them, because no common set of interests united them. "In point of manners, Religion and the other circumstances, which sometimes beget affection between different communities, they were not more assimilated than the other States," Madison observed. More to the point, their economic interests were fundamentally different. In Massachusetts the dominant interest was fish, in Pennsylvania flour, and in Virginia tobacco.

On its merits this should have been a winning argument, and Madison and his allies—Hamilton, James Wilson, and Rufus King—hammered away at their opponents: Roger Sherman and Oliver Ellsworth of Connecticut, William Paterson of New Jersey (another Princeton man), and John Dickinson, now representing Delaware. Madison was soon disabused of the idea that he could prevail by appeals to reason, justice, and necessity. The small-state champions held their ground—or rather they shifted their lines of defense with little regard for consistency. Sometimes they argued that the integrity of their states would be lost if the equal vote was not retained. Sometimes they suggested that the state *governments* deserved protection against federal encroachments, and since those governments were equal in stature, they deserved an equal vote in one house at least. There was a simpler argument still. The equal state vote had been the rule since 1774, and the small states should not be forced to give it up, especially when the larger states seemed intent on increasing their own power. If this was unjust, well, the Confederation was meant to be perpetual, not perfect.

In mid-June the small states countered with the New Jersey Plan, drafted by Paterson. It too proposed a new structure of three independent departments. But Congress would remain a single chamber in which each state cast one vote, and its measures would remain resolutions instructing states, not laws binding citizens. This was a fatal weakness, and one reason why the delegates barely discussed the New Jersey Plan on its merits. Its true purpose was to demonstrate that the small states meant to preserve their equal vote in one house. Lest the point be missed, Dickinson took Madison aside right after Paterson spoke. "You see the consequences of pushing things too far," he warned. Many small-state delegates favored Madison's reforms, "but we would sooner submit to a foreign power" than "be deprived of an equality of suffrage, in both branches of the legislature." Madison ignored the warning. On June 19 he opened the debate by arguing that the New Jersey Plan would not "provide a Governmt. that will remedy the evils felt by the States both in their united and individual Capacities." Most delegates concurred, and when Madison finished, they handily rejected the New Jersey Plan, seven states to three. Given a choice between Madison's sweeping agenda and Paterson's modest one, the convention clearly agreed that the project of reform must remain comprehensive.

The field was now cleared for impasse. As Madison observed on June 19, "the great difficulty lies in the affair of representation, and if this could be adjusted, all others would be surmountable." But *adjustment* still meant that one side had to prevail, the other yield. As the convention entered its second month, Madison still hoped his arguments would wear down the opposition. Yet no matter how hard they were pressed to explain how the large states could ever unite against them, the small-state spokesmen insisted their constituents would face real danger if they were denied an equal vote in one house. "Altho' no particular abuses could be foreseen by him," Ellsworth grudgingly admitted on June 30, "the possibility of them would be sufficient to alarm him." Madison thought he was dealing in probabilities based on the real diversity of interests among the largest states. Ellsworth favored a lower standard of proof: it was enough that "the danger of combinations among them is not imaginary."

To answer such claims, Madison resorted to one further argument. Let us admit that one purpose of representation was to secure "every peculiar interest whether in any class of citizens, or any description of States," as much as possible, he observed on June 30. "Wherever there is danger of attack there ought to be given a constitutional power of defence." But because the mere size of a state did not create a real or permanent interest, the small states had no reason to fear the large states simply because they were more populous. There were, however, other ways to distinguish the states that did identify lasting divergent interests. These differences, Madison continued, "resulted partly from climate, but principally from the effects of their having or not having slaves." The true "division of interest in the U. States" thus "lay between the Northern and Southern" states.

This was a dangerous gambit, Madison admitted, because it risked multiplying the number of conflicts the convention had to resolve.

Worse still, this conflict between two great regions did not resemble the milling array of interests that Madison evoked when he explained why large republics would be more resistant to factious divisions than smaller ones. Instead, it identified one overarching division that tracked a dangerously neat geographic line. Yet bringing the question of slavery into play did have a potential intellectual advantage. It confirmed that the supposed conflict between large and small states was only a passing threat to the politics of the convention, not a permanent one threatening the stability of the Union.

The appeal proved unavailing. On July 2 the convention deadlocked on Ellsworth's motion to give each state an equal vote in the upper house. Over Madison's objection it then appointed a "grand committee" to explore a compromise. When such committees were chosen, the whole convention elected a member from each state, and its choice indicated that the advantage was tilting away from Madison. Paterson, Ellsworth, and Luther Martin were named for the smaller states, whereas the large states would be represented by the relatively conciliatory trio of Benjamin Franklin, Elbridge Gerry, and George Mason.

The committee met on July 3, joined their colleagues in celebrating the eleventh anniversary of independence on the Fourth, and reported the next day. As Madison had warned, the committee could discover no magical solution that could not also be proposed to the convention as a whole. But once appointed, its members felt obliged to propose something, especially with Franklin, the world's greatest fount of practical advice, encouraging action. Accordingly it produced two resolutions. One would allocate seats in the lower house on the basis of population and require all appropriations of public funds to "originate" there, and not be subject to amendment in the upper house. The second resolution simply stated that "each State shall have an equal vote" in the upper chamber. Delegates from the smaller states were quick to label this a "compromise," but Madison and his allies were not fooled. The limitation on the upper house's power to *amend* money bills meant nothing, Madison noted, so long as it could simply *reject* them. In exchange for this meaningless legislative privilege, the large states were being asked to accept the equal state vote in the upper house. For Madison this was defeat, not compromise.

On this question the impasse remained intact and there was little left to say. Instead the convention turned to representation in the *lower* house. Here it faced the troubling issue Madison had raised when he used the lasting division between slave and free states to trump the passing dispute between small and large ones. The delegates agreed that the lower house should be "the most exact transcript of the whole Society." The sentiment that John Adams first voiced in 1776—that such an assembly should be a "miniature" of society—was now a commonplace of American thinking. But where in that miniature could one place the hundreds of thousands of slaves who provided the labor for the south's plantation economy? If they were solely chattel property, lacking rights of any kind, they were not really part of society at all—however much southern society depended on their toil from dawn to dusk.

So most northern delegates reasoned. In its origins, they argued, representation was only a means of collecting the sentiments of the community when it had grown too numerous to deliberate as a body of citizens. If slaves were never part of that community, why count them to determine the allocation of representation? Southern delegates did not see the issue that simply. When an elected assembly legislated, they responded, it acted on property in all its forms. That peculiar form of property called slaves grew the commodities on which much of the American economy depended. In that sense they *were* part of society, indeed a unique part deserving special recognition precisely because of their heavy concentration in one region. Leave that property unrecognized, apportion representation only on the basis of a state's free population, and southern interests would be dangerously impaired.

The issue came to a head after July 5 when the convention successively appointed two committees to set the initial distribution of seats in the lower house. Disappointed by their reports, southern delegates pressed the committees to justify their estimates, then grew more apprehensive with the vague answers they received. They were also troubled when Gouverneur Morris and Nathaniel Gorham proposed that the existing seaboard states should unite to guarantee themselves a permanent majority of seats, thereby preventing the balance of political power from shifting westward as the population surged across the Appalachians and into the interior of the continent. That might make sense if the states from New Hampshire to Georgia shared one common interest. But Madison had already explained why they did not. Expecting future migration to narrow the population gap between north and south, southern delegates insisted on a regular reapportionment of seats among the states. Rather than leave the basis for this reapportionment to future legislatures, as Morris suggested, they believed a rule should also be locked into the text of the new articles of union.

Here was forged the convention's first (and arguably more genuine) compromise over representation. To keep the mirror of representation brightly polished, there would be a regular census of the population. To prevent the populous north from perpetuating its current advantage, a constitutional rule for reapportionment had to be fixed now. The existence of slavery would be recognized through a clever formula apportioning representation and direct taxation among the states on the basis of population, with slaves (obliquely called "all other Persons") valued as three fifths of "the whole Number of free Persons." Never inclined to moderation in the defense of slavery, the South Carolina delegates held out for counting slaves equally. But the Virginia delegates conceded that three fifths—the ratio Madison first proposed in 1783 as a formula for dividing national expenses among the states—would do.

To our way of thinking this fateful compromise was both a moral blot on the Constitution and a crude index of the perceived racial inferiority of African Americans. To northern delegates at Philadelphia it was merely a political bargain, the price of placating a minority region. By accepting slavery as a dominant institution of southern society, this bargain confirmed that Madison was right to make its presence or absence the true fault line of the Union. But sealing this bargain carried a cost. If the interests of one minority of states (the south) could be protected through a deal over representation, why not those of another (the smaller states)?

That was the challenge Madison still faced as the convention returned to the issue of the upper house. True, the interests of citizens of small states did not differ in kind from those of their countrymen elsewhere. Pick three farmers on the eastern shore of the Chesapeake, let each raise tobacco and wheat and take communion in the Methodist church that would soon reap souls by the thousands across the Delmarva Peninsula. Would it matter that one lived in populous Virginia, the second in moderate-size Maryland, and the third in minuscule Delaware? But the smaller states all had a century and more of history on their side and, because they were smaller, an acute sense of their communal integrity. Perhaps they could not refute Madison's logic. But neither would they allow his reasoning to govern their political identity.

After one last round of debate and a Sabbath for contemplation, the convention took its decisive vote on Monday morning, July 16. Few minds had changed, and the basic division of states stood fast—with one exception. Elbridge Gerry and Caleb Strong of populous Massachusetts broke with their colleagues King and Gorham, splitting their delegation and allowing the equal state vote in the upper house to carry by the narrowest margin possible, five states to four, with one divided. Had either Gerry or Strong stood fast, the deadlock would have remained.

A revealing interlude followed. The convention moved on to the next item on the agenda, the key article describing the legislative powers of the Union. Madison's strategy presumed that a decision on this point would depend on a satisfactory solution to the representation question. Everything the Virginia Plan proposed, Randolph noted, was "founded on the supposition that a Proportional representation was to prevail in both houses of the Legislature." Having lost that point, its supporters were unsure how to proceed. Perhaps the convention should adjourn, Randolph suggested. The large states could "consider the steps proper to be taken in the present solemn crisis of the business," while the small states discussed "the means of conciliation." This remark drew a strident response from William Paterson. Indeed, "it was high time for the Convention to adjourn," he exclaimed, revoke "the rule of secrecy," and let their constituents know what was going on. The small states were not to be intimidated into yielding the point they had just gained. Randolph immediately complained that "his meaning had been so readily & strangely misinterpreted." He did not want an adjournment sine die, only overnight. The point clarified, Paterson seconded his motion, and tempers cooled.

What the delegates said to one another that afternoon and evening, at dinner or in their lodgings or strolling along the city's wharves, is lost to history. The conversations must have been intense. Both blocs had spent six weeks arguing and bluffing, threatening and cajoling, presuming and not merely pretending that everything hinged on one decision. Now, his bluff called, Madison had to concede that one of his main goals had fallen victim to two mercurial Massachusetts men voting against their state's interests.

The next morning the large state delegates caucused to weigh their options. Several members from the small states sat in, to learn that their opponents had no fallback strategy to conceal. "The time was wasted in vague conversation," Madison noted, "without any specific proposition or agreement." A few delegates wanted to fight on, even to consider reconvening as a rump body representing a majority of the American people. But others "seemed inclined to yield to the smaller States," even if that meant recommending an "imperfect & exceptionable" constitution that might enjoy the support of a majority of states representing only a minority of the people. Nor did they even agree about the "importance" of the central point in dispute: that representation could only be about the people themselves, not the legal entities of the states.

Here lay the basis for the "compromise" the convention presented to those same people two months later. For the moment Madison rightly recognized it as a defeat, not only for his agenda but also for basic principles of political equality and majority rule. He could never share the reaction this decision soon drew from Jefferson, who claimed to be "captivated by the compromise of the opposite claims of the great & little states." When the time came in *The Federalist* to discuss the equal state vote, Madison dismissed it with the faintest praise possible as "a lesser evil."

This great defeat was immediately followed by another. When the convention met on July 17, it spent a few minutes fencing over the article defining the powers of the legislature. Then it turned to another of Madison's key ideas: the national veto on state laws. Though Madison had not persuaded the Virginia delegation that this draconian power should operate "in all cases whatsoever," it still covered "all laws passed by the several States contravening in the opinion of the National Legislature the articles of Union." Dispirited as he felt, Madison still defended his proposal as "the most mild & certain means of preserving the harmony of the system" and one vestige of British rule Americans could well adopt. But the vote to represent the states equally was a reminder of their residual status as the original members of the Union. However badly their legislatures had acted, did they really merit the kind of scrutiny Madison's favored proposal would require? How would the national legislature even manage to examine the volumes of laws that Madison himself predicted the states would produce? The negative had survived an early test back in June. But now it was rejected, seven states to three.

These two reverses, coming on consecutive days, clearly repudiated the dual assault on the sovereignty of the states that was the core of Madison's strategy. He had argued that states, as such, did not deserve representation in the national government, nor could they remain sole judges of their own legal authority. That did not mean that he imagined the states would wither away into hollow jurisdictions. Indeed, it was precisely because they would continue to legislate so much that their acts needed to be subject to national review. But just as Madison believed the Union could no longer rely on the states to execute its decisions, he also wanted to deny them any direct influence on national affairs. Nothing said in rebuttal of his arguments had convinced him he was wrong. But the convention had ruled against him.

Two months of deliberation remained, but after mid-July, the tenor of debate shifted markedly. For the next week and a half, the convention focused on the executive, a subject on which Madison's ideas were far more tentative. He agreed with the vote of June 1 to vest executive power in a single person, rather than a council or ministry. He was less certain of the wisdom of giving the executive a limited veto on legislation, as the convention also agreed to do in early June. His objection was not that the legislature should go unchecked, but rather that the veto would be better wielded, as the Virginia Plan proposed, by a joint executive-judicial council of revision. This proposal also failed. In rejecting it his colleagues clearly indicated that the proper time for judges to review an act of legislation was not *before* its passage but *after*, in a properly presented legal case that would allow them to gauge whether an act passed constitutional muster. Here was the concept of judicial review of legislation, the distinctive American practice that is often associated with the famous 1803 case, *Marbury v. Madison*, but which was an idea that the framers already grasped and accepted.

These two early decisions came relatively easily. But asking how the executive would be appointed plunged the delegates into "tedious and reiterated discussions." There were three basic problems. First, the dominant models of executive power in the eighteenth century were either monarchical or ministerial—and Americans had already repudiated both. Second, it was difficult to imagine the political dimensions of executive power in a republic, where the people's own representatives were supposed to bear their trust. Or rather, it was easy to envision an executive aspiring to be either a tyrant or a demagogue, but difficult to conceive how a republic resting on the judgment and consent of the many could ever place its confidence in the authority and will of one.

Third, and most vexing, each scheme the delegates considered for appointing this officer met strong objections. An executive chosen by the legislature could become a toady to its will—unless barred from seeking reelection, which would deny the public the benefit of allowing an able executive to continue serving. Election by popular vote faced two objections. One was that voters would prefer candidates from their own state, making it difficult to fashion a majority for any candidate other than the first likely incumbent, Washington. The other was that a popular election conducted in a single national constituency would disadvantage the south, because only free citizens could vote and the three-fifths rule would not apply. Then there was the novel idea of creating an elite body of citizen electors. Since they would meet for the sole purpose of appointing the executive, they could not control their choice's conduct after he assumed office. But who would these electors be? Even if they were better informed than ordinary voters, they might not "be men of the 1st. or even the 2d grade in the States."

Suitably perplexed, the delegates spent ten days cycling through these alternatives, only to end where they began, with an executive chosen by the legislature for a single seven-year term. Then they rested while a five-member committee of detail set about making a working draft of a constitution from the two dozen general resolutions adopted thus far. Though not a member, Madison stayed in Philadelphia while the committee met, and it's likely that Randolph and Wilson consulted him as the new text took shape. Madison had his own project to pursue: turning his daily copious notes into a fuller record of the debates. The result, though not a complete transcript, carefully traced the flow of debate and the changing moods of the convention. Serving the needs of history was proving to be "drudgery," as he described it to Jefferson, but a task to which he was committed.

Fulfilling its task, the committee added detail after detail to the skeletal framework of the convention's resolutions. The most important change in its report of August 6 was to convert the Virginia Plan's broad grant of legislative authority into a list of particular powers. Some scholars see this change as a virtual coup, suggesting that the committee deflected the convention from endorsing a version of parliamentary sovereignty and instead led it to adopt a distinctly American notion of government as a set of specific delegated powers, each requiring its own purpose and justification. If the committee really was departing from its instructions, however, it seems surprising that no one noticed at the time. A better explanation is that Madison intended the open-ended language of the Virginia Plan to serve as a placeholder. Before July 16 he repeatedly argued that agreement on the powers to be delegated hinged on how the questions of representation were resolved. But that position was tactical, not strategic, designed to focus his colleagues' attention on the issue he deemed most important. As a matter of principle, Madison believed that legislative powers *should* be enumerated, not inferred from a vague grant that would embolden legislatures to act as they wished.

After the committee reported, the pace of deliberation quickened and Madison's "drudgery" eased, much to the regret of scholars and jurists eager to recover the framers' "original intentions." On August 17, for example, Madison and Gerry proposed amending the clause authorizing Congress "to make war" to read "declare" instead, "leaving to the Executive the power to repel sudden attacks." Judging from Madison's notes, the ensuing debate may have lasted all of fifteen minutes, leaving us wondering what the framers really thought about a host of situations in which the president might order armed forces into limited combat without prior congressional approval.

The most important developments of the final weeks of debate involved the executive. Here two concerns converged. One was the continuing perplexity over the mode of election. The other was a reaction against the Senate (as the committee of detail christened the upper house). Well into August the delegates assumed that the upper house would be responsible for diplomacy and war. But a senate elected by the state legislatures, with states voting equally, seemed to resemble the Continental Congress, with its unhappy history of divisive debates over diplomacy and inefficient management of the war effort. As that perception took hold, the delegates came to see advantages in augmenting the power of the executive, giving this least republican officer the opportunity to act as something more than a faithful servant of the legislative will.

Even as this shift took place, the mode of electing the executive remained a puzzle. In a test vote on August 24 the convention deadlocked over trusting the appointment to a corps of special electors or the legislature. With other matters pressing for attention, it referred the issue to another grand committee (sometimes called the committee on postponed parts). Its report of September 4 contained the key innovations that gave the presidency (as the office was now called) its final form. Acting with the advice and consent of the Senate, the president would have the power to make treaties and appointments to high offices, including the Supreme Court. The committee also abandoned the idea of a single-term president appointed by Congress in favor of a scheme of electors who would meet in their separate states, ballot once, and forward the results to Congress. If no candidate won a majority, the decision would go to the Senate. Since electors would be apportioned among the states on the basis of total membership in Congress, the large states would have the advantage in promoting candidates (though small states and southern states would receive more electors than they deserved on the basis of free population alone). But the small states would have the greater say whenever the election went to the Senate—a result that would happen nineteen times out of twenty, Mason predicted, because the people would know so little about national characters.

This proposal had one advantage and two flaws. The advantage was that it built upon the key decisions of July by incorporating both the three-fifths clause and the equal state vote into the allocation of electors. The defects were that the system seemed likely to produce finalists, not winners, leaving the ultimate decision to the institution with which the president was now linked through the treaty-making and appointments clauses. In that case the president might still be a tool of an "aristocratic" Senate rather than the independent official the framers wished for.

It took three days for the convention to stumble on a solution, and even then only *after* it tentatively approved the committee's recommendation. First, Hugh Williamson was inspired to suggest that the "eventual choice" be made by Congress "voting *by states*"; then Roger Sherman had the better idea of shifting the election to the House, again voting by states. This preserved the bargain between large and small states while solving the separated-powers problem of making the president politically dependent on the Senate. But for all its ingenuity the key advantage of this scheme was that it avoided the killer objections against election by the people or Congress. Rather than specify how electors would be appointed, the convention dumped that responsibility on the state legislatures—the institution of American governance that Madison dreaded most.

The same day the convention solved this problem, Madison sent Jefferson a brief sketch of the Constitution. Adjournment was a week or two away, so there was no danger of violating the basic rule of secrecy that the delegates had scrupulously honored. He closed his sketch on a dispirited note. "The extent" of the changes the convention would propose "may perhaps surprize you," Madison wrote, in a wonderful understatement. "I hazard an opinion nevertheless that the *plan should it be adopted* will neither effectually *answer* its *national object* nor prevent the local mischiefs which every where *excite disgusts* agst the *state governments.* The grounds of this opinion," he promised, "will be the subject of a future letter."

Before that letter could be written, the convention still had to finish its work and launch the process of constitutional ratification. In the eleven days left, the delegates wrapped up loose ends, turned the Constitution over to a committee of style for a final rigorous editing, and made a feeble effort to persuade three wavering members—Mason, Randolph, and Gerry—to swallow their doubts and sign the completed document. Knowing Mason's sense of rectitude and Gerry's reputation as a maverick, Madison suspected that these two proud men would go their own stubborn way. But the defection of his "dear friend" Randolph was painful. Like Mason, Randolph professed to worry that southern interests would be hurt by "navigation acts" that could be passed by a simple majority.

Randolph had another concern, however, and one that threatened to undermine Madison's entire strategy for ensuring that the Constitution would be ratified by the unequivocal voice of the American people and then operate as supreme law, binding on states and citizens alike. Madison had mapped that strategy before the convention, once again combining principled analysis with prudential calculation. With the exception of Massachusetts, all the constitutions Americans had written since 1776 had been approved not by the people themselves but only by their representatives, meeting either in the provincial conventions of 1775–1777 or, in the special case of the Confederation, in the state legislatures. In legal terms these constitutions thus had no greater authority than ordinary statutes. Just as one legislature could repeal the acts of its predecessors, so a legislature choosing to violate some provision of a constitution was, in theory, perfectly free to do so. Call a constitution what you will, Jefferson observed in his *Notes on the State of Virginia*, if it had only legislative approval, it could not bind later legislatures. It would not really be a constitution at all—if by constitution one meant a fundamental law that a legislature could not alter.

Following Jefferson's reasoning and the Massachusetts precedent, Madison concluded that a new constitution must be ratified by the people, not the state legislatures. That meant bypassing the amendment rules of the Confederation. Although a few of the framers expressed qualms about this departure, most agreed that "first principles" or "revolution principles" should trump legalistic scruples. The same consideration freed them from adhering to the unanimity rule that had thwarted every previous effort to amend the Confederation. The new constitution would take effect upon the approval of nine states, not all thirteen. Once that threshold was crossed, dissenters and holdouts would have a strong incentive to rejoin the reconstituted Union.

In effect, the convention was appealing to the Lockean right of a people "to alter or abolish governments" as they wished. But it was one thing to vindicate that right in theory, another to guarantee its prudent exercise. For a sovereign people to approve the whole process of constitutional reform, Madison believed, their decision must be limited to a simple acceptance or rejection of the Constitution in its entirety. To do anything else—to allow the state conventions to accept some articles, reject others, or make their approval contingent on securing desired amendments—would doom the document to endless renegotiations. But that was exactly what Randolph seemed to desire. "The State Conventions shd. be at liberty to offer amendments," he insisted, which should then "be submitted to a second General Convention, with full power to settle the Constitution finally." Madison knew this was political madness. What was to stop the states from instructing their delegations to gain or defend some special point at all costs, thus making impasse far more likely than compromise? One great advantage of Madison's unique preparations for Philadelphia was that they enabled him to set the agenda the convention followed, even if it rejected the points he treasured most. There would be no such element of surprise the next time. Everyone would arrive forewarned and forearmed.

The other delegates understood this, and Randolph was left in dissent. So were Mason and Gerry. Their last complaint was voiced on September 12, when they moved to add a bill of rights to the Constitution. "It might be prepared in a few hours," Mason noted—meaning the convention could copy his Virginia declaration of 1776 and perhaps add a few other articles appropriate for national governance. This suggestion too was dismissed out of hand, without the support of a single state. Madison's notes indicate that only one delegate, Roger Sherman, thought the motion even merited comment.

Five days later the three dissenters resisted Benjamin Franklin's last appeal for unanimity and watched as the thirty-nine other members present signed the Constitution. In its final form it looked very different from the material referred to the committee of style a week earlier. The skillful quill of Gouverneur Morris had grouped these twenty-three resolutions into seven well-constructed articles, with numbered sections and distinct clauses, giving the completed text a more polished and organized character than any previous draft. Whether this new structure was intended to carry important clues and keys to the meaning of the text, as many interpreters assume, is hard to say. The delegates admired the committee's handiwork. But in their eagerness to adjourn, they did not ask, much less discuss, whether some of the changes Morris had silently made in the text would carry significance for later interpreters. Ratification had to come first; interpretation could wait its turn.

From the State House the members strolled east to the City Tavern for the last of many dinners they had consumed there. September 17 was a cool day, so perhaps they ate heartily and drank modestly. From there Washington returned to his lodgings at the home of Robert Morris (who sat through the convention without speaking once—but his friend Gouverneur Morris talked enough for both of them), "and retired to meditate on the momentous w[or]k which had been executed." Three doors down, Madison was also lost in thought. Not being a diarist, he did not record whether his equanimity as a veteran legislator outweighed his regret over crucial points lost. He had thought enough about the nature of political passion to realize that those who had just been engaged in the process of constitution making could not be "the best judges" of the Constitution's "merits or its faults." That assessment would better be made, he wrote to Edmund Pendleton, by those who could join "knowledge of the collective & permanent interest of America" to "a freedom from the bias resulting from a participation in the work."

Madison's own bias in the following weeks carried his thoughts in two opposing directions. On the one hand, he was dispirited over having lost the points he valued most. He devoted nearly half of the seventeen-page promised letter he finally wrote to Jefferson on October 24 to the "immoderate digression" of defending the congressional negative on state laws as he first conceived it, as a device to operate "in all cases whatsoever," with the paramount goal of protecting individual rights against the unjust laws of the states. "The evils issuing from these sources contributed more to that uneasiness which produced the Convention, and prepared the public mind for a general reform, than those which accrued to our national character and interest from the inadequacy of the Confederation to its immediate objects. A reform therefore which does not make provision for private rights," he concluded, "must be materially defective."

Yet however keenly he felt this defeat, Madison was far more impressed by the Constitutional Convention's devotion to completing its task. Given the "difficulties" the framers faced, he wrote to Jefferson, "it is impossible to consider the degree of concord which ultimately prevailed as less than a miracle." He made the same point in *Federalist* No. 37, one of his most brilliant contributions to the ratification debate and an essay uniquely devoted to asking Americans to ponder just how difficult the enterprise of constitution making on this scale must have been. "The real wonder is that so many difficulties should have been surmounted," he observed, "and surmounted with a unanimity almost as unprecedented as it must have been expected." This was not hyperbole meant to gull a credulous populace, but a reflection and confession of Madison's deepest convictions.

That conviction guided his conduct throughout the ratification campaign that began as soon as the convention adjourned. That campaign necessarily had two dimensions, one subject to careful management, the other volatile and impossible to cabin; one concerned with the decisions of the state conventions, the other with the public debate that would take place in newspapers and pamphlets, taverns and churches, and everywhere Americans exchanged political opinions. For the same reasons that he opposed a second convention, Madison and other Federalists planned to restrict the state conventions to a straight up-or-down vote on the Constitution as a whole, unencumbered by amendments or conditions. But in the battle for public opinion, no simple formula could predict how Americans would view the Constitution. What Edmund Burke said of Americans in 1775 was still true: "They augur misgovernment at a distance; and snuff the approach of tyranny in every tainted breeze."

The first challenge to ratification, however, occurred in Congress. That was where the framers first sent the Constitution, with the expectation that Congress would quickly ask the state legislatures to call elections for the prescribed conventions. Eager to see this process begin without a hitch, Madison left Philadelphia by September 21. He would have liked to linger in Princeton, where he was to receive the honorary degree of Doctor of Laws on the twenty-sixth. But the business was too urgent, and so it was on to New York and Congress. That was a wise decision. When he arrived on the twenty-fourth, he learned that his Virginia colleague Richard Henry Lee wanted Congress to add its own amendments to the Constitution before transmitting it to the states. If it did, Madison worried, the Constitution would become an act of Congress, not of the convention, and that might well make it more difficult to evade the amendment rule of the Confederation, with its impossible hurdle of obtaining the unanimous consent of all thirteen legislatures. To avoid that pitfall, Madison accepted a compromise under which Congress would send the Constitution to the states unendorsed but also unaltered, leaving Lee to circulate his amendments privately to other Anti-Federalists.

This was a critical point to gain, because the Constitution's Anti-Federalist opponents could have argued that the ratification procedure the convention had devised was illegitimate and illegal. Remarkably, twelve states accepted the new rules without protest. Only Rhode Island dissented, and even it went the convention one better by holding a statewide referendum that roundly rejected the Constitution. Neither the convention nor Congress, however, set a timetable for action. That depended on when the legislatures next met and how much time they believed their states would need to reach a decision.

Madison stayed in New York, attending Congress and monitoring early reports of the Constitution's reception. "The newspapers here begin to teem with vehement & virulent calumniations of the proposed Govt.," he wrote to Washington on October 18. Judging from the papers alone, he informed Randolph three days later, "one wd. suppose that the adversaries were the most numerous & the most in earnest." But the evidence was scattered and inconclusive, and Madison remained "far from considering the public mind as fully known or finally settled on the subject."

How to settle the public mind was the great challenge Federalists faced as the ratification campaign proceeded on its twelve separate fronts (counting Rhode Island as a lost cause from the outset). Few issues better illuminate the complexity of Madison's thinking than his approach to the question of public opinion. A constitution could not be imposed against the judgment and desire of the people, nor could it become a constitution in the new meaning of the term without their consent. Yet his entire analysis of American politics rested on grave doubts about the capacity of ordinary citizens to treat public measures with self-restraint, prudence, and concern for the true good of the community. Interest and ambition, opinion and passion were, he believed, the dominant forces governing ordinary political judgment.

When it came to considering how public opinion was actually formed, Madison remained more skeptic than cynic. He reserved his deepest doubts on the subject for Edmund Randolph, whom he hoped to woo back to the Federalist fold. Just after the New Year, Madison wrote to Randolph at length, couching his opposition to the second convention his friend was still advocating in a revealing discussion of public opinion. As profoundly as he respected "the rights of private judgment" in principle, Madison observed, "there can be no doubt that there are subjects to which the capacities of the bulk of mankind are unequal." The Constitution was one. "The great body of those who are both for & against it, must follow the judgment of others not their own." Had exactly the same document been framed "by an obscure individual, instead of the body possessing public respect & confidence," no one would take it seriously. Had Randolph, Mason, Henry, and Lee not proclaimed their dissent, "Virginia would have been as zealous & unanimous as she is now divided on the subject." From this assessment of the dynamics of public opinion Madison drew one controlling lesson. The success of the constitutional project required "a fortunate coincidence of leading opinions, and a general confidence of the people in those who may recommend it." A second convention would threaten both, releasing a flood of "various and irreconcileable" opinions and giving "opportunities to designing men" to exploit the ensuing cacophony.

These concerns illuminate what Madison hoped to accomplish in his own contributions to the public debate, the twenty-nine essays he wrote for *The Federalist.* That project was the creation of Alexander Hamilton, who shared with John Jay the leadership of Federalist forces in the uphill ratification struggle in New York. Had Madison not remained in New York City, attending Congress, Hamilton would not have asked him to join this ambitious project, which involved nothing less than a broad defense of the basic purposes of Union and a comprehensive, article-by-article exposition of the Constitution. Fortunately for the history of American political thinking, Madison was available to join Hamilton and Jay under their shared pen name, "Publius," and he either assumed or was given the primary task of expounding those aspects of American constitutionalism that had most deeply engaged his own reflections since 1780: federalism, legislation, and representation. In discussing these topics, Madison did not always disclose the full range of his ideas. There was no need, for example, to mention the veto on state laws, since that was not part of the document his countrymen were being asked to judge. This was the Constitution he was defending, not an abstract political theory. Yet his essays, like Hamilton's, are remarkable not for how much they hide, but for what they reveal of their author's concerns, insights, and way of thinking.

Madison's first contribution was the essay that scholars still study most closely: *Federalist* No. 10, with its public restatement of the theory of the extended republic first formulated in the memorandum on "The Vices." He pursued that theme in *Federalist* No. 14, an essay which reminded Americans that it was their "glory" not to have "suffered a blind veneration for antiquity, for custom, or for names, to overrule the suggestions of their own good sense, the knowledge of their own situation, and the lessons of their own experience." Three other essays, summarizing the lessons he had derived from his reading on ancient and modern confederacies, followed in December. But the bulk of the early writing was done by Hamilton and was devoted less to the Constitution itself than to the imperative need to make adequate provision for national defense and revenue.

That emphasis shifted when Madison's sixth essay, *Federalist* No. 37, was printed the day after he wrote to Randolph. From this point on, the subject was no longer the advantages of union but the substance of the Constitution, and Madison chose to open this second, longer half of the entire series by explaining just how difficult a task the convention had faced. (Fittingly, when *The Federalist* was published as a book later that year, this essay opened the second volume.) That difficulty took multiple forms. The framers had to reconcile two principles of government that were not easily balanced: the "inviolable" attachment to liberty and republicanism, which was the nation's political birthright, and "the requisite stability and energy" essential to good government. So too, assessing their efforts to divide power between the Union and the states, or among the three branches of national government, required considering how much harder it was to classify the phenomena of political life than it was to identify the elements of the natural world (which he called "the district of vegetable life," "the neighboring region of unorganized matter," and "the animal empire"). In nature one could assume that "all the delineations are perfectly accurate"—that is, really exist—even if mankind lacked the tools to detect them. But in politics no such certainty applied. "Questions daily occur in the course of practice, which prove the obscurity which reigns in these subjects, and which puzzle the greatest adepts in political science"—of whom, of course, he was one. Any effort to resolve these problems would be further vexed, Madison added, by the "unavoidable inaccuracy" of language itself. This had become a central concern of philosophy since Locke wrote *An Essay Concerning Human Understanding*, but one whose implications Americans, in their provincial simplicity, had never applied to the science of politics.

To read these observations and trace their application in later essays is to realize how little interest Madison had in catering to mass opinion. That was not his intended audience. He aimed higher, appealing to men of standing and influence who, though wavering on the Constitution, were open to persuasion, and who could carry others along in their wake. That is why it is foolish to label *The Federalist* as propaganda. Properly used, that term describes sophisticated efforts to manipulate base feelings and prejudices, and hardly covers the careful reasoning and prudent distinctions that mark these essays.

Perhaps Madison's most revealing essay—because it confirms his anxiety about the role of the same public opinion to which he was appealing—was *Federalist* No. 49. Having just cited Jefferson's *Notes on the State of Virginia* to support the idea that the greatest threat to the separation of powers would come from the "impetuous vortex" of Congress, Madison took a curious digression. Now he quoted another passage in which Jefferson proposed to correct violations of this separation by calling popularly elected conventions whenever two departments agreed that the third was engaged in improper "encroachments." That proposal, he conceded, had one sovereign advantage. Because "the people are the only legitimate fountain of power," they had an undoubted right to declare the "true meaning" of a constitution. But what was justifiable in theory might prove dangerous in practice. "Frequent appeals" to the people would deprive the government "of that veneration which time bestows on every thing, and without which perhaps the wisest and freest governments would not possess the requisite stability." Worse "was the danger of disturbing the public tranquility by interesting too strongly the public passions" in constitutional matters. Notwithstanding the justifiable pride Americans could take in their constitutional enterprises, they needed to recall that such "experiments are of too ticklish a nature to be unnecessarily multiplied." In 1776 a whole array of special circumstances had favored the first round of constitution making, and by implication similar conditions were at work in 1788. "Future situations" of constitutional disagreement would be less favorable to rational discussion.

Madison saved his "greatest objection" for last. The most likely source of encroachments would be the legislature, he predicted, the branch of government with the greatest "personal influence" over the people. The idea that the people could umpire a politically charged struggle between the branches was therefore naive. Legislators would have the best chance of being elected to the conventions, becoming "parties to the very question to be decided by them." Even when the legislature's misbehavior was so "flagrant" as to swing the advantage to the other departments, the people would be ill equipped to play the part Jefferson assigned them. Why? Because such disputes would inevitably rouse a spirit of partisan acrimony and thus summon the people's "passions," not their "reason," to render judgment.

Nothing better captures the ambivalence of Madison's position in 1787–1788 than the correspondence between the private view of public opinion he expressed to Randolph and the public argument of *Fed-*

eralist No. 49. Together they provide a portrait of an applied theorist of the science of politics who believed that a true constitution required the free assent of the people it would govern, but who also worried about engaging the public too strongly in matters it was not fully competent to debate. Madison was at once a constitutional radical, celebrating the capacity of his countrymen to rethink basic questions of republican government, and a political conservative who never underestimated the risks they were taking. That too was part of his political genius.

Madison's last essay as Publius appeared on March 1, 1788. Three days later he was Virginia-bound. He stopped first in Philadelphia, then at Mount Vernon, before arriving home on March 23, the day before the election for the state's ratification convention. Originally he did not intend to serve, fretting it would "involve me in very laborious and irksome discussions." But entreaties from friends in Virginia, including Washington, convinced him otherwise. Privately Madison preferred not to appear at the election, "to avoid apparent solicitude on the subject." But friends insisted his presence would be "indispensable." His stature in Orange County being what it was, he would have been elected regardless. As it was, he outpolled the runner-up, 4–1.

After the intense activity of the past year, Madison enjoyed a leisurely spring, traveling about the county while he waited for the Richmond convention to assemble in June. He kept up his active political correspondence and closely monitored how the Constitution was faring in other states. By May, eight had ratified it and only Rhode Island had dissented, leaving New York, New Hampshire, Virginia, and North Carolina still to act. The two northern states would also meet in June—the New Hampshire convention for a second time, after an abortive session in February. There and in Virginia the prospects for ratification were a toss-up. In the other two states Anti-Federalists held solid majorities. To court wavering Anti-Federalists in closely divided states, Federalists now agreed that the conventions could *recommend* amendments for future consideration, so long as their subsequent adoption was not made an express condition for a state's ratification.

Madison could claim one success even before the convention opened on June 2. Randolph has "thrown himself fully into the federal scale," he wrote to Washington shortly after his arrival in Richmond, while Henry and Mason, the two Anti-Federalist leaders, "made a lame figure & appeared to take different and awkward ground" in their first remarks. Following a suggestion from Richard Henry Lee, Mason insisted that the convention examine the Constitution clause by clause, and was surprised when the Federalists happily agreed. That agreement rested on the recognition that the best path to success in a closely divided convention was to answer the opponents' every objection. To that end Madison had already asked Hamilton to ship multiple copies of *The Federalist,* the better to arm the orators on his side. A different kind of patience was needed for Henry, who was just too freestyle an orator to follow a script. Much to Mason's irritation and Madison's frustration, Henry ranged widely, even wildly, from issue to issue, often playing to the audience's worst fears. One was that a national government dominated by northern states would literally sell the south's special interest in the Mississippi down the river. Another was that the new Congress could use its power to organize, arm, and discipline the militia to deny the state the arms it would need as its main line of defense against slave rebellion.

The convention met for three full weeks. As in other states, its galleries were open to the public and its debates partially reported in the *Virginia Independent Chronicle.* In addition, a lawyer named David Robertson took copious shorthand notes while trying to ignore the constant comings and goings of those seated around him. Recording Madison was almost as difficult as following the dazzling shifts of Henry's oratory. Madison had never learned to project his voice as easily as Henry did, and he was also suffering from a "bilious indisposition which confined me for some days" early on and sapped his strength after he returned. Still, he was the leading Federalist speaker and strategist, and there was much strategizing to do because "the parties continue to be nicely balanced," each allowing its wishes to influence its nose-counting, and each nervous about the outcome. He worried that the Anti-Federalists would try to postpone a decision, hoping that some Federalists might drift away from fatigue, or that favorable news from New York, where the opponents had a decided majority, would encourage Virginia to hold out.

In the end Madison had to make one critical concession. The convention had to choose between two alternatives. Henry moved that Virginia first submit a list of forty amendments—half a bill of rights, half structural changes to the Constitution—to the other states and delay ratification until they responded. The Federalist alternative tracked the strategy first used in Massachusetts: ratify unequivocally now, and propose amendments to be considered by the first Congress to meet under the Constitution. In the climactic debate, Madison gained the critical support of Randolph in reiterating all the reasons why any scheme of ratification contingent on the states first vetting one another's amendments would lead to a procedural morass so dense as to defy unknotting. In the decisive vote of June 25, the convention first rejected Henry's motion, 88–80, then approved the Federalist resolution, 89–79. Amendments would be proposed, not required.

The same day New Hampshire provided the decisive ninth vote needed for ratification. Virginia took another two days to draft its amendments and complete its act of ratification. Madison hoped that Anti-Federalists would show moderation in the changes they sought. That was not the case. Many of the "recommendatory alterations" were "highly objectionable" but "impossible to prevent," he informed Hamilton. Still, the news that Hamilton and Jay really wanted was simply that the other two states had ratified the Constitution. With that report in hand, they were able to break the dominant Anti-Federalist majority in the New York convention into two blocs, one willing to see the Constitution ratified if their state imitated Virginia by proposing both a declaration of rights and structural amendments. It took several weeks of Jay's best diplomacy to conclude a deal by which these amendments would be merely recommended. On July 25 New York raised the eleventh pillar of the new federal edifice by the narrow vote of 30–27.

By then Madison was back at Congress. He first spent three days at Mount Vernon, celebrating the twelfth anniversary of independence and the success of the constitutional project, and no doubt discussing future plans in the moments when Washington was not besieged by visitors. Whether they discussed the obvious issue of the general's willingness to serve as first president is not known. Madison was now his closest political adviser, but Washington cherished the idea of retirement no less in 1788 than he had before, and Madison would have been cautious about pressing the point. Then again, perhaps no urging would really be needed. "There is no doubt that Genl. Washington will be called to the Presidency," he wrote to Jefferson on October 8, and if called, how could he refuse?

The question of the vice presidency was more delicate, however, and

Madison turned to it in his next letter to Jefferson, a week later. If the president came from the south, the "next in rank" must hail from the other part of the continent. But the two obvious candidates seemed "objectionable": John Hancock, as "a courtier of popularity given to low intrigue," and John Adams—well, Madison hardly knew where to begin. In addition to the old rumor that Adams had been part of an anti-Washington "cabal" back in 1777, there were his personal failings: his "extravagant self importance" and an "impatient ambition" that "might even intrigue for a premature advancement."

These political musings were only a prelude to another topic: the merit of a bill of rights. Even from Paris, Jefferson had been an influential advocate on this point, which Madison and other Federalists viewed as a Trojan horse masking their adversaries' hopes for more radical amendments. Still, endorsing the idea of a declaration of rights was the easier pill to swallow in the final days at Richmond, and once taken, Madison felt bound to honor that pledge. Yet what he felt obliged to do in public did not sway his private skepticism about the value of a bill of rights. "I have never thought the omission a material defect," he told Jefferson, "nor been anxious to supply it even by *subsequent* amendment, for any other reason than it is anxiously desired by others." If well-meaning but misguided citizens felt it was necessary, if this was the price of persuading moderate Anti-Federalists to accept the Constitution, Madison would not only support it but take the lead in its adoption.

That did not mean, however, that he expected it to do much good once adopted. Madison had thought too long and hard about the problem of protecting rights in a republic to believe that the clamor for a bill of rights proved his prior analysis wrong. Madison ticked off three quick objections. Following an argument first made by James Wilson in a controversial public speech a year earlier, he noted that the limited nature of the authority delegated to the national government might make the adoption of a bill of rights dangerous by implying that more power had been granted than in fact was the case. Second, if one had to express a right in a form generally acceptable to the public, "there is great reason to fear that a positive declaration of some of the most essential rights could not be obtained in the requisite latitude." A broad statement of the sacred rights of conscience, for example, might not be possible to attain throughout the states. Third, the federal division of power between two levels of government would "afford a security" unknown to other governments.

Madison saved for last the argument that mattered most. "Experience proves the inefficacy of a bill of rights on those occasions when its controul is most needed," he observed. "Repeated violations of these parchment barriers have been committed by overbearing majorities in every State." Those majorities, he made clear, were *popular*, not legislative, in nature. They would act by passion, not reason, and their will would prevail over any mere statement of principles. Traditionally, bills of rights were thought to operate as a restraint on government by providing the people with a basis for knowing when their rulers were overstepping their power. But that function no longer fit the political life of a republic. "Wherever the real power in a government lies, there is the danger of oppression," Madison observed. "In our Governments the real power lies in the majority of the community, and the invasion of private rights is *cheifly* to be apprehended, not from acts of Government contrary to the sense of its constituents, but from acts in which the Government is the mere instrument of the major number of the constituents." From his distant vantage point in France, which was tottering on political upheaval, Jefferson could cling to the traditional view. But writing from America, Madison knew better after confronting "facts, and reflections suggested by them" that Jefferson, in his European exile, could overlook.

Facts and reflection, experience and theory: once again Madison was displaying his facility for translating his involvement in political life into the propositions of republican constitutionalism. Once again he could not avoid pursuing his thinking by posing a question that led to a fresh hypothesis. "What use then it may be asked can a bill of rights serve in popular Governments?" Two answers suggested themselves. One was the traditional one of providing standards by which the people could act politically against "usurped acts of the Government." He doubted such usurpations would be likely in a republic, but the prospect could not be ruled out. The more arresting thought could be stated in a single sentence. "The political truths declared in that solemn manner acquire by degrees the character of fundamental maxims of free Government, and as they become incorporated with the national sentiment, counteract the impulses of passion and interest." Bills of rights would work best not by establishing *legal* claims that injured parties could pursue, but by modifying the *political* behavior of the dominant majority.

The great irony of this discovery was that the most dangerous majorities would exist precisely where the Constitution did not yet reach. At the national level of government, the multiplicity of interests should work well enough to prevent those majorities from forming. But within the states they could still coalesce to work their mischief, beyond the reach of the federal umpire his colleagues at Philadelphia refused to create. In the meantime, though, he was committed to adding rights-protecting articles to the Constitution. Alone among the new congressmen who met in New York in March 1789, he took that promise seriously enough to make his colleagues do the right thing and swallow what he privately called "the nauseous project of amendments."

The adoption of these amendments was not by itself a pivotal moment in American history. Well over a century passed before the interpretation and enforcement of the Bill of Rights would move to the very center of constitutional controversy. When it did, that was only because the Fourteenth Amendment adopted after the Civil War ultimately gave the national government the authority with which Madison had hoped to arm it four score years earlier: the power to intervene within the states individually, the better to protect minorities from the unjust acts of factious majorities. Yet in political terms the adoption of these original amendments provided a critical postscript to the constitutional settlement of the late 1780s, assuring Americans that their exercise of the right to alter and abolish governments had been legitimate. That act began with the bold but desperate initiative that the Annapolis commissioners had launched at Mann's Tavern in September 1786, but its completion required turning the genie of popular sovereignty into a workable constitutional practice. If theory and practice meshed as well as they did in these years, it was not least because Madison creatively thought his way through a set of problems no lawgiver before his time had been fortunate enough to confront, much less managed to solve.

9

The State Builder

A FEW WEEKS before Madison sent Jefferson that long October account of the convention, Alexander Hamilton drafted a very different set of "Conjectures About the New Constitution." Madison was still ruing his setbacks at Philadelphia, but Hamilton was already thinking about what it would take for the Constitution not only to be ratified but effectively implemented. Atop the "circumstances" favoring its adoption, he placed "a very great weight of influence of the persons who framed it, particularly in the universal popularity of General Washington." Next came "the good will" of merchants, "most men of property," and "the hopes of the Creditors of the United States," by which he meant all those individuals holding the various paper certificates that Congress had issued to pay for the war. Then Hamilton turned to the consequences of rejection, down to the remote chance of a "reunion with Great Britain," with a son of George III occupying an American throne and "a family compact" tying the two nations. Hamilton ended his "Conjectures" on an upbeat note. Should the Constitution pass, Washington would probably become the first president, ensuring "a wise choice of men to administer the government and a good administration." That in turn would "conciliate the confidence and affection of the people and perhaps enable the government to acquire more consistency than the proposed Constitution seems to promise for so great a Country."

Hamilton's "Conjectures" are as revealing as Madison's defense of his negative on state laws. Madison believed that the Constitution embodied a set of principles to which Americans must adhere for its promise of a more perfect Union to be realized. In his view, maintaining the Constitution was an end in itself, and recognizing its "landmarks" became more important as he came to suspect that Hamilton viewed it in very different terms. Hamilton, by contrast, was far more willing to allow each branch to test and push the limits of its power, particularly if doing so would expand the authority of the national government at the expense of the states. He treated the Necessary and Proper Clause of Article I as a discretionary license for Congress to legislate.* In *Federalist* No. 78 he offered a robust statement of the emerging idea that courts could properly strike down unconstitutional laws. But Hamilton is best known, and remains most controversial, for his broad view of executive power. He was the one founder most willing to restore a measure of royal prerogative to the novel office of the presidency. If Hamilton was an equal-opportunity constitutionalist when it came to encouraging each branch of government to press its claims, that did not mean that all branches were created equal. The executive had a unique capacity for action. In the right hands it could set the agenda to which Congress, its political rival, would have to react. Madison expected the House of Representatives, the branch most responsive to the voice of the people, to be the dominant institution of government. Hamilton had other ideas.

That is what makes Hamilton's "Conjectures" so prescient. It opens and closes on the same note: the key to ratifying the Constitution and launching the government was Washington, with the right men around him. He counted himself high among those Washington would choose, and the federal Treasury the obvious site for his talents. He had made public finance a pet subject of study during off hours as the general's aide-de-camp and staunchly supported the ambitious program that Superintendent of Finance Robert Morris tried to impose on Congress in 1782–1783.

An incident in June 1789 reveals how obvious a choice he was. When the House took up the bill creating the Department of State, it faced its first constitutional puzzle: would the removal of high officials need the same consent of the Senate required for their appointment? In *Federalist* No. 77 Hamilton had suggested it did, and that passage was cited by

* In full, the clause empowers Congress "to make all laws which shall be necessary and proper for carrying into execution the foregoing powers, and all other powers vested by this Constitution in the government of the United States, or in any department or officer thereof."

William L. Smith, a young South Carolinian who soon became a loyal Hamilton supporter. The next day one of Hamilton's New York associates passed Smith a note, revealing "that upon mature reflection" the writer known as Publius "*had changed his opinion* & was now convinced that the President alone should have the power of removal at pleasure." Then Smith added his own aside: "He is a Candidate for the office of Secretary of Finance!"

It seems surprising, though, that Hamilton ever thought the Senate would have a role in the removal of executive officers. In his famous convention speech of June 18, 1787, he answered his own question, "can there be a Govt. without a good Executive?" by boldly declaring that "the English model was the only good one on this subject." The plan of a constitution he read that day proposed vesting sole "appointment of the *heads or chief* officers of the departments of finance war and foreign affairs" in the "Supreme Executive." A president who needed the Senate's consent to remove key subordinates could hardly be master of his own government. But a shrewd chief executive also knew when to defer to the talented ministers he had recruited. That was Washington's style. The general had not been an imperious commander. Trusted subordinates were given room to operate, and councils of war were moments for genuine discussion. On matters beyond his ken Washington was happy to take counsel. At the Treasury Hamilton expected to play the ministerial role he outlined in an intriguing aside in *Federalist* No. 35. "Nations in general, even under governments of the more popular kind, usually commit the administration of their finances to single men or to boards composed of a few individuals," he wrote, trusting "Inquisitive and enlightened Statesmen" to determine "the objects proper for revenue" before submitting their refined plans to "the sovereign or Legislature" for approval.

When "enlightened Statesmen" of the age considered the subject of public revenue, the model they studied most closely was Britain. When Hamilton told the delegates at Philadelphia that "the British Govt. was the best in the world," his praise went beyond "the excellence of its constitution." What he admired even more was the British *state*—its system of laws and public administration. Hamilton happily quoted another of those great ministers whom he hoped to emulate: Jacques Necker, the Geneva banker who had tackled the heroic task of reforming the finances of America's wartime ally. It was Necker, Hamilton reminded the convention, who called Britain the only nation "'which unites public strength with individual security.'" Security meant the personal liberty that the British constitution famously preserved. But "public strength" referred to the capacity of the British state to unite revenues from its customs duties and excise taxes with the private wealth that the Bank of England mobilized for public purposes. This was the great resource that enabled Britain to maintain naval superiority over its European rivals and military parity with far more populous France.

Hamilton's unrepublican affection for the unreformed British constitution could shock his fellow Americans, as it did the framers at Philadelphia. It did so again at a cabinet dinner Thomas Jefferson hosted in April 1791, replete with the usual libations from his amply stocked cellar. "'Purge that constitution of it's corruption,'" Jefferson recalled John Adams declaiming, "'and give to it's popular branch equality of representation, and it would be the most perfect constitution ever devised by the wit of man.'" Do that, Hamilton replied after a moment's thought, and "'it would become an *impracticable* government: as it stands at present, with all it's supposed defects, it is the most perfect government which ever existed.'"

This candid admiration for the virtue of corruption was not original to Hamilton. Decades earlier, David Hume had argued that if the Crown could not use its patronage to manage Parliament, the vaunted British constitution could never work in practice. "Give to this influence what name we please," Hume wrote, "call it by the invidious appellations of *corruption* and *dependence*," the basic truth remained that without it, the balanced constitution could never survive. Hamilton had quoted Hume's essay in his earliest political writings, published while he was still at King's College (now Columbia University). He never experienced the contrition that befell Jefferson, who "devoured" Hume's *History of England* in his youth, only later to repent how hard he had to work "to eradicate the poison it had instilled into my mind."

An American who retained a fondness for Hume could also be capable of fresh thinking. So it was with Hamilton. No one did more to set the agenda of American politics after 1789. It was to counteract his commanding influence over policy that Jefferson and Madison developed the constitutional arguments on which they founded their opposition Republican party. But Hamilton was more than a master of public policy. He was also a great risk-taker, given to gambits that sometimes paid off and sometimes made him his own worst enemy. Although he had his moody, even brooding moments, Hamilton was never indecisive. That was why he was most at home in the executive branch, why his grasp of the uses of presidential power helped make the control of that office the decisive fact of American politics, and why his personal relationship with Washington mattered so much.

As concerned as Washington had been with establishing his own "character," he was no less observant of the character of his aides, subordinates, and rivals. That judgment allowed him to overcome his initial bias against New Englanders to make Henry Knox and Nathanael Greene his two most trusted senior commanders. It also kept him from allowing Hamilton's occasionally immature or rash behavior to lessen his appreciation of the younger man's talent, genius, and desire to establish a "character"—a reputation—that might rival, if not quite equal, his own.

It was Hamilton's peculiar destiny, however, to have his own quest for fame intimately bound up with the overriding objective of securing the national character and reputation of the United States—that is, its public credit. Among all the ambitious projects and policies he launched in the 1790s, by far the most important were his several reports on that subject. Public credit, in the simplest sense, was the capacity of a government to devise an efficient and productive system of taxation, in part to defray its immediate revenue needs, but also, more important, to enable it to enter domestic and foreign credit markets and obtain vital loans from its own citizens or investors elsewhere. A nation with public credit could mount enterprises that would be far more difficult to sustain if it had to rely solely on the forms of onerous taxation that would leave its citizens groaning and grumbling, and its government wary of testing its subjects' loyalty and support. The most important of those enterprises was the ability to wage war, and the officials most likely to understand the value of public credit would be those who had witnessed firsthand the suffering of armies in its absence. Here was a deeper bond between Washington and Hamilton that the Treasury secretary's two chief opponents, Secretary of State Thomas Jefferson and Representative James Madison, could not sever.

Even the Constitution has its urban legends. One says that the clause requiring the president to be a natural-born citizen was conceived to bar the Nevis-born Hamilton from the nation's highest office. Alas for the constitutionally gullible: the full clause extends eligibility for the presidency to anyone who was "a citizen of the United States, at the time of the Adoption of this Constitution." As a subject of King George who immigrated to New York in 1773, Hamilton did not require naturalization to become a citizen of the new republic or to serve in Congress or the New York legislature. In 1787 his citizenship was no more in question than his patriotism.

His parentage was another matter. Their origins were respectable, their own reputations less so. James Hamilton was a younger son of a Scots laird, a status that led him into a feckless career as a merchant in the West Indies. Rachel Faucett was the daughter of a physician-planter. In 1745, while living with her mother on the Danish island of St. Croix, Rachel made an ill-advised marriage to Johan Michael Lavien, who dressed well but acted badly. Rachel jilted Lavien after he had her briefly jailed for unspecified "indecent and very suspicious" behavior, and then returned to her native Nevis. There she and James Hamilton became an item and remained so after Lavien divorced her. They never formally married but kept a stable common-law marriage for a decade and a half. One son, James, was born in 1753. Alexander came along either in 1755, as a legal document from 1768 suggests, or 1757, the date he claimed and his family dutifully honored. Nothing hinges on the year, for Hamilton was just as precocious in either case.

In 1765 James Hamilton took his family to St. Croix but soon abandoned them and moved on to St. Kitts. Three years later Rachel died, taken by one of the fevers that spurred luckier Europeans in the West Indies to make their fortunes quickly, then return home. Rachel had no fortune to leave and what remained of her estate as a storekeeper went to her firstborn son, Peter Lavien. His half-brothers faced further downward mobility and were soon apprenticed, James to a carpenter, Alexander to the merchants Beckman & Cruger. A cousin who became their guardian committed suicide in 1769. Alexander's first extant letter, written that year, blasted the "grov'ling" status to which "my Fortune &c. condemns me" and ended with its young author saying "I wish there was a War" to offer an escape from clerical boredom.

So there would be—though not the usual imperial struggle over sugar and trade that Hamilton would then have imagined. Meanwhile the countinghouse where he drudged came under the direction of Nicholas Cruger, the offspring of a prominent New York mercantile family, and Hamilton took on greater duties. He began submitting pieces to the local newspaper, poetry (of a sort), and in 1772 a vivid account of the hurricane of August 31. Lacking other prospects, it is easy to imagine Hamilton eventually becoming a trader in his own right. But commerce appealed to Hamilton more as a subject for analysis than a vocation. That autumn, opportunity did come when a group of local notables, alert to his manifest intelligence, clubbed together to send him off to the mainland for study, perhaps in the hope that he would return to St. Croix as a physician. For a time Hamilton attended the Elizabethtown Academy in New Jersey, lodging in the home of William Livingston, John Jay's future father-in-law. His hope to go on to the nearby College of New Jersey was thwarted, probably because President Witherspoon balked at his proposal to be allowed to advance at his own quickened pace. He turned instead to Jay's school, King's College, first as a private student, then as a member of the class of 1776.

When the revolutionary crisis broke in 1774, Hamilton quickly aligned himself with the colonists. Unverified accounts have him addressing city crowds during that first turbulent summer. His real political debut came when he published two hefty tracts rebutting Tory attacks on the First Continental Congress. In the summer of 1775, while still a student, he joined a volunteer company of artillerymen and helped liberate royal cannon from the fort at the Battery. Here Hamilton had his baptism of fire when the hovering warship *Asia* opened fire on the colonists. In the spring of 1776 he received his first commission, captaining a company of New York artillery. Hamilton and his ever-diminishing band fought throughout the disheartening New York and New Jersey campaigns and saw action at Trenton and Princeton.

Most of what is known about Hamilton's early service comes from later reminiscences, which are always subject to the coloring of time and partiality. But it seems clear that intelligence and bravery quickly marked him as a promising young officer. Washington's general orders for March 1, 1777, announced that Hamilton "is appointed Aide-De-Camp to the Commander in Chief, and is to be respected and obeyed as such." Though the last phrase sounds like a formula, it bespoke the confidence that Washington placed in his young aides. Washington never relaxed the habit of wanting to oversee every aspect of the administration of the army. But he also had to maintain an extensive correspondence with civilian authorities in Congress and the states as well as other officers he commanded. Within the guidelines he laid down, he trusted better-educated aides like Hamilton and Jack Laurens to couch his recommendations and orders in the most appropriate form.

In this Hamilton proved a prodigy, and he soon became a nearly indispensable part of Washington's staff. That was hardly the path to glory that Hamilton sought. Yet so long as the middle states remained a central theater of action, he could subordinate his desire for a combat command to the recognition that he was doing essential service and would still be at the general's side whenever battle offered. This was the period when the officers immediately around Washington were keenest about protecting his reputation against the pretensions of his rivals. It was Hamilton who received the sensitive mission of asking Horatio Gates, triumphant at Saratoga, to detach units to strengthen Washington's central force near Philadelphia. Hamilton was present with Charles Lee when a wrathful Washington stemmed the American retreat at Monmouth. There he fought as valiantly and recklessly as Jack Laurens, when for a second time Laurens's horse was shot beneath him. It was Laurens who dueled Lee over the resulting recriminations, with Hamilton as his second, but the roles could as easily have been reversed.

After Monmouth, however, the northern war became a prolonged sitzkrieg, removing the best chances for the glory combat could uniquely bring. If not quite wallowing in self-pity, Hamilton began indulging the gloomier side of his temperament. In December 1779 Laurens advanced Hamilton as a candidate for the posting he had just spurned as secretary to Benjamin Franklin. Though Hamilton appreciated the gesture, he reckoned he had no chance of gaining appointment. Six years an American, he remained "a stranger in this country. I have no property here, no connexions." Even if he had the "talents and integrity" that Jack saw in him, "these are justly deemed very spurious titles in these enlightened days." Then came a postscript. "I have strongly sollicited leave to go to the Southward," as Laurens would do, but his request was refused on grounds he also found spurious but which "gave law to my feelings." Washington valued his services too highly to send him off to combat. Hamilton closed on a Hamlet-like note. "I am disgusted with every thing in this world but yourself and *very* few more honest fellows, and I have no other wish than as soon as possible to make a brilliant exit. 'Tis a weakness, but I feel I am not fit for this terrestreal Country."

With no real home of his own, Hamilton made a select circle of fellow officers his own family. As in all wars, the intimacy of camp quarters and the intensity of combat forged close friendships unknown in civilian life. When Laurens headed south in 1779, Hamilton opened his first letter to him by professing his love for Jack in the passionate terms with which young officers sometimes addressed each other. But he then asked Jack to scout a wife for him in Carolina and went on to list the qualities he sought. "She must neither love money nor scolding, for I dislike equally a termagent and an oeconomist," wrote the future proponent of a national bank. "But as to fortune, the larger stock of that the better." Hamilton instructed Jack that *"size"* mattered in one other respect: "mind you do justice to the length of my nose," he ordered, using the same euphemism for the male anatomy with which Jefferson befuddled Maria Cosway.

Having abandoned his own wife, Laurens was no marriage broker, and Hamilton in any case was about to make a notable "connexion" of his own. One advantage of the northern stalemate was that it enabled him to go courting—or rather, brought courting his way, as dinners and balls went on even amid the heavy snows of 1780. In early February he renewed his acquaintance with Elizabeth Schuyler, General Philip Schuyler's daughter, who was in camp visiting her aunt. Hamilton waged an intense siege, and within two months they were engaged. The Schuylers were one of New York's most eminent families, members of the Dutch patroon elite whose estates dominated the Hudson Valley. Yet far from treating Hamilton as a fortune-seeking upstart, the Schuylers welcomed him into their family, knowing his intelligence and reputation were the best tokens for the future.

Marriage to Betsey (or Eliza) gave Hamilton a place in the New York elite, sealing the political connections he began forming when he was still a novice artilleryman. As early as December 1775, for example, he wrote Jay a lengthy letter, presumptuously citing lessons "confirmed to me both by reading and my own experience" to warn the "political pilots" against allowing the mass of the people "to grow giddy" and "run into anarchy." Soon after joining Washington's staff, a committee of the legislature asked him to act as their correspondent at headquarters. Hamilton accepted, noting only his opinions must be read as "private sentiments, and are never to be interpreted as an echo of those of the *General.*" Hamilton thus formed direct contacts with New York's leaders, who pegged him as a rising talent worth cultivating. As Washington's aide, he met regularly with the committees that Congress dispatched to camp whenever the army's logistics lurched toward collapse. He was, in short, a thoroughly political officer whose views older men respected.

One such occasion occurred in September 1780, after Congressman James Duane stopped at camp and solicited his thoughts on "the defects of our present system, and the changes necessary to save us from ruin." The request must have been open-ended, for Hamilton's reply was more treatise than letter. From his opening premise that "The fundamental defect is a want of power in Congress," he wrote as an unabashed nationalist who believed that the proper gauge of its authority was not the unratified Articles of Confederation but rather its duty *"to preserve the public from harm."* The superior political resources available to the states made it ludicrous to worry about the danger of vesting effective power in Congress. Pages of analysis of the problems afflicting the war effort culminated in his ideas for establishing a national bank, modeled on the Bank of England, to attract the capital of the country's "monied interest" with the promise of dividends funded by national taxes collected by national officials.

These were not impromptu reflections, dashed off to answer a congressional request. The topics Hamilton examined—military logistics, soldiers' pay, revenue, public credit, money—were staple subjects of discussion in 1780, when the war effort was faring even worse than it had during the Valley Forge saga two years earlier. But Hamilton was unusual in how much thought he had already devoted to these topics. Pressed as he was with the countless duties of his position, he still found time to pursue the education the war had interrupted. In effect Hamilton designed his own course of independent study, much of it devoted to mastering the great topics of political economy: finance, taxation, banking, and commerce. It was a project destined to have a profound impact on his adopted nation.

Shortly before Christmas 1780 the young couple were wed at the Schuyler mansion in Albany. Awake or asleep, he found his thoughts "monopolized by a little *nut brown maid* like you," he wrote to Betsey in October, "and from a soldier metamorphosed into a puny lover." A general's daughter knew better than to think that her charms could distract Hamilton from his military ambitions. Just before the wedding, he again asked Washington to give him "a conspicuous part in some enterprise that might perhaps raise my reputation as a soldier above mediocrity." When a proposed assault on British posts on Manhattan was shelved, Hamilton and Lafayette concocted another plan to wangle an appointment as adjutant general. This too came to naught. Lafayette assured Hamilton that the general's "friendship and gratitude for you" were "far greater than you imagine." But after four years of personal service, Hamilton felt he was owed more.

The break between them came on February 16, 1781, in a trivial incident at Washington's headquarters at Windsor, New York. Meeting on the staircase, the commander asked Hamilton to wait upon him; Hamilton went downstairs to deliver a letter, then was detained by Lafayette as he returned. He was away barely two minutes, Hamilton wrote to his father-in-law. He found the general seething at the head of the stairs, having waited, he reckoned, a good ten minutes. "'I must tell you Sir you treat me with disrespect,'" the general snapped. "'I am not conscious of it Sir,'" his aide replied, "'but since you have thought it necessary to tell me so we part.'" Within an hour Washington sent another aide with "Proposals of accommodation." But Hamilton stayed "inflexible" even after Lafayette, Schuyler, and others urged him to relent.

He explained himself most fully to Schuyler, whom he rightly expected to question his decision. The truth was that "for three years past I have felt no friendship for him and have professed none." When Washington offered some measure of the personal affection he notably extended toward Lafayette, Hamilton made it plain that "I wished to stand rather upon a footing of military confidence than of private attachment." You can imagine, he told Schuyler, "how this conduct in me must have operated on a man to whom all the world is offering incense." Why then had he persisted in Washington's service? Because Hamilton had subordinated private inclinations to the greater good. "His popularity has often been essential to the safety of America, and is still of great importance to it."

If Hamilton thought Washington would repair the breach by offering him the combat command he long sought, he miscalculated. In April 1781, when the newlyweds were living just across the Hudson from headquarters, Hamilton asked Washington "how you will be able to employ me in the ensuing campaign." An "embarrassed" Washington replied promptly but offered nothing. The usual objections applied: line officers in existing units would resent any departure from the usual seniority-based rule of gaining command. Hamilton tried again, noting that "my early entrance into the service" during "the campaign of 1776, the most disagreeable of the war," would surely have led to a higher command had he not joined the general's staff. But for now Washington still had no assignment to proffer.

With his military career in limbo, Hamilton's thoughts—or at least those not devoted to romantic bliss—returned to finance and politics. Just after Washington's first rebuff he sent the new superintendent of finance, Robert Morris, another mini-treatise akin to his letter to Duane, this one discussing issues of revenue and taxation and the advantages of a national bank. Writing as "The Continentalist," he began a set of newspaper essays exposing the inadequacy of the Articles of Confederation. The opening sentence set the tone: "It would be the extreme of vanity in us not to be sensible," he wrote, "that we began this revolution with very vague and confined notions of the practical business of government." The final installment in this series concluded with what was already becoming a familiar theme. Until permanent revenues were assigned to Congress, its public credit would never be established. In a closing tribute to Morris, Hamilton sounded a note to which he would later return. It is "impossible, that the business of finance could be ably conducted by a body of men, however well composed or well intentioned."

Returning to camp just before "The Continentalist" essays went to press, Hamilton sent Washington a letter making a last bid for command and enclosing his commission in case it was denied. Had he not known Hamilton so well, a man with Washington's temper could have accepted the resignation on the spot and rid himself of this petulant youth. Instead he sent an encouraging message that he would "endeavour by all means to give me a command." Three weeks later the wish was fulfilled when Washington gave him a newly formed battalion of New York infantry. Three weeks after that, Hamilton wrote "My Dearest Angel" with the news that he was Virginia-bound, with no time for a farewell—or rather, that "it would be thought improper to leave my corps at such a time and upon such an occasion." The battalion marched to Head of Elk, Maryland, where Sir Billy Howe had debarked almost exactly four years earlier, and then sailed down the Chesapeake to join the siege at Yorktown. He would be home by late fall, he promised, in time for the birth of the firstborn he rightly knew would be *"a boy."*

Hamilton's moment of glory finally arrived on October 14—but to attain it, he had to make one more plea to Washington. The objective was the two redoubts—man-made hillocks—at the far left of the British defenses. The British troops packed into this constricted space could smell the York estuary lapping beneath them, a few hundred yards from the great bay, a salty reminder that escape would be easy, had the Royal Navy broken the French blockade. Lafayette was assigned Redoubt 10, and he initially placed his own aide, a Frenchman, in charge. Hamilton protested, and to assuage his feelings and reclaim the honor for his own countrymen, Washington gave him the command. Hamilton would lead the main part of his force in a frontal assault, while Jack Laurens took eighty men around to the rear. The attack went as planned, and after a sharp but brief combat that left eight dead on each side, the redoubt was taken.

Two days later Hamilton let Eliza know that "honor obliged me to take a step in which your happiness was too much risked." Whether it was honor, ambition, or glory that drove him to seek combat is a fair question, but he could now reassure her that his war was over. "To-morrow Cornwallis and his army are ours," he wrote two days later, and within days he would start for Albany. Still peeved "over the difficulties I met with in obtaining a command," he deemed it "incompatible with the delicacy due myself" to solicit a new posting, should active combat resume. But neither was he ready to resign his commission. If Washington still wished his services, he could come calling. Once home, Hamilton threw himself into becoming a good husband, father, and provider, resuming the legal studies he had just begun before war led him to another profession. "You cannot imagine how entirely domestic I am growing," he wrote to an army friend in March, soon after Betsey delivered the son he had predicted, named Philip for his grandfather. "I lose all taste for the pursuits of ambition, I sigh for nothing but the company of my wife and baby." It was probable his "public life" was over.

That resolution lasted all of a few weeks. In early May 1782 Robert Morris asked him to fill the vacant position of collector of continental taxes in New York. Hamilton hesitated. True, "my military situation has indeed become so negative that I have no motive to continue in it." Morris would have understood that Hamilton did not regard Washington as a patron. Still, the position would not pay enough to justify the distraction from his legal studies. After Morris reassured him that he would have both time and income enough to allay his concerns, Hamilton accepted, on the condition that he would be authorized to lobby the assembly to support the revenue program Morris was pushing Congress to adopt. This was exactly what Morris wanted. He regarded his collectors not as mere cashiers or functionaries but as a potential political network of his own. As Hamilton observed in his final "Continentalist" essay, "The reason of allowing Congress to appoint its own officers of the customs, collectors of the taxes, and military officers of every rank is to enable it to create in the interior of each state a mass of influence in favor of the Foederal Government." Substitute Morris for Congress and the analysis still holds.

Hamilton promptly set off for Poughkeepsie and did all he could to answer expectations. He lobbied the relevant committees and even managed, with the aid of his senatorial father-in-law, to have a resolution adopted calling for a general convention to revise the Articles of Confederation. Once back in Albany, he prepared for his bar examination under a special waiver releasing military officers from the usual three-year reading period. He still found time to send Morris a detailed dispatch about New York politics, replete with pointed sketches of its leaders, such as the superintendent's sworn opponent, Abraham Yates, "whose ignorance and perverseness are only surpassed by his pertinacity and conceit."

Morris already knew one other outcome of Hamilton's visit: his election to Congress to fill a seat Schuyler had declined. Hamilton felt some pangs about venturing to Philadelphia. He doted on his son, who at seven months already "has a method of waving his hand that announces the future orator." Like son, like father. For when Hamilton took his seat, he quickly entered the continuing debate over the lamentable condition of congressional finances, joining James Wilson and the Philadelphia merchant Thomas Fitzsimons in promoting the comprehensive program for funding the public debt that Robert Morris had presented in July. But they could not rely on oratory alone. To attain his objectives, Morris first had to persuade Congress to endorse his program of land, poll, and excise taxes. But under the Confederation only the states could levy taxes, and Congress would still need their further approval for Morris's program to take effect. To muster political support at both levels of governance, Morris could turn to two interest groups. One was the national community of public creditors—individuals holding the various paper instruments that Congress had issued during the war. The other was the army, with its arrears in salary, and particularly its officer corps, who had been promised half-pay pensions for life—an offer that clashed with republican principles of selfless service but also promised compensation for the real sacrifices the officers had endured.

With his close ties to the officers, Hamilton became a crucial link between the debates in Congress and the rumblings of discontent, and perhaps even mutiny, at the main army camp at Newburgh. This was the first occasion when Hamilton exposed the political adventurism that has led both critics and admirers to detect a deeply Machiavellian strain in his political character. Presumably acting in consultation with Robert Morris and his assistant, Gouverneur Morris, Hamilton attempted to make Washington the vehicle for persuading Congress to adopt the superintendent's program. In mid-February 1783, Hamilton wrote Washington a remarkable letter, effectively urging the commander *"to take the direction"* of the army's protests over wages and pensions. You should act decisively but discreetly, Hamilton advised, the better to "preserve the confidence of the army without losing that of the people. This will enable you in case of extremity to guide the torrent, and bring order perhaps even good, out of confusion." A week later, Hamilton attended a key congressional caucus held at the Philadelphia home of the merchant-delegate Thomas Fitzsimons. There he dissented from the compromise that James Madison and others present fashioned, one that fell short of what Morris insisted, and Hamilton believed, was necessary. After informing his fellow delegates that Washington was "becoming extremely unpopular among all ranks" at camp, Hamilton suggested that their aim should still be to make Washington "the conductor of the army in their plans for redress."

At first Washington ignored Hamilton's letter, to its author's anxiety. When he finally replied, he politely but firmly declined the quasi-Caesarean part cast for him. He intended to "pursue the same steady line of conduct which has governed me hitherto," Washington wrote on March 4. A week and a half later, he put an end to the continuing ferment among his officers by appearing unannounced at a fresh meeting chaired by none other than Horatio Gates. His old nemesis could hardly refuse Washington's request to address the restive officers. Washington began by reading the speech he had personally drafted, recalling the officers to the first principles of republican duty. Then, pausing to read a letter from a member of Congress, he performed perhaps the single most famous physical gesture in American history (at least until Harry Truman held up the *Chicago Daily Tribune* that wishfully proclaimed his defeat in 1948). Reaching into his jacket for never-worn-in-public eyeglasses he had just received, Washington observed "that he had grown gray in their service"—that is, the service of the officers—"and now found himself growing blind." The gesture and remark, "so natural, so unaffected," met the highest aesthetic standards of the day and touched feelings in his audience deeper even than their desire for justice. There the protest ended, and with it the fleeting prospect of mutiny.

Now Hamilton had to backtrack. He and Washington had not been in regular contact since 1781, yet now they resumed a candid correspondence that effectively renewed their alliance. Hamilton smoothed over the dangerous role he had urged Washington to play by emphasizing that his aim had been to "confine" the protests of the army "within the bounds of duty." Hamilton admitted that he was not wholly reconciled to the republican principles that Washington insisted on honoring. "I confess could force prevail I should almost wish to see it employed," he wrote on March 25. "I have an indifferent opinion of the honesty of this country, and ill-forebodings as to its future system." Washington shared those forebodings, and that common opinion outweighed whatever suspicions he felt about exactly what Hamilton may have had in mind for him—or the army he had labored eight years to shape in his own image. Rather than keep Hamilton at arm's length, Washington invited him to fulfill his promise to provide his further thoughts on the state of politics. The two men thus effectively restored the confidential relationship they had enjoyed until their displays of temper at Windsor had come between them. By April, Washington was addressing his former aide as "my dear Colo. Hamilton" while noting that "I write to you unreservedly." Though their correspondence again lapsed when both returned to private life, the trust they had reestablished did not.

This revival of personal trust had one other source. For both men the great lesson of the war was that Americans had to acquire "Continental views," to learn to "think continentally," to transcend "local prejudices." They had to recognize that the world remained as dangerous after independence as it had been before, that European powers would find fresh incentives to meddle in their politics, that Americans had to develop modes of governance superior to those with which they had scraped through to independence. Both men had provincial loyalties: Washington to the dominion of his birth, Hamilton to the state whose elite had adopted him. But those attachments were now superseded by another. Though the word *nationalism* had yet to be coined, both men were nationalists in every sense of the term. Their presumptuous belief that they were thinking *continentally* reveals how expansive their ambitions for the struggling federal Union could be.

Hamilton was eager to return to private life. "There is so little disposition either in or out of Congress to give solidity to our national system," he complained to Nathanael Greene, "that there is no motive to a man to lose his time in the public service." As soon as "New York is evacuated," he added, "I shall set down there seriously on the business of making my fortune." Meanwhile he dutifully attended Congress, taking on the numerous assignments that came his way. The foremost was chairing the committee charged with drafting the plan for a postwar military establishment, a task that again required consulting Washington. When an unruly band of Pennsylvania soldiers marched on Congress to demand their back wages, he headed the committee sent to ask the state's executive council to call out the militia for its protection. Speaking for the council, John Dickinson declined the request, doubting the militia would come if called. A confused scene soon unfolded. Mostly the mutineers stood around, Madison reported, "individuals only occasionally uttering offensive words and wantonly pointing their muskets to the Windows" of the State House, as nervous delegates peered out at them. After a while "spirituous drink from the tipling houses" nearby produced a different mood. Leaving the State House, the delegates faced "a mock obstruction" before the soldiers let them pass. With the council declining to restore order, Congress voted to decamp to Princeton. Hamilton was the lead negotiator throughout this episode. For his troubles, it was "insinuated" that he was secretly contriving to get Congress to relocate to New York.

Hamilton left Congress in late July. Among his last acts was to draft, circulate, but then withhold resolutions calling for a general convention to remedy the defects of the Confederation. Much like his letter to Duane of 1780, his resolutions provided a sweeping assessment of why "the United States have been too often compelled to make the administration of their affairs a succession of temporary expedients." Once he set out to explore a subject, no rule of literary economy could bar its exhaustive analysis. But the constitutional conclusions that flowed from that assessment were no match for another underlying political reality. "The road to popularity in each state is to inspire jealousies of the power of Congress," he wrote to John Jay, even "though nothing can be more apparent than that they have no power" of which to be jealous.

Taking "the road to popularity" was the last thing on his mind as his young family set up housekeeping at 57 Wall Street in the fall of 1783. Hamilton was bent on establishing his legal practice and averse to any public engagement that could distract him. When his name appeared in a list of candidates for the state assembly, he let the public know he would "decline public office." But that did not mean swearing off politics. Hamilton greeted the new year of 1784 with his latest political tract, *A Letter from Phocion*, which vigorously challenged a spate of anti-Tory legislation that the New York assembly had just enacted. Hamilton soon served as lead attorney in *Rutgers v. Waddington*, one of the early cases that set a precedent for the emerging idea of judicial review, the novel idea that the acts of a supreme legislature representing a sovereign people could still be subject to judicial scrutiny.

Supporting the lenient treatment of loyalists was Hamilton's chief public cause during the first years of peace. But he was preoccupied with profession and family. His legal business and reputation boomed. In September 1784 Betsey delivered a daughter, Angelica, named for the sister with whom her husband carried on a correspondence so flirtatious that biographers wonder whether an affection deeper than family ties was involved. Hamilton finally returned to politics in the spring of 1786, with election to the state legislature. He entered the assembly as an opposition leader to the dominant political coalition of Governor George Clinton. A nationalist during the war, the popular Clinton, poised to win his fourth statewide election, now harbored "anti-federal" attitudes. He was particularly disillusioned with Congress over its failure to support New York's effort to recover the "revolted" territory of Vermont. With the ample revenues it collected from duties on the imports flowing along the wharves of Manhattan, the state needed Congress less than Congress needed New York.

Even in opposition, however, Hamilton was an obvious choice to represent New York at the Annapolis Convention of September 1786. Along with Egbert Benson, he rode south on September 1 to join the ten other commissioners who met at Mann's Tavern eleven days later. By Madison's recollection, Hamilton drafted the letter that this rump meeting then issued, proposing a general convention for Philadelphia in May. Whether this *coup de marche* was his stratagem is less easy to ascertain. He had suggested such a convention to Duane in 1780 and again during his last weeks in Congress in 1783. It was just the kind of bold stroke that he relished. In March 1787 the New York assembly named him one of the state's three delegates, along with two Clinton lieutenants, Robert Yates and John Lansing. Hamilton agitated to expand the delegation to five, with the special aim of enabling John Jay, a man of "acknowledged abilities, tried integrity and abundant experience," to join the deliberations. But the senate held firm at three, which meant that his colleagues could always outvote him. Still, he arrived early in order to attend the concurrent general meeting of the Society of the Cincinnati.

Hamilton attended the convention for a month, then left for New York on June 29. Yates and Lansing, incipient Anti-Federalists, followed two weeks later, leaving the state unrepresented. Once home, Hamilton wrote to Washington to report on "the public sentiment" he had sounded on his trip home. "The prevailing impression among thinking men," he reported, "is that the Convention, from a fear of shocking popular opinion, will not go far enough." The delegates should not "allow too much weight to objections drawn from the supposed repugnancy of the people to an efficient constitution." Hamilton then added a revealing comment. "Not having compared ideas with you, Sir, I cannot judge how far our sentiments agree." Taken literally, this suggests that the two men had *not* actively consulted during their six weeks together in Philadelphia. Nor was it evident that another opportunity would soon arise. Hamilton intended to stay home another fortnight. Then, if he found reason to think his attendance "will not be mere waste of time," he might deign to return. Washington replied a week later, sharing Hamilton's doubts about the tenor of the convention but urging him to come back, even if his vote would have no effect. Hamilton briefly went back to Philadelphia in August, then returned to New York before retaking his seat on September 6, in time for the key debate on the presidency. Once back he opened his first speech by noting "his dislike of the Scheme of Govt. in General." His last remarks demonstrated little more generosity to what had been done in his absence. "No man's ideas were more remote from the plan than his [own] were known to be," Hamilton observed, even as he urged the non-signers Mason, Randolph, and Gerry to abandon their dissent.

Given this attitude as well as his extended absence, one can hardly call Hamilton a major framer of the Constitution. Yet it is equally difficult to imagine the Constitution being implemented without him. His effort began shortly after the convention adjourned, when he launched *The Federalist*, the one truly comprehensive analysis of the Constitution that ratification produced. Scholarly commentary gives Madison's contributions pride of place because they lay out general theories of federalism and separation of powers. But Hamilton's more numerous essays, well over half the total, are filled with striking insights and rendered in a voice that gave no quarter to the Constitution's detractors. Take his assessment of national priorities in the realms of revenue and defense. Far from conceding the danger of allowing the national government to command and supply the country's armed forces, Hamilton argued that no other solution was possible. Once it was agreed that defense was one of the great "objects" of government, logic alone prohibited imposing any constitutional limit on its powers of taxation. He was equally candid in dismissing the hackneyed notion that a republic should rely upon a militia drawn from the mass of the citizenry, the better to avoid the political evil of a standing army of mercenaries. For the militia to be "well regulated" (meaning effective), it had to be a "select corps of moderate size," receive the serious training a mass body could never get, and thus be "ready to take the field whenever the defence of the State shall require it."

Hamilton's early essays as Publius were a tribute to the power of the national idea in his thinking. National government was first and foremost about national security; national security required maintaining public credit; and commerce was important not least because its easy taxation would provide the revenues needed to sustain that credit. By contrast, mere domestic concerns seemed a matter of near indifference. Where Madison reminded readers of all the advantages the states would retain in any competition with the national government, Hamilton almost imagined a day when the states would cease to be major institutions of governance. Over time "the wants of the states will naturally reduce themselves within *a very narrow compass*," he predicted. Next to the duties of the national government, the costs of "the mere domestic police of a state" would be "insignificant."

The most revealing passages that Hamilton wrote as Publius came with his discussion of the presidency. After devoting several essays to a prosaic summary of its duties, Hamilton took a deep breath and opened *Federalist* No. 70 by pronouncing those truths about executive power that misguided republicans needed to hear. The axiom "that a vigorous executive is inconsistent with the genius of republican government" was plain wrong, he argued. Just as good government requires energetic administration, so national security depends upon energetic leadership—and to be effective such leadership must be unitary. A multitude of voices and time for deliberation are essential for the legislature, Hamilton conceded. But in the executive, which often has to act with "decision, activity, secrecy and dispatch," divided counsels are dangerous and unity vital.

This praise of the executive rested on a profoundly psychological understanding of the incentives that would attract individuals to public service. That insight was best captured in *Federalist* No. 72, where Hamilton's defense of allowing presidents to be reelected led him to meditate on "the love of fame, the ruling passion of the noblest minds." To speak in praise of passion, as Hamilton did, was to take the wrong side in a familiar debate. Madison spoke for the reigning orthodoxy when he closed *Federalist* No. 48 by noting that "the passions ought to be controlled and regulated by the government." For Hamilton, by contrast, the impassioned love of fame would encourage just the kind of behavior he wanted to nurture in the executive. Only "the prospect of being allowed to finish what he had begun" would "prompt a man to plan and undertake extensive and arduous enterprises for the public benefit." Remove that incentive, and the best one could expect "from the generality of men" would be "the negative merit of not doing harm." That might work for physicians, but not for statesmen laying a foundation for national greatness.

One American, of course, had already proved that it was possible to pursue fame without sacrificing republican principles. Hamilton had written to Washington in October 1787, enclosing the inaugural installment of *The Federalist.* Then their correspondence again lapsed. Hamilton next wrote in mid-August 1788, enclosing the just-published McLean edition of *The Federalist.* Hamilton went straight to the point. "I take it for granted, Sir, you have concluded to comply with what will no doubt be the general call of the country in relation to the new government," he wrote. "It is to little purpose to have *introduced* a system, if the weightiest influence is not given to its firm *establishment* in the outset." The general's reply was affectionate but noncommittal. True to character, Washington worried that if he returned to public life, "the world and Posterity might *accuse* me of *inconsistency* and *ambition.*" Having firmly resolved to resume his private station when he resigned his military command, Cincinnatus fretted that he would risk his reputation even by acceding to his countrymen's just call.

This was enough of an opening for Hamilton to send off a lawyer's brief of a letter that turned the general's concern for reputation back upon itself. Washington had no choice in the matter, he argued. "It would be inglorious in such a situation not to hazard the glory however great" that he had already earned. Washington's "signature to the proposed system" was a further pledge on both his honor and the reputation of the convention. Should the new government fail for his lack of participation, all of the framers could then be faulted for having "pulled down one Utopia" (the Confederation) only "to build up another" (the Constitution). By withholding "your future aid to the system," he warned, you would risk "greater hazard to that fame" that is so "dear to you" than he would by honoring his stated intention of 1784.

Washington replied once more, wavering still, and Hamilton answered with a brief note, repeating his belief that Washington's service was "indispensable." There the exchange ended. In the end, Washington hardly needed the counseling of Hamilton to accept the presidency when the unanimous electoral vote he garnered in the first federal elections offered the best proof of his reputation. But enough had been said to demonstrate that their mutual trust was instantaneous and genuine. The letters were couched in terms of mutual respect, confidence, and affection. Each man had become, in that revealing eighteenth-century usage, "my dear Sir," not in their formal salutations, but at those points where they wrote with the greatest candor. No longer the impetuous officer who had contrived an exit from Washington's family in

1780, then pestered him for the combat command he coveted, Hamilton was once again the sage and precocious counselor his superior had always valued.

The remarkable candor of these letters extends to Hamilton's direct appeal to the theme of fame. He could not have struck that chord so boldly had he not long ruminated on the subject himself. The reference to fame as "the ruling passion of the noblest minds" was not some throwaway literary reflection. It expressed Hamilton's own passion, evidence that his prominent place in the New York bar and in state politics was not the height of his ambition. By appealing to Washington's sense of fame, he was laying a foundation for his own.

Fame was again on Hamilton's mind when he eulogized Nathanael Greene at the July 1789 meeting of the Cincinnati. Amid "those great revolutions which occasionally convulse society," he reminded his fellow officers, "human nature never fails to be brought forward in its brightest as well as blackest colors." The image of Benedict Arnold must have flashed through the minds of all present. Yet among the "advantages which compensate for the evils" these convulsions produce, Hamilton went on, was this: "they serve to bring to light talents and virtues which might otherwise have languished in obscurity or only shot forth a few scattered and wandering rays." Many of those present could ask how well that lesson applied to themselves. None could doubt it fit Hamilton as perfectly as the man he honored. Greene "descended from reputable parents, but not placed by birth in that elevated rank, which under a monarchy is the only sure road" to preferment, his ability would have gone unrecognized, and Greene would have remained "scarcely conscious of the resources of his own mind." The Revolution had intervened, allowing Greene to discover "the vigor of his genius."

More than a tribute to a friend, this passage was also a meditation on the experience of a generation and an autobiographical reflection that everyone in the room immediately recognized. Yet his contrast between monarchical patronage and republican opportunity masked a more complex aspect of Hamilton's thinking. Not long after Washington arrived in New York on April 23—still irked that Congress had taken so long to muster the quorum needed to confirm his election—the two men passed an evening together, and Washington asked his advice about "the etiquette proper to be observed by the President." Hamilton replied with a thoughtful memorandum regretting that American "notions of equality" were "too strong to admit of such a distance being placed between the President and other branches of the government" as he might wish. Still, one could try. The "door of access" to the president should be limited, he proposed, to "heads of departments," foreign ambassadors, and senators, because they were "in a degree his constitutional counsellors." Mere members of the House—the people's own representatives—did not merit this access because they lacked senators' constitutional connection to the executive.

Exactly when Washington fixed on nominating Hamilton to the Treasury is not certain. "Hamilton is most talked of" for the position, Madison noted in late June. On September 2, Washington signed the bills creating the departments of state, treasury, and war. He sent Hamilton's nomination to the Senate on Friday, September 11. He was confirmed that afternoon and went to work that weekend. Perhaps he attended nearby Trinity Church that Sunday to seek divine support for the "arduous trust" he had assumed. But it is certain that he visited the comte de Moustier, the French minister to the new government, and launched his official correspondence with a letter to Thomas Willing, Robert Morris's old partner, asking for a loan of $50,000 from the Philadelphia-based Bank of North America. "It will always give me pleasure to promote the interest of the Institution over which you preside," Hamilton noted in closing.

While Hamilton eagerly took up his duties, members of the new Congress were looking forward to their first recess. They had been sitting since April, and in peacetime spending six months away from home was a lot to ask. Congress had done its chief financial business in the spring, haggling over the schedule of duties on imports. That was a stopgap measure, designed to raise quick revenue from the spring shipments of European goods in order to cover the operating expenses of government. A different kind of haggling preoccupied Congress during the summer: the location of a permanent national capital. When it adjourned, it had done nothing about the massive public debt accrued during the war, much less considered a plan for establishing public credit. Instead the House of Representatives instructed Hamilton to "prepare a plan" to provide "for the support of the public credit." As part of his charge he was asked to correspond with the governors of the states to determine the size of their own debts and the amount of continental securities they held in their treasuries.

Hamilton had three months to complete this task, and the result was the first great state paper—arguably the greatest state paper—in the nation's history. It was a task for which he had been preparing a good decade, going back to his wartime studies of public finance and his active involvement in the politics of revenue reform in the early 1780s. It was also a task for which he was superbly qualified, much as Madison was the right person to lead the great constitutional project of 1787. Both men understood the value of seizing the political initiative by setting the agenda to which others—less enterprising, industrious, or brilliant—would have to respond.

From the outset Hamilton made the free flow of British goods into American harbors a foundation for his entire policy. The best thinkers on political economy agreed that duties levied on imports were the most productive sources of revenue and the easiest to collect. Unlike land or poll taxes, they did not bring hardworking, property-owning citizens into direct and unpleasant contact with the state. Instead they silently added a marginal cost to the prices consumers would pay for those extra things that made life more pleasant. For the foreseeable future, Hamilton rightly understood that Britain would have the only economy capable of supplying American wants and tastes. Rather than encourage patriotic Americans to spurn goods coming from their recent enemy, the restoration of public credit required them to consume the objects British manufacturers were eager to provide.

This strategy, however, brought Hamilton into direct conflict with Madison, who strongly favored a plan of *discrimination*, imposing higher duties on goods coming from nations that had not negotiated treaties of commerce with the United States. That category included Britain, which still barred American ships from imperial ports. Madison wanted to use discrimination to pry commercial concessions from Britain. Hamilton hoped the concessions would come as normal relations were restored. In the meantime the establishment of public credit came first.

Madison failed to carry his point in Congress, but his potential opposition remained a problem. Hamilton missed a chance to court his support when he did not learn that Madison had stayed on in New York after Congress recessed, editing his notes from the Constitutional Convention while hoping to catch Jefferson when he debarked from France—the better to persuade him to accept his nomination as secretary of state. Once Madison left town, Hamilton sent a catch-up note after him, asking his thoughts on additional sources of revenue and the public debt. But Hamilton left a more revealing mark of his opinion of Madison in a visit he paid to George Beckwith, a former British officer acting as his country's unaccredited emissary to the United States. His purpose was to suggest that the time was ripe "to form a Commercial treaty with you to Every Extent to which you may think it for Your interest to go." True, the United States would retain its ties to France. But British "productions" were more to American tastes, "nor do our raw Materials suit her so well as they do you." Hamilton made his own preferences clear. "I have always preferred a Connexion with you, to that of any other Country," he declared. "*We think in English*, and have a similarity of prejudices, and of predilections." Why could they not therefore share interests as well?

The conversation naturally turned to the discrimination issue. "I was much surprized," Beckwith noted, to find among those "decidedly hostile to us in their public conduct" a man "whose Character for good sense" would predict "a very different conduct." "You mean Mr. Maddison," Hamilton answered, confessing he was surprised as well. Madison was "uncorrupted and incorruptible," he conceded. But he added a gratuitous observation meant to curry Beckwith's favor at his former ally's expense. "The truth is, that although this gentleman is a clever man, he is very little Acquainted with the world."

Whatever this disparaging remark meant, it was not lack of acquaintance with the world that explained Madison's views, but a plausible belief that Britain would never offer the United States a favorable treaty except under duress. As Hamilton prepared his report on public credit, he began with the very different assumption that he had conveyed to Beckwith. Maintaining the unhindered flow of British goods to American consumers would provide the critical revenues upon which his plan of public credit depended—if he could persuade Congress to accept it.

Hamilton was ready when the second session of Congress convened in early January 1790. His report, dated the ninth, went to the House five days later. That it was *his* report was clear from its opening paragraphs, which spoke of "the anxieties" its author naturally felt "from a just estimate of the difficulty of the task, from a well-founded diffidence of his own qualifications for executing it with success, and from a deep and solemn conviction of the momentous nature of the truth" that the House had pronounced by asking him to make "an *adequate* provision" for the public credit. Diffidence was the last quality that anyone ever associated with Hamilton. It was precisely because he was so well prepared for this task that the House had given it to him to begin with. He had not let them down. The report began with a characteristic statement of "plain and undeniable truths" about public credit and revenue. Here was an echo of his clarion voice in *The Federalist*, a voice that could sound more like Jefferson, with his penchant for the bold statement, than Madison, with his ingrained love for distinctions and nuance. He hoped to use that voice to present his ideas in person, and his good friend Elias Boudinot made just that suggestion when the House learned the report was ready. "It was a justifiable surmise," Boudinot gently noted, "that gentlemen would not be able clearly to comprehend so intricate a subject without oral illustration." But after a half-comic debate, which revealed that some members seemed to think they had to choose between an oral or written report, Hamilton lost his opportunity.

The "Report on Public Credit" was a landmark document for three reasons. The first and most obvious is its testimony to the breadth of its author's vision, which extended well beyond the great challenge of resolving the debt problem, the Confederation's chief bequest to the new government. For as Hamilton's admirers and detractors then and since have recognized, the report was the inaugural step in a larger program that could be either heralded for boldly imagining the formation of a modern nation-state or faulted for badly misjudging the temper of his countrymen. Over the next two years other reports followed, most notably the proposal for chartering a national bank, which Hamilton presented to Congress in December 1790.

Together these papers constituted a state-building project that could never have emerged from the serial debates of a body like the new Congress, which for decades would remain an assembly of amateur lawmakers who typically served a term or two before returning home and swearing off national office altogether—unless a nice sinecure as a customs collector came their way. Indeed, the very idea of a legislative *program* was still a novelty in the eighteenth century, when legislative calendars overflowed with petitions from communities and individuals seeking legal authority to resolve some parochial problem. Hamilton was right to assume that the pursuit of so visionary an idea of the national interest required the initiative that only an ambitious statesman could launch. But that perception in turn exposed the two other great legacies of the "Report on Public Credit" and its sequels. One was that his vision ran well ahead of what the Federalist movement of the late 1780s had needed to pursue its agenda of constitutional reform. That did not make it unrepresentative of Federalist thinking. But neither did it ensure that his program would command unequivocal support. The most troubling dissenter would be Madison, a legislative force in himself, now braced in his emerging opposition by a renewed alliance with Hamilton's cabinet rival, Thomas Jefferson. Hamilton's program was a work of creative statesmanship, but its Anglophilic aspects also sparked intense opposition, which gave surprising coherence to the national political alignments that began to form by 1792.

To overcome that opposition, Hamilton had to recruit a corps of loyal supporters in Congress, and that challenge defined the final great legacy of his project. In drafting his reports, Hamilton always acted as the servant of Congress—or more directly, the House of Representatives—responding to its requests and instructions. As secretary of the Treasury, he was uniquely obliged to do so because the long association in Anglo-American history between representation and taxation meant that his department was not wholly like other departments. (That was why the act establishing the U.S. Treasury immediately created the distinct positions of comptroller, treasurer, register, and auditor, each accountable to Congress as much as to the secretary.) But as Hamilton actively advocated for his program, his relation to Congress changed as well, and with it, the untested relation between the executive and the legislature. Hamilton was at once a servant of Congress, a specially trusted subordinate to the president, and a potential prime minister in terms of both his relation to Washington and to the Congress, over which the Treasury secretary was acquiring his own political influence. As his quarrel with Madison and Jefferson escalated from disputes over policy into disagreements over the Constitution, Hamilton's greatest asset—the trust of the president—was implicated as well.

For as Washington was reluctantly drawn into this controversy, the presidency, not Congress, became the focal point of American politics. Hamilton's great ambition was to give the country the apparatus of a nation-state, applying British lessons to American conditions. But his great achievement was to nationalize American politics, making his program and influence objects of public controversy while spurring the formation of national political parties. In this process, Jefferson and Madison ultimately proved greater innovators and better strategists. But they had to act as they did because they were confronting someone of Hamilton's genius.

At the outset, however, the dispute seemed to be a simple matter of public policy. The main point in controversy was Hamilton's proposal to combine the separate public debts of the nation and states into one consolidated mass. Beyond relieving the states of their leftover obligations from the Revolution, this federal *assumption* of state debts would strengthen the case for securing the permanent revenues that would accustom Americans to the novel act of paying taxes. But sensible as those objectives might be as public policy, the assumption plan faced two objections. Some states had already discharged much of their debt; for example, in Massachusetts, debt-reducing policies had provoked a popular uprising, Shays's Rebellion, which alarmed so many budding Federalists in 1787. Assumption would therefore not benefit the states equally. Then there was the matter of speculation. Since war's end, and especially since ratification, there had been an active market in the various instruments Congress and the states had issued. Should speculators who had bought at bargain prices really be paid as generously as Hamilton's report proposed to do?

For Hamilton the imperative of establishing public credit trumped the question of justice. The larger the public debt, the stronger the rationale would become for levying serious taxes and the more confidence future lenders would have in the government's capacity to pay its creditors. But why, Madison objected, should Americans pay higher taxes to reward current debt-holders who had acted for profit, not patriotism? Hamilton had reason to be surprised by the Virginian's opposition. In April 1783 the two had served on the committee charged with drafting a public address to accompany the financial resolutions Congress had just passed. Madison was its chief author, and near its close he condemned the idea of discriminating among different classes of creditors. "If the voice of humanity plead more loudly in favor of some than of others," he wrote, "the voice of policy, no less than of justice pleads in favor of all." Hamilton could not have said it better. In 1790 no less than 1783, "the voice of policy" dictated treating all creditors equally, to ensure that others would lend to the United States in full confidence. Another memory further inclined him to treat Madison's opposition as a personal betrayal: "a long conversation" they had held "in an afternoon's walk" during the Constitutional Convention, during which they agreed on "the expediency and propriety" of exactly the measure Hamilton was now proposing.

The two men conversed again before the House took up Hamilton's report. Madison did not dispute Hamilton's recollection but insisted that "the very considerable alienation of the debt" since their earlier conversations "essentially changed the state of the question." His full array of motives for opposing Hamilton's scheme was more complex, ranging from high principle to doubts whether assumption would benefit Virginia at all. "I go on the principle that a Public Debt is a Public curse," he wrote to Henry Lee ("Light-Horse Harry" of wartime fame), "and in a Rep[ublican] Govt. a greater than in any other." Perhaps Hamilton would not have gone so far as to call the debt a public blessing, worth perpetuating for its own sake. He certainly believed that a national debt could be instrumentally useful. By justifying the taxes required to fund it, it would prove that the United States would act in good faith whenever it had to enter credit markets, whether at home or abroad. Moreover, by regularly paying interest to its current holders, the debt would create a powerful constituency—an interest group in a literal sense—with a stake in supporting the national government. Regarding the states more as a historical nuisance than a tribute to the beauties of federalism, Hamilton saw assumption as an opportunity to teach influential citizens that they could derive greater benefits from a nation-state than a nation of semi-sovereign states.

The first converts he had to make, though, were in the House, and here Hamilton faced an uphill struggle. Test votes in the committee of the whole revealed how closely the House was divided. In early April Madison rightly predicted that the fate of assumption might turn on a single vote. A week later the House rejected this key plank in Hamilton's program, 31–29. Even had the vote been reversed, Madison thought, "It would certainly be wrong to force an affirmative decision on so important and controvertible a point—by a bare majority." Yet he did not extend his doubts about the policy to malign the motives of its author. "The Novelty and difficulty of the Task he had to execute form no small apology for his errors."

Far from dooming assumption, the April vote only gave way to further discussion. At this point, the politics of funding the debt intersected with another issue that was just as divisive. This was the location of a permanent capital, a question that split both houses into regional and provincial blocs, based on obvious calculations of local interest dressed up in noble rhetoric, which fooled no one. The Pennsylvania delegation, led by Senator Robert Morris, hoped to restore Philadelphia as the capital. The Virginians, led by Madison, Jefferson, and not least important, the president, were equally intent on creating a new federal city along the Potomac, which they still fancied could become a major avenue of commerce into the interior.

As painstaking historical detective work has established, the deal making that allowed the assumption plan to pass the House in exchange for a decision to build a capital on the Potomac did not occur solely because Jefferson, on or about June 20, invited a despondent Hamilton and a wary Madison to a conciliatory dinner at his new lodgings at 57 Maiden Lane. Some kind of bargain between Virginia and Pennsylvania, the nation's most populous states, was already afoot, under which the government would temporarily return to Philadelphia while work on a Potomac site began. But dinner chez Jefferson significantly advanced the negotiations.

Jefferson left two accounts of how that famous meal came about. The first, probably written in the 1790s, was almost certainly the basis for the second, set down in 1818 in the introductory passage to *The Anas*, his memoir of the disputes of the 1790s. The two accounts agree in most respects. Jefferson recalled that he was on his way to Washington's residence when he bumped into Hamilton, looking "dejected beyond description" and even unkempt. Either they "stood in the street" or tramped "backwards & forwards before the President's door for half an hour." Hamilton unburdened himself over the looming fate of his plan for public credit and asked for Jefferson's support. Though they managed "separate departments, yet the administration and it's success was a common concern," Hamilton observed, and "we should make common cause in supporting one another."

In *The Anas* Jefferson embellished this point in an intriguing way. Now he recalled Hamilton stating that they had a common duty to "rally around" the president, who "was the center on which all administrative questions ultimately rested," and to "support with joint efforts measures approved by him." If that was indeed how Hamilton posed the issue, the logical course of action was for both men to knock on Washington's door and take the matter up with him directly. That was not what happened. Instead, Jefferson extended his dinner invitation to meet Madison and seek "a compromise which was to save the union."

Perhaps the glimmer of difference between these two accounts should not be pushed too far. Yet they do implicate an intriguing aspect of this key episode. Was the president more a head of state, chiefly responsible for executing the laws enacted by Congress, or the head of an administration with its own program to promote? There is no doubt that Washington fully supported Hamilton's plan for public credit or that he was privy to the details of the project that would place the capital a few miles from Mount Vernon. But the relation between the political wishes of the president and his constitutional duty was yet to be drawn. If Hamilton did indeed appeal to Jefferson on the basis of official loyalty to Washington, he was invoking a model of presidential leadership that had yet to be clearly established. A year earlier Hamilton had justified denying presidential access to members of the House on the premise that the president had no special connection to the House, the chamber with first call on all revenue measures. The influence that Washington was quietly trying to wield on the location of the capital owed less to his official authority as president than to his personal stature. It was Hamilton's administration, not the president's, that was in jeopardy, which was why he told Jefferson that he "was determined to resign" if Congress rejected his plan. Washington could hardly have made the same threat.

It took another month of maneuvering to convert this dinnertime understanding, whereby Madison would relax his opposition to assumption in exchange for Hamilton's support for the Potomac, into the firmly sealed Compromise of 1790. To satisfy the Virginians, Hamilton offered to engage in some creative bookkeeping that would adjust the sum credited to Virginia in the final settlement of accounts between the Union and the states. The Pennsylvanians still hoped that Congress, once settled in urbane Philadelphia, would balk at the expense of removing to the mosquito-infested Potomac marshlands. The whole affair of locating the capital, Madison noted, was "a labyrinth for which the votes printed" in the newspapers "furnish no clue."

Yet the national government had escaped this maze intact, and this was no small feat. Just as important, this first fabled compromise left Hamilton free to pursue his broader agenda. Before recessing in August 1790, the House instructed the secretary to report on "such further provision as may, in his opinion, be necessary for establishing the public credit." His reports were due when Congress reconvened at Philadelphia four months hence. This was, again, Hamilton's charge alone. Washington not only meant to decamp to Mount Vernon as soon as possible, but he also desired "to have my mind as free from public cares during my absence." Working right up to his December 13 deadline, Hamilton sent his two supplemental reports to the president only at the same time as he conveyed them to the House.

The first report proposed to raise additional revenues through excises on spirits brewed at home and duties on imported liquors—such as the hundred cases of Champagne, Sauternes, Bordeaux, and Frontignan that Jefferson, in his secondary role of sommelier of state, had ordered from France on the president's behalf. The original "Report on Public Credit" contained similar proposals that Congress had not accepted. Hamilton was certainly right to think that liquor taxes could prove highly productive. One of the great social changes that the Revolution somehow inspired was a growth in the per capita downing of spirits so prodigious that one modern scholar has rightly labeled the new nation the "alcoholic republic." The catch with an excise lay in its collection. Here Hamilton used this supplemental report to explain that its success would require effective enforcement by the revenue officers whose conduct he would oversee. "As long as men" possessed "unequal portions of rectitude," Hamilton noted, "the most conscientious will pay most; the least conscientious, least." The virtue of paying taxes was a lesson Hamilton ardently wanted his countrymen to learn, not one they already knew.

Another lesson needed teaching, and that was the subject of the paper Hamilton filed the next day. This was the "Report on a National Bank," the subject he had been pondering and advocating since his wartime reading on public finance. Like the original "Report on Public Credit," it began on a didactic note. "Public opinion being the ultimate arbiter of every measure of Government," it was essential "to accompany the origination of any new proposition with explanations, which the superior information of those, to whom it is immediately addressed, would render superfluous." This was a bit of flattery to congressmen who Hamilton suspected could also profit from a serious explanation of the advantages a bank could bestow. Banks were still a novelty to Americans. As Hamilton noted, the nation as yet had only three, all chartered under terms wholly inadequate to the government's present and future needs. That these were *needs*, and not merely optional benefits, was one of the great themes of his report. While ordinary citizens could certainly learn from the lessons he laid out, there was a deficit of congressional information to correct as well.

Most congressmen welcomed Hamilton's tutorial. While the House took up the excise first, discussion of the bank bill began in the Senate. The measure passed quickly with few changes and went to the lower chamber on January 20, 1791. There too its prospects seemed good—until Madison opened a surprise challenge to the bill, on grounds not of policy but constitutionality. Nothing in the Constitution, he argued, empowered Congress to issue charters of incorporation. Madison well knew, and now informed his fellow congressmen, that just such a power had been proposed to the convention in its last days, but rejected. That motion had come from him, and he took its rejection to mean that the convention had denied Congress that power. Granting charters of incorporation was not, in his view, a trivial or incidental act. Corporate privileges were quasi-monopolistic in nature. They gave substantial advantages to their holders, including the right to adopt bylaws with legal force of their own. The power to incorporate was consequential enough to merit explicit recognition. It could not simply be inferred.

Neither Hamilton nor most members of Congress found this objection persuasive. Everyone knew that Madison had been a dominant presence at the convention and that he had kept detailed notes of its debates. Yet there was no precedent for his appeal to these notes, which remained unpublished, as did the official journal of motions and votes that was in Washington's safekeeping. Madison could ascribe one meaning to the convention's rejection of his motion. But how could Congress rely on his impressions or interpretation? After all, there was another plausible explanation for the convention's action. The delegates might have rejected an explicit power of incorporation simply because other clauses already covered it by implication or *construction.* The obvious reply to Madison was that any combination of the taxing and spending powers vested in Congress and the discretionary sweep of the Necessary and Proper Clause could enable it to charter an institution from which, as Hamilton's report ably explained, so many benefits would flow.

In the House, Madison's objections proved unavailing. The bank bill passed easily on February 8 and was sent to the president for his signature. "The prudence of the President is an anchor of safety to us," Jefferson wrote the next day. This terse remark suggests that he and Madison already hoped that Washington would veto the act on constitutional grounds. The president was certainly open to that idea. Respecting Madison as he did, especially in matters constitutional, Washington first asked the two Virginians in his cabinet, Attorney General Edmund Randolph and Jefferson, for their opinions. Both sided with Madison. Jefferson ended his memorandum on a tempered note, however. Unless the president firmly believed the measure to be unconstitutional, he should defer to the judgment of Congress. "It is chiefly for cases where they are clearly misled by error, ambition, or interest," Jefferson concluded, "that the constitution has placed a check in the negative of the President."

This final fillip was as much a matter of interpretation as the textual analysis that formed the bulk of Jefferson's memorandum. The Constitution says nothing about the purposes of the veto. If polled, the framers would probably have said that they intended the president to be able to protect his office against "encroachments" from Madison's "impetuous vortex" of Congress. Many would have agreed that the president's sworn duty to "preserve, protect, and defend the Constitution" required him to veto acts he deemed clearly unconstitutional, while allowing the collective judgment of a congressional supermajority to override his individual scruples. But Jefferson's reference to "error, ambition, or interest" insinuated the existence of other, avowedly political criteria of judgment. To say that the veto was "chiefly" designed to make the president a judge of the motives for legislation was to make him a monitor of the conduct of legislators—and perhaps even of ministers who misled Congress into "error." There could be little doubt whom Jefferson wished to hold accountable.

There was, however, one more voice to be heard. Just as it made sense to let the critics speak first, so Hamilton deserved a right of rebuttal. Even though it was meant for an audience of one, Hamilton's "Opinion on the Constitutionality of an Act to Establish a Bank" is better read as a state paper than a simple memorandum of advice. Running thirty-eight pages in the modern edition of his papers, it dwarfs the productions of Randolph and Jefferson. In nuance, detail, and rhetorical pitch, it is also far more compelling and intellectually sophisticated. Like his other papers, it begins by confessing its author's "uncommon solicitude" for his task, driven in part by personal concern for "the successful administration of the department under his particular care," but also by "a firm persuasion" that the "principles of construction espoused" by his cabinet colleagues "would be fatal to the just & indispensible authority of the United States." The principle at stake seemed plain: "every power vested in a Government is in its nature *sovereign*, and includes by *force* of the *term*, a right to employ all the *means* requisite, and fairly *applicable* to the attainment of the *ends* of such power." In his view, this was a sufficient proof for a congressional power of incorporation. If Hamilton deigned to answer the Virginians' other objections, it was only because the president deserved full reassurance on these points. So the lion of the Wall Street bar set to work, knowing that the president needed not so much to be persuaded on every point as shown that each objection could be plausibly countered.

Batting last in the order of advisers gave Hamilton an edge he fully exploited. But his real advantage lay in the common lessons that he and the president drew from the war. Reading Jefferson's carping objections surely reminded Washington of the feeble responses his wartime pleas for support had elicited from Congress and the states. Jefferson wrote as if the great task was to maintain a precious semantic distinction between what was truly necessary—that is, indispensable—and what was merely "convenient." So long as the government could find other means of managing its financial needs, the fact that a bank would be convenient could not satisfy the stricter criterion of necessity. But necessity could also take a more pointed meaning drawn from the bitter lessons of war that Hamilton and Washington had never forgotten. On February 25 Washington signed the bank bill into law, closing one dispute within the cabinet and unknowingly opening a broader one beyond it.

Three days later Jefferson wrote to Philip Freneau, Madison's college classmate and one of the nation's leading poets. Might he be interested, Jefferson asked, in "the clerkship for foreign languages" in his department? "The salary indeed is very low," but there was "so little to do as not to interfere with any other calling the person may chuse." This sounded as if Jefferson the aesthete wanted to become a patron of poetry. But in fact he and Madison had other literary labors in mind. They had hatched a plan to turn Freneau, who was known to have journalistic aspirations, into a political editor and place him in charge of a new newspaper, eventually known as the *National Gazette*. In support of his policies, Hamilton had enjoyed the consistent aid of John Fenno's *Gazette of the United States*. Now his opponents proposed to pursue the dispute within government by appealing to public opinion outside it.

When Jefferson, Randolph, and Hamilton wrote their dueling memoranda on the bank, they aimed to sway the opinion of one. Washington, in turn, had exercised his constitutional power to require the secretaries to give their "Opinion, in writing . . . upon any Subject relating to the Duties of their respective Offices." It was a very different opinion—public opinion at large—that Jefferson and Madison wanted Freneau to help them sway. In one sense, there was nothing new about this. American politicians had been using the press in this way for decades. Yet something had changed since the late 1780s. When the great question of constitutional ratification broke upon the nation in 1787, publication of legislative debates was still a novelty. But the debates in the state conventions were covered in something like real time, and that set a precedent for the new government. While the Senate kept its doors closed until 1795, the House opened its galleries from the start, and those attending included reporters who did their best to record speeches as they were delivered. The opportunity now existed for citizens to react to measures as they made their way through Congress, and thus for public opinion, however partially expressed or crudely measured, to influence the legislative process.

It was, therefore, a new and dynamic conception of public opinion that the emerging party leaders of the early 1790s planned to exploit. That it was something to reckon with they all agreed. Hamilton called it "the ultimate arbiter of every measure of Government." In hoping that the president might veto the bank bill, Jefferson also recognized that "the opinion of the public, even when it is wrong, ought to be respected to a certain degree." Madison would soon open a new series of essays that he contributed to Freneau's paper with this declaration: "Public opinion sets bounds to every government, and is the real sovereign in every free one."

It took some wheedling by the Virginians to enlist Freneau in their cause. While those discussions were beginning, Jefferson and Madison took a late-spring jaunt through New York and New England, sailing north up the Hudson and then returning along the Connecticut in the well-traveled Jefferson phaeton. Though their travel journals suggest that they really were going as tourists, Hamilton heard that politics was on their agenda too. In New York City, his friend Robert Troup reported, the two men had closeted in "pasionate courtship" with Aaron Burr and Robert Livingston, known foes of Hamilton. "Delenda est Carthago I suppose is the Maxim adopted with respect to you," Troup surmised. "They had better be quiet, for if they succeed they will tumble the fabric of the government in ruins to the ground." Still, if their animosity toward Hamilton was so fierce, it seems surprising that they accepted the hospitality of his in-laws when they reached Albany and toured the Saratoga battlefield with his brother-in-law John Schuyler.

Yet Troup was right to suspect that principled questions about "the fabric of the government" were coming into play. Just as Jefferson suspected Hamilton of monarchical leanings, so Hamilton regarded Jefferson as a closet Anti-Federalist who had never abandoned his initial misgivings about the Constitution. By the summer of 1791 the rift between the two men was irreparable, but for the moment it affected only the rivals and a few confidants. Then Freneau, coaxed by Madison, finally accepted Jefferson's offer of employment. The first issue of his *Gazette* appeared on the last day of October. Jefferson never wrote for it, but Madison was an early contributor, starting with a brief article, "Public Opinion," on December 19. He returned to that subject a month later in the essay "Charters." It began by reminding readers of the Revolution's most lasting achievement. "In Europe, charters of liberty have been granted by power," meaning that governments sometimes doled out rights to their subjects. "America has set the example, and France has followed it, of charters of power granted by liberty," meaning that free peoples could now write constitutions vesting their governments with such power as they chose to delegate. This achievement made the present age "the most triumphant" in history. But its preservation, Madison went on, ultimately depended on "public opinion." Indeed, "the stability of all governments and the security of all rights may be traced to the same source." In the United States this principle had a particular meaning. The "public opinion" of the American people "should attach itself to their governments as delineated in the *great charters*," the written constitutions they had adopted.

If these essays read like a civics primer, it was because Madison was laying down the rudiments of a party platform. It was to be a party defined by principles rather than a program. It would take a great deal of reading between the lines to determine what, if anything, the author of these pieces wanted government to *do*. The real actors appeared to be the people themselves. Theirs was the duty to enforce "the partitions and internal checks of power." Their "eyes must ever be ready to mark, their voice to pronounce, and their arm to repel or repair aggressions on the authority of their constitutions."

None of these dire warnings would need utterance now unless the Constitution and the republic were under some unspecified threat. Still, these essays and other pieces that Freneau initially published hardly mounted a frontal assault on either Hamilton or his policies. He had other, more personal worries as the election year of 1792 opened. Not long after the government moved to Philadelphia, with his family remaining at New York, Hamilton began an affair with Maria Reynolds, a young, attractive married woman who called one day seeking the largesse he was known to dispense. Some frisson of desire passed between them, and their affair began that night at her lodgings. Late in 1791 the absent James Reynolds, whom Maria initially portrayed as an abusive husband, returned to the city. Soon Hamilton was receiving visits and notes from the ostensibly offended spouse, some seeking his patronage, others threatening blackmail. Maria wrote too, in the pathetic voice of the neglected lover. A halfhearted effort to break off the affair gave way to lingering desires, while fear for the loss of reputation and concern for family left Hamilton in a state of inner torment.

Not so tormented, however, that he could ignore the political challenge he now faced. From its moderate beginnings, Freneau's paper was becoming a partisan rag, repeatedly attacking Hamilton and his measures. Madison was clearly the leader of an opposition faction in Congress. In late February the president finally learned of the animosity between the two secretaries. On Leap Year Day, Jefferson visited his lodgings early, to continue a subject they had begun discussing the day before, when Jefferson had suggested moving the post office business from the Treasury to his own department. He was not seeking power, Jefferson noted, for he meant to stay in office no longer than the president. As a matter of policy, the move made sense. If the purpose of the post was not to raise revenue but to disseminate information across the republic—and thereby inform public opinion—then it belonged among the domestic duties of the State Department. But his "real wish was to avail the public of every occasion during the residue of the President's period to place things on a safe footing."

"Safe" meant this: not under Hamilton's control. When Jefferson returned the next morning, perhaps after Washington had finished one of his 5 A.M. rides, they breakfasted, then retired to his study. The conversation turned to Jefferson's hint that he might retire, and they vied over which of them wanted to be back in Virginia more. Washington hoped that Jefferson would stay on after his own return to Mount Vernon—thereby indicating uncertainty as to whether a new president would be free to replace his predecessor's ministers with his own. He tried to mollify Jefferson by assuring him that its multiple duties made his office "much more important" than the Treasury, with its "single object of revenue." That consolation overlooked one key fact: control of the post office would give Jefferson a source of patronage to rival Hamilton's closely supervised network of revenue agents. The president also worried that Jefferson's leaving would aggravate the recent "symptoms of dissatisfaction" with the government.

That remark gave Jefferson an opening to air his real misgivings. "There was only a single source of these discontents," Jefferson answered, launching into a litany of complaints that staggered Washington with their breadth and intensity. Jefferson made the wild charge that Hamilton was a one-man engine of corruption. He was destroying the morals of "our citizens," seducing them into speculating with paper money, tempting them away "from the pursuits of commerce, manufacturing, buildings & other branches of useful industry." He was corrupting members of Congress, encouraging them to invest in the public debt so they would support his program. His latest report, proposing subsidies for manufacturers, was more dangerous than the report on the bank. If adopted under the doctrine that Congress could tax and spend merely to promote the general welfare, it would destroy the "limited" nature of the federal government.

Washington said nothing to Hamilton about Jefferson's charges. But Hamilton did not need to have the president disclose what his friends were telling him, what Freneau's *Gazette* regularly revealed, and what his dealings with Jefferson further confirmed. The evidence "comes through so many channels and so well accords with what falls under my own observation," he wrote to Edward Carrington in late May, that he could "entertain no doubt" that Jefferson and Madison were leading *"a faction decidedly hostile to me and my administration"* and doing so in accordance with ideas *"subversive of the principles of good government and dangerous to the union, peace, and happiness of the Country."* Carrington, a former army officer whom Hamilton had known since 1780, was now federal supervisor of revenue for Virginia. But he also corresponded with both Jefferson and Madison, and in giving Carrington pages of analysis of the words and deeds of his foes, Hamilton was striking behind their lines, trying to impugn their motives in their native state. "If I were disposed to promote Monarchy & overthrow State Governments," he concluded, "I would mount the hobby horse of Popularity—I would cry out usurpation—danger to liberty &c." He would, that is, act just as Jefferson (or his surrogates) was acting now. Yet Hamilton could not yet bring himself to conclude that the Virginian, though "'A man of profound ambition & violent passions,'" was quite *that* demagogic.

Hamilton might have escaped this last ounce of doubt had he read the letter Jefferson had written Washington three days earlier. A "corrupt squadron" (whose unnamed captain could only be Hamilton) meant to change "the present republican form of government, to that of a monarchy" modeled on Britain, Jefferson warned. But he did not ask Washington to remove his rival; rather, he hoped Washington would simply defer his known desire for retirement. Yet again the president had to bear the unique flattery he attracted. "There is sometimes an eminence of character on which society have such peculiar claims as to controul the predilection of the individual for a particular walk of happiness"—that word again! That is "your condition," a "law imposed on you by providence in forming your character." Still, the president could take heart from the naive hope Jefferson voiced: that the coming congressional elections would break Hamilton's influence and enable him to retire to Mount Vernon well before his second term expired.

That was where Jefferson sent this letter, hoping that Washington would ponder it all the more seriously away from the press of business. But the letter missed its recipient on his way back to Philadelphia. Washington delayed taking up the charges with Jefferson until July 10. He found the whole subject "painful," he apologized, because of his own agonizing over retirement. The president was worried that "a decay of his hearing" might foretell that "other faculties might fall off & he not be sensible of it." But he also made it clear as the meridian sun (as he liked to say) that Jefferson's letter had neither fallen on a deaf ear nor drawn the response desired. The president could regard public attacks on the policies of the government only as attacks on his own stewardship. If these critics "thought there were measures pursued contrary to his sentiment," Washington warned, "they must conceive him too careless to attend to them or too stupid to understand them." Jefferson renewed his charge that the decisive votes in Congress were being cast by members under Hamilton's sway and repeated his concern that the public debt was being used to fix excessive taxes on the people. But "finding him really approving the treasury system," Jefferson backed off.

Washington waited three weeks to seek reassurance that he had answered Jefferson correctly. The president was back on the Potomac in late July when he wrote to Hamilton "in strict confidence, & with frankness & freedom," laying out twenty-one numbered concerns. These began with the size and disposition of the public debt but soon moved to the charges of corruption and crypto-monarchism. Washington did not mention Jefferson but instead intimated that he was reporting opinions he had heard on his travels, both "from sensible & moderate men" and former Anti-Federalists, including "my neighbour, & quandom friend Colo M[ason]." But it is unlikely that Hamilton was deceived. With its list of charges, Washington's letter implied that it was a response to some bill of particulars forced on his attention, not a public opinion survey he had compiled on his own.

Hamilton took a fortnight to reply. Once again, as with his opinion on the bank, he was devastatingly thorough, mingling a close defense of his program with ridicule of the hyperbolic notion that there might exist a covert plan "of introducing a monarchy or aristocracy into this Country." The whole idea was "utterly incredible," he scoffed; "its absurdity refutes itself." Such a "plot" would take over a lifetime to complete. "A people so enlightened and so diversified" as ours could never be "cajoled" in this way—unless "popular demagogues" arose to flatter "their prejudices," spark "their jealousies and apprehensions," and bring on the "anarchy" that anyone schooled in the history of ancient republics would recall was ever the seedbed of tyranny. Without naming names, Hamilton alluded to one potential demagogue both of them knew: not Jefferson but his newfound ally, Aaron Burr, a man "known to have scoffed in private at the principles of liberty" while also possessing "the advantage of military habits."

The task of answering Washington's letter reinforced Hamilton's determination to mount a public defense of his policies and conduct. He would have to do so anonymously, or rather pseudonymously, for even in fashioning the new phenomenon of national public opinion, the partisans preserved the eighteenth-century convention of the pen name. In the summer and fall of 1792, writing under at least nine pseudonyms, he dashed off essay after essay. Freneau was a particular object of scorn. Hamilton linked the poet-translator's public employment to his role as the surrogate mouthpiece of the secretary of state. Jefferson himself was a man who "arraigns, the principal measures of the Government" to casual acquaintances, and "with *indiscreet* if not *indecent* warmth." Yet he was the only officer of state to escape Freneau's censure. "Even the illustrious patriot who presides at the head of the Government has not escaped your envenomned shafts."

As to the illustrious patriot who wanted to play Cincinnatus returning to his plow one last time, the rivals were united on one point. That part was not ready for him yet. Were Washington to step aside, the most likely successor was Adams. Hamilton and Madison disliked the vice president profoundly. Jefferson's friendship with his old congressional and diplomatic colleague was already waning. Rather than risk the uncertainty of a contested election, all three united in urging the president to stay on. Acceding, Washington asked the rivals to make a sustained effort to replace their "wounding suspicions, and irritable charges" against each other with "liberal allowances—mutual forbearances—and temporising yieldings on *all sides*." If these two remarkably talented men shared the same commitment to the public weal that they demanded of him, the president must have thought, surely they could subordinate their "differences in political opinions" to the cause of maintaining the Constitution itself.

Washington asked more than either man was prepared to give. On September 9 both secretaries sat down—Hamilton in Philadelphia, Jefferson at Monticello—to answer his pleas for civility and cooperation. Writing with the "greatest frankness," Hamilton admitted that he had joined in the "retaliations which had fallen upon certain public characters." But he also declared that *"for the present"* he could not "recede" from the literary combat. After all, he and his policies had been the initial object and victim of Jefferson's assault, an assault which aimed at "the subversion of measures, which in its consequences would subvert the Government."

Jefferson was no more conciliatory. He also apologized for his "mortification at being myself a part" of the "internal dissentions" within the government. But his "deepest regret," he immediately added, was over having been "duped" by Hamilton "and made a tool for forwarding his schemes, not then sufficiently understood by me." The plain reference was to the bargain of 1790 over the assumption of state debts, a deal that Jefferson had freely initiated, and from which he, Madison, and Washington all thought their state had gained an advantage in the decision for the Potomac. Jefferson then returned to his familiar brief against Hamilton's system and its treacherous implications for constitutional government, tossing in a lengthy defense of the whole Freneau business for good measure. The letter closed on a bitterly personal note. He was resolved to honor his plan to retire at the moment when the Constitution "contemplated a periodical change or renewal of the public servants," Jefferson reminded the president. When he returned to Monticello—the home he now meant to rebuild completely—he would "not suffer my retirement to be clouded by the slanders of a man whose history, from the moment at which history can stoop to notice him, is a tissue of machinations against the liberty of the country which has not only received and given him bread, but heaped it's honors on his head."

But the chief source of those honors was Washington, and the "history" that Jefferson mocked covered those long years of military service and sacrifice that the president and Hamilton had shared. Jefferson had spent those same years first on his visionary project to revise the Virginia legal code and then in a governorship that illustrated the evils that Hamilton's system—and Madison's constitutional project—were designed to prevent. The president had already cautioned Jefferson not to assume that he was too stupid to understand the measures he had signed into law. In the end Hamilton's program was his own, not perhaps in every detail, but in the central objective of creating an American state that could surmount the problems Washington had encountered during the war. Next to the fanciful, even fantastic evils against which Jefferson railed, those that the president and his Treasury secretary recalled were real and vivid. Men had died needlessly because of them, and the war had dragged on years longer than it might have, making Washington a Fabius by circumstance, not choice.

As president, though, Washington could choose his course, and the letters of September 9 made such a choice unavoidable. Two more able or brilliant men have never served in the cabinet, much less at the same time. No doubt that was part of the problem. Their disputes over policies and principles fused with intense personal suspicions, which doomed Washington's sensible gesture at accommodation from the start. Jealousy and betrayal, gossip and honor were all elements of this story. Jefferson's bitter remarks on his rival seem little less petty than Hamilton's sniping against Jefferson and Madison. *"They have a womanish attachment to France,"* he complained to Carrington, *"and a womanish resentment against Great Britain,"* as if the Virginians' ideas of foreign policy were determined by flighty impulse. So too Hamilton's condescending jibe at Madison as a man "very little Acquainted with the world" looks less impressive coming from a loving husband and father who was still being blackmailed by the husband of the mistress he could not yet bring himself to abandon. Perhaps one could know the world too intimately.

Hamilton certainly knew the president more intimately than either of the two Virginians. If Washington had to choose between them, as Jefferson insisted he must, he would take the side of his wartime aide rather than his native countrymen. Making that choice also required Washington to act as something more than a presiding officer, coordinating and occasionally mediating the activities of subordinate ministers. By asking Washington to veto Hamilton's pet project of a bank, and then by pressing him to repudiate the secretary's program, Jefferson and Madison turned the president into the chief decision maker of his government. To control the government, one had to capture the presidency. Whenever the first incumbent retired—whether in 1796 or sooner—it would be essential to compete for that office.

That contest was still four years off, and September 9, 1792—the day the dueling letters were written—cannot be singled out as the birthday of America's first political parties. Other events, notably the foreign policy crises that would begin the next year, would still have to occur before the rivalry within the cabinet would develop into real competition between organized political coalitions. Jefferson remained in the cabinet for another year, postponing his retirement until 1794. "I am then to be liberated from the hated occupations of politics, and to remain in the bosom of my family, my farm, and my books," he wrote to one of the women whose charming company had most delighted him in Europe—ironically, none other than Hamilton's sister-in-law, Angelica Schuyler Church. "I have my house to build, my fields to farm, and to watch for the happiness of those who labor for mine." He meant his slaves, in what is surely the strangest use Jefferson ever made of his magical word. For what measure of happiness could a slave ever feel, the author of *Notes on the State of Virginia* had already asked, who was "born to live and labour for another," to "lock up the faculties of his nature" without the benefit of the public education Jefferson wanted to bestow on all free citizens, and to "entail his own miserable condition on the endless generations proceeding from him"?

Jefferson's dependence upon slavery was hardly lost to his political nemesis, who held very different notions of how the American economy should develop. To Hamilton it was only one more example of the gap between Jefferson's starry vision of the American future and the worldly realities he failed to confront. But there, perhaps, is where we should leave these two rivals over the meaning of the Revolution. Jefferson's vision had formed in advance of independence, from his education and reading, and was best conveyed in his precocious realization that "the whole object of the present controversy" was not to reclaim endangered rights but to institute new and idealized forms of government and the social arrangements—equitable land distribution, public education, an enlightened approach to religion—that would make them work. Hamilton's ideas too depended significantly on education and reading, sometimes in the works of authors, such as Hume, whom Jefferson despised. But they were tempered as well by lessons of experience that Jefferson was loath to absorb, lest they disrupt the vision he had already formed. If Hamilton can be regarded as America's Niccolò Machiavelli, as one recent biographer has suggested, Jefferson might be called its Thomas More—a utopian without the religious faith that sent More to the stake.

There was, however, one trait the two men shared with others of their generation, or rather, the overlapping cohorts who made the Revolution: those who, like Adams, Jefferson, and Washington, brought the nation to declare independence, and those who came of age with the struggle to secure it. Whatever their social origins, whether eminent like Jefferson's or Madison's, modest like John Adams's, or doubtful like Hamilton's, all of them shared that one characteristic that Hamilton memorialized in Nathanael Greene. "Those great revolutions which sometimes convulse society," Hamilton reminded his brother officers of the Cincinnati, had also this merit: "that they serve to bring to light talents and virtues which might otherwise have languished in obscurity or only shot forth a few scattered and wandering rays." How these unlikely provincial revolutionaries discovered the talents they did remains an enduring puzzle of our history, easier to ponder than finally to explain, yet one ever worth exploring.

ACKNOWLEDGMENTS

NOTES

SOURCES AND FURTHER READING

INDEX

Acknowledgments

Some time in 2001, before the catastrophe that will forever mark that year, my Stanford colleague David Kennedy offered me some useful advice. "You should talk to my agent," David said, and so I did. That turned out to be Don Lamm, who had been lured out of retirement from his former career as a book publisher to go over to the dark side of literary agency. I am extraordinarily grateful to Don for helping me to develop the ideas that led to *Revolutionaries*, for doing the other nice things that agents can do, and for regularly popping into my office to share his vast stock of anecdotes and, even better, his friendship. Along the way, some free contractual advice from my brother-in-law, Stephen Scharf, helped me understand why he is a player in the Southland.

Amanda Cook generously took this project on when its original editor fled to another firm. She has been the toughest editor I have ever worked with, repeatedly reminding me that the historical knowledge I assume everyone just knows may be a bit more obscure than I realize. She has a ruthless eye for redundancy and a good ear for the hollow note.

As writers go, I am rather a lone ranger about sharing my work, but versions of various chapters were tried out on willing audiences and readers whose comments I have always valued. These include my colleagues at the Department of History's faculty workshop; another workshop organized by Mark Brandon at Vanderbilt Law School; and my good buddy Ron Hoffman and his band at the Omohundro Institute of Early American History and Culture at the College of William and Mary. A very special thanks goes to my Evanston High classmate Cindy Huber Major and her reading group from Gunston Hall, that jewel of colonial architecture. Peter Onuf, my chief amigo in Charlottesville, read the chapter about his own special founder while our time in Oxford overlapped, and deemed it acceptable. Edie Gelles was always happy to respond to my queries about the Adamses, whom she knows so well.

Over the years, I benefited from the research assistance of a terrific group of students: Sonia Mittal, Piotr Kosicki, Adam Cooke, and Eric Zimmerman.

A sabbatical at the Center for Advanced Studies in the Behavioral Sciences was everything I was always told it would be, thanks to its former director, Claude Steele, and his wonderfully supportive staff. Thanks too to Deans Sharon Long and Arnold Rampersad of the School of Humanities and Sciences at Stanford for helping to make that year possible.

When I began working on this book, our sons Rob and Dan were just completing the Stanford and Palo Alto–based phases of their education; as I finish it, they have both recently received their graduate degrees and begun their careers as budding diplomatist and diplomat. After three decades on the coast, I like to believe that some kind of serendipitous synergy is at work here, especially when I recall how their grandfather began his academic career with similar interests. To Helen, my bride of so many years ago, I can only repeat how thankful I am for the good luck of having met her at the Freshers' Fair that first long-ago day in Edinburgh.

This book is dedicated to my mentor, Bernard Bailyn. I wandered into his graduate seminar in 1969, knowing little more about early American history than I had learned as editor of our "class book" on the Revolution in Mrs. Bingham's fifth-grade at the J. J. Finley School in Gainesville, Florida. For me, as for so many others, the understanding of history that I began to acquire in that seminar (and what followed) merely proved to be the decisive intellectual event of my life. Not only is Bud a truly creative historian; in his sheer enthusiasm for historical study he also remains the youngest scholar I know.

Notes

ABBREVIATIONS

AFC Lyman H. Butterfield et al., eds., *Adams Family Correspondence* (Cambridge 1963–)

AH: Writings Joanne Freeman, ed., *Alexander Hamilton: Writings* (New York, 2001)

BF: Writings J. A. Leo Lemay, ed., *Benjamin Franklin: Writings* (New York, 1987)

D&A Lyman H. Butterfield, ed., *Diary and Autobiography of John Adams* (Cambridge, 1961)

DPDC Ronald Hoffman et al., eds., *Dear Papa, Dear Charley . . .* (Chapel Hill, 2001)

FC Philip Kurland and Ralph Lerner, eds., *The Founders' Constitution* (Chicago, 1987)

GW: Writings John Rhodehamel, ed., *George Washington: Writings* (New York, 1997)

JCC Worthington C. Ford, ed., *Journals of the Continental Congress, 1774–1789* (Washington, D.C., 1904–1937)

JJ Richard B. Morris, ed., *John Jay: The Making of a Revolutionary* (New York, 1975–)

JM: Writings Jack N. Rakove, ed., *James Madison: Writings* (New York, 1999)

LDC Paul H. Smith, ed., *Letters of Delegates to Congress, 1774–1789* (Washington, D.C., 1976–2000)

PAH Jacob Cooke and Harold Syrett, eds., *The Papers of Alexander Hamilton* (New York, 1961–1979)

PBF Leonard W. Labaree et al., eds., *The Papers of Benjamin Franklin* (New Haven, 1959–)

PGM Robert Rutland, ed., *The Papers of George Mason, 1725–1792* (Chapel Hill, 1970)

PGW: CS W. W. Abbot et al., eds., *The Papers of George Washington: Colonial Series* (Charlottesville, 1983–1995)

PGW: RWS Philander Chase et al., eds., *The Papers of George Washington: Revolutionary War Series* (Charlottesville, 1985–)

PHL Philip Hamer, ed., *The Papers of Henry Laurens* (Columbia, S.C., 1968–)

PJA Robert Taylor et al., eds., *Papers of John Adams* (Cambridge, 1977–)

PJM William Hutchinson, William Rachal et al., eds., *The Papers of James Madison* (Chicago and Charlottesville, 1961–1991)

PMHB Trevor Colbourn, ed., "A Pennsylvania Farmer at the Court of King George: John Dickinson's London Letters, 1754–1756," *Pennsylvania Magazine of History and Biography,* 86 (1962), 241–86, 417–53

PNG Richard K. Showman et al., eds., *The Papers of Nathanael Greene* (Chapel Hill, 1976–)

PTJ Julian Boyd et al., eds., *The Papers of Thomas Jefferson* (Princeton, 1950–)
RDC Francis Wharton, ed., *Revolutionary Diplomatic Correspondence of the United States* (Washington, D.C., 1889)
RFC Max Farrand, ed., *Records of the Federal Convention of 1787* (New Haven, 1966)
TJ: Writings Merrill Peterson, ed., *Thomas Jefferson: Writings* (New York, 1984)
WSA Harry Alonzo Cushing, ed., *The Writings of Samuel Adams* (New York, 1904–1908)

page

Prologue

1 "great Chair . . . parasite": *D&A*, I, 13–14.
2 "Abrahams Faith . . . no farther": *D&A*, I, 28.
"last year . . . means of Happiness": *D&A*, I, 35–36.
"future Happiness": *PJA*, I, 13.
3 "not without Apprehensions": *PJA*, I, 19.
"Engines . . . Liberty to think": *PJA*, I, 17.
"my Loss . . . Stamp Act": *D&A*, I, 263.
"my Geers": *D&A*, I, 264–65.
"at a loss": *D&A*, I, 266–67.
5 "heard or seen": *PMHB* 86: 254.
"much inferior . . . August place": *PMHB* 86: 259.
6 "gay assembly . . . disgustful": *PMHB* 86: 262, 274, 254.
"young fellow's . . . inferiors": *PMHB* 86: 269–70.
"cheerful": *PMHB* 86: 263.
"eyes on the ground . . . blue ribbond": *PMHB* 86: 429–30.
7 "pride": *PMHB* 86: 278.
8 "'Easy to be bought'": *PMHB* 86: 268.
"old-fashioned religion . . . all empires": *PMHB* 86: 421–22.
"greatest men" and "brutal, mean": *PMHB* 86: 257, 265.
"resolved to remember": *PMHB* 86: 428.
9 "Bulletts whistle": *GW: Writings*, 48.
12 "frequently tumbling": *GW: Writings*, 43.
13 "violent illness . . . rivulets": *PGW: CS*, I, 331, 336–37, 339–40.
18 "lawgivers of antiquity": Adams's *Thoughts on Government* is reprinted in *FC*, I, 110.
"speech-making": Hannah Arendt, *On Revolution* (New York, 1963), 26.
21 "supreme, irresistible": Sir William Blackstone, *Commentaries on the Laws of England*, Bk. I, §2, 49.
22 "hushed into silence": *D&A*, I, 324.

1. Advocates for the Cause

29 "Salvation of their Country": L. F. S. Upton, "Proceedings of the Body Respecting Ye Tea," *William and Mary Quarterly*, 22 (1965), 297.
31 "the Medes and the Persians": *Collections of the Connecticut Historical Society*, 2 (1870), 140.
37 "his Duty": *Wm. & M. Quarterly*, 22: 297.
43 "Labours and Drudgery": *D&A*, I, 38–39.
"What is the End": *D&A*, I, 337–38.
"a fixed Resolution . . . declared Enemies": *D&A*, II, 63.
"desirous to avoid": *D&A*, II, 73–74.
44 "gallant, generous": *D&A*, II, 79.
"Brace of Adams's": *D&A*, II, 54–55.
"Bone of our Bone": *D&A*, II, 81.
"a Part in public Life": *D&A*, II, 82.
"a Dignity . . . our Calamities": *D&A*, II, 85–86.
45 "a double share": *WSA*, III, 105.
"not Spirit enough": *PJA*, II, 83.
"patience, and Industry": *D&A*, II, 95.
48 "vengeful Stroke": *WSA*, III, 114.
49 "must suffer Martyrdom": *AFC*, I, 107.
50 "grand Scene": *D&A*, II, 96.
"Court of Ariopagus": *PJA*, II, 99.
"furbishing up": *AFC*, I, 115.
"American Senator" and "New England Educations": *PJA*, II, 109.
51 "prepared an entertainment": *D&A*, II, 97.
"all the Bells": *D&A*, II, 100.
solid Lead . . . as neat": *D&A*, II, 103–4.
"talk very loud": *D&A*, II, 109.
"is a Champaign": *D&A*, II, 111.
52 "Dirty, dusty": *D&A*, II, 114.

"hanging [of] the Quakers": *D&A*, II, 107.

53 "Phrenzy" and other expressions: Clarence E. Carter, ed., *The Correspondence of General Thomas Gage with the Secretaries of State, 1763–1775* (New Haven, 1931), I, 380, 367, 368, 371.
"constitutional death": *D&A*, II, 121.
"excellent Library": *D&A*, II, 122.

54 "Saml. Adams": *D&A*, II, 115.
"Field of Controversy": *D&A*, II, 123.
"Fleets and Armies": *D&A*, II, 124.
"unprepared with Materials": *LDC*, I, 31.

55 "fired on the People": *LDC*, I, 32.
"indignation . . . Applicable": *LDC*, I, 34.

56 "By g-d": *LDC*, I, 70.
"arbitrary will . . . no longer": *JCC*, I, 32–37.

57 "happiest days": *D&A*, II, 134–35.
"figurative Panegyricks": *LDC*, I, 130.
"Each others Language": *AFC*, I, 163.
"intemperate . . . attended to": *LDC*, I, 100.

58 "live wholly without": *LDC*, I, 106.
"Southern Gentlemen": *LDC*, I, 59.
Galloway offered: *LDC*, I, 117–18.

59 "no means remarkable": *LDC*, I, 120.
"instantly relieved": *LDC*, I, 133.
"cheerfully complied": *LDC*, I, 134.

60 "able, willing . . . Our power": *LDC*, I, 158–61.
"safe asylum": *LDC*, I, 161.
"necessary expenses": *JCC*, I, 54.
"the horror": *JCC*, I, 60–61.

61 "a great Man": *AFC*, I, 164.

62 "cheerfully consent": *JCC*, I, 68–69.
"whole art of government" . . . over another: *PTJ*, I, 134–35.

63 "may it please . . . our complaints": *JCC*, I, 115–20.
"the happy, the peacefull": *D&A*, I, 157.

64 "fears and apprehensions": *AFC*, I, 172.
"missed Fire," "a vagabond," and "Country People": *Correspondence of Gage*, I, 380.
"The People would cool": *Correspondence of Gage*, I, 382.
"prison of Boston": *LDC*, I, 552.

65 "pressing Invitations": *D&A*, II, 159.

66 "beat up a Breeze": *WSA*, III, 206.

68 "All Protestantism": Quotations from Burke's speech on conciliation in this and following paragraphs are taken from the excerpt in *FC*, I, 3–6.

2. The Revolt of the Moderates

72 "want of some Person . . . sick of the subject": *JJ*, I, 138–39.
"plunge[d] into the midst": *LDC*, I, 280.
"Trade of Patriotism": *LDC*, I, 173.

73 "decisive & determined": *LDC*, I, 515.
"inexpressible Pain": *LDC*, I, 301–2.

74 "bask in the sunshine": *FC*, I, 343.
"Men of property": *LDC*, I, 454.

77 "*Christian* Sparta": *WSA*, IV, 238.

81 "America's first hit song": Calvert, *Quaker Constitutionalism*, 211.

82 "Coach and four": *D&A*, II, 117.
"beauty full Prospect" and "Sweet Communion": *D&A*, II, 133, 137.

83 "scrawls": *JJ*, I, 125.

84 "horrid Opinion": *D&A*, II, 151.
"ministerial plan": *JCC*, I, 87–89.

86 "human Policy . . . first Act": *LDC*, I, 332.
"Persons appearing": Quoted in Rakove, *Beginnings of National Politics*, 71.

87 "Glorious Appearance . . . risk your life": *LDC*, I, 339–41.

88 "proposals for raising . . . Regulating of Trade": *LDC*, I, 351–52.
"act on the defensive . . . insult and injury": *JCC*, II, 52.
"has been converted" and "a more total revolution": *LDC*, I, 343, 367.
"exposed to the ravages": *LDC*, I, 394.
"Dishonor to their Arms": *LDC*, I, 374–75.

89 "that Bitter Cup": *LDC*, I, 377.
"single Iota . . . Parent State": *LDC*, I, 378.

90 "suited as nearly": *LDC*, I, 384.
"Bill of Rights . . . Congress withdraw'· *LDC*, I, 401–2.
"measures be entered into": *JCC*, II, 66.

91 "an honour": *LDC*, I, 528.
"Pride and Pomp": *LDC*, I, 537.
"strong Jealousy": *LDC*, I, 497.
"piddling Genius": *LDC*, I, 658.

92 "Chesnut Street": *D&A*, II, 173.

93 "most happy effect": *LDC*, II, 285–86.
"never to have Salt Petre": *LDC*, II, 226.

94 "sole purpose": *JCC*, III, 392.
"The rebellious war": The king's speech is quoted in Merrill Jensen, *The Founding of a Nation* (New York, 1968), 649.

95 "Small islands not capable": Eric Foner, ed., *Thomas Paine: Collected Writings* (New York, 1965), 28.
"eating is the custom": *Paine: Writings*, 22.
"interested men": *Paine: Writings*, 25–26.
"ungenerous & groundless": *LDC*, III, 175–77.
96 "If Parliament": *LDC*, III, 33.
97 "all our Vessells": *LDC*, III, 244.
"limited to terms" and "can only Treat": *LDC*, III, 269, 366.
"very long, badly written": *LDC*, III, 232.
98 "Where the plague . . . path": *LDC*, III, 495.
"intestine division": *LDC*, V, 412.
99 "I confess": *LDC*, IV, 146–47.
"tollerably well": *LDC*, III, 600–2.
100 "Novelty of the Thing": *LDC*, III, 529.
101 "equitable Terms": *JCC*, IV, 143.
"thank herself": *LDC*, IV, 146–47.
102 "Every Post . . . deceitfull Expectations": *LDC*, IV, 40–42.
103 "utterly reject": *LDC*, II, 319–20.
"Even brutes": *Paine: Writings*, 23.
"finishing Blow": *LDC*, IV, 352.
"no youthful Lover": *New-York Historical Society Collections*, 11 (1878), 29.
104 "Purity of my Intentions . . . little Power": Dickinson to an unknown recipient, draft, Elizabethtown, N.J., Aug. 25, 1776, R. R. Logan Collection, box 1, folder 12, Historical Society of Pennsylvania.
"caused division": *LDC*, IV, 511.
105 "interest & inclination . . . Judgements & wishes": *LDC*, IV, 511–12.
"masterly Understanding": *LDC*, III, 587.
106 "Our people . . . a Blow": *LDC*, V, 624.
107 "veriest Slave . . . Estates to loose": *LDC*, VI, 87–88.
"very usefull . . . about the matter": *LDC*, VI, 216.
"If the Congress": *LDC*, V, 609.
108 "actual imprisonment": *LDC*, II, 514.
"Sensible part": *LDC*, IV, 174.
"so long deprived": *LDC*, V, 225.
"horrid consideration . . . the Question": *LDC*, V, 224.
109 "passed the Rubicon": *JJ*, I, 348.
"spend the remainder . . . Mother of Slaves": *JJ*, I, 352–53.
"King of Heaven": [John Jay], *Address of the Convention of the Representatives of the State of New-York to Their Constituents* (Fish-Kill, 1776), 10.
"the roads will permit": *JJ*, I, 229.
"lazy he is": *JJ*, I, 500.
110 "little party politicks" and "such rugged Times": *JJ*, I, 550, 557.

3. The Character of a General

112 "long been known": *LDC*, V, 648.
"vain to ruminate": *PGW: RWS*, VII, 439–41.
"ravages in the Jersies": *PNG*, I, 375.
113 "all of Opinion . . . so much Freedom": *PGW: RWS*, VII, 414–16.
"For heaven's sake": *PGW: RWS*, VII, 423.
114 "existence of our Army . . . a ploughshare": *PGW: RWS*, VII, 382.
115 "perfect reliance": *JCC*, VI, 1040, 1043–46.
"Congress never thought": *LDC*, VI, 541.
"Happy it is": *PGW: RWS*, VII, 495.
"We rejoice": *PGW: RWS*, VII, 485.
116 "Oh, Heaven!": *D&A*, II, 265.
117 "gladly give": *PGW: CS*, X, 23–24.
"situation & Circumstances": *PGW: RWS*, VII, 466.
118 "lordly Masters": *GW: Writings*, 130.
"Not that we approve": *PGW: CS*, X, 96.
"anything to be expected": *PGW: CS*, X, 109–10.
119 "countenance": *LDC*, I, 61.
"sundries": *PGW: CS*, X, 166–67.
"This Appointment": *LDC*, I, 497.
"firmly Cements": *LDC*, I, 499–500.
120 "trust too great": *PGW: RWS*, I, 3–4.
"Imbarkd . . . political motive": *PGW: RWS*, I, 12–13, 15–17, 19.
"dainty food . . . Repenting Sinners": *PGW: RWS*, IV, 412; III, 288, 569–70; V, 381, 424.
121 "often thought . . . own Conscience": *PGW: RWS*, III, 89.
"not sufficient": *PGW: RWS*, III, 569.
"flower of the British troops": *PGW: RWS*, III, 24.

122 "dirty and nasty": *PGW: RWS*, I, 335, 372.
"dearth of Publick Spirit . . . accept this Command": *PGW: RWS*, II, 446, 449; III, 287.
"His Excellency": *PNG*, I, 163–64.
"abuses in this army": *PGW: RWS*, I, 99–100.
123 "not only Soldiers": *PGW: RWS*, I, 119.
"while the Enemy": *PGW: RWS*, I, 128.
"now the Troops": *PGW: RWS*, I, 50, 54.
"This day giving": *PGW: RWS*, III, 1.
"nor can they": *GW: Writings*, 543–44.
124 "Raw, and undisciplined . . . curry favour": *PGW: RWS*, III, 275, 238.
125 "Life, Liberty": *PGW: RWS*, III, 13–14.
"A Soldier reasoned": *PGW: RWS*, VI, 394–96.
"Gentlemen, and Men": *PGW: RWS*, VI, 395–96.
"Sloth, negligence": *PGW: RWS*, VI, 464.
126 "shameful inattention": *PGW: RWS*, VI, 462.
"fruitless experiments": *PGW: RWS*, VI, 463–65.
"free correspondence": *PJA*, IV, 427–28.
127 "unless two or three": *PJA*, IV, 498.
128 "the General was enraged": John Rhodehamel, ed., *The American Revolution: Writings from the War of Independence* (New York, 2001), 220.
"How much better": *PGW: RWS*, VI, 553.
129 "The enemy wait": *PGW: RWS*, VII, 412.
130 "the game": *PGW: RWS*, VII, 370.
"the impossibility of serving . . . to loose my Character": *PGW: RWS*, VI, 441–42; VII, 105.
"be done with the subject": *PGW: RWS*, VI, 442.
131 "boldly stepping forth": *PGW: RWS*, VII, 497.
"scatters our Armoury . . . gone tomorrow": *PGW: RWS*, VIII, 126, 439.
132 "total depression": *PGW: RWS*, IX, 128–29.
133 "Howes present Plan": *PJA*, V, 221.
"Robbing, Plundering": *PGW: RWS*, X, 130.
"long enough": *LDC*, VII, 532.
135 "distress": *PGW: RWS*, XI, 103.
"extreamly well armed . . . cock their Hats": *LDC*, VII, 533–34, 538–39.
"that fop": *LDC*, VII, 565–66.
136 "his true Designs": *LDC*, VII, 565–66.
137 "a place of safety": *PAH*, I, 326–28.
"active masterly Capacity": *D&A*, II, 265.
"further Reinforcement": *PGW: RWS*, XI, 339.
"This Army": *PGW: RWS*, XI, 373.
138 "stretched in the doorways": *PGW: RWS*, XI, 399.
"to Genl. Gates": *PGW: RWS*, XII, 60–61.
"depredations & ravages": *PAH*, I, 353–54.
139 "immortaliz'd himself": *PNG*, II, 195.
"rendered Phila.": *PGW: RWS*, XII, 426.
140 "ignorant & impatient populace": *PGW: RWS*, XII, 377–79, 414–17.
141 "felt myself greatly embarrassed": *PGW: RWS*, XII, 606.
142 "infinite pleasure": *PGW: RWS*, XII, 669–70.
"great and capital change": *PGW: RWS*, XII, 683–84.
"lodging complaints . . . as we have done this": *PGW: RWS*, XII, 686–87.
144 "gaming is again creeping": *PGW: RWS*, XIII, 171.
"reforming the abuses": *JCC*, X, 40.
145 "important alterations": *PGW: RWS*, XIII, 377ff.
"see a defeat": *PNG*, II, 247.
"glow of patriotism": *PGW: RWS*, XIII, 424–25.
146 "interest is the governing principle . . . greatest merit": *PGW: RWS*, XIII, 377–78; XIV, 573–74.
"numerous bickerings": *PGW: RWS*, XIII, 388.
"Line of splendor": *PNG*, II, 307.
147 "He told me": *LDC*, X, 27 n.7.
148 "our Affairs": *LDC*, VIII, 329.
"the *jealousy* . . . and Fortitude": *PGW: RWS*, XIV, 577–78.
"fatal Consequences" and other quotations in this and next paragraph: *LDC*, VIII, 57–61.
150 "poison the minds": *PGW: RWS*, XIV, 547.
"secure, fortified Camp": *PGW: RWS*, XIV, 567.
"remaining on the defensive": *PNG*, II, 384.

151 "strange kind of entertainment": *American Revolution: Writings*, 428–29.
152 "The commander in Chief": *PNG*, II, 451.
153 "Folly, caprice": *PAH*, I, 425–27.

4. The First Constitution Makers

157 "Knowing how little": *PGM*, II, 488.
159 "real constitution": *PJA*, I, 165–66.
160 "No government . . . *per bookum*": Jack P. Greene, *Peripheries and Center: Constitutional Development in the Extended Polities of the British Empire and the United States* (Athens, Ga., 1986), 65.
"Conquest or usurpation": "Of the Original Contract," in David Hume, *Essays Moral, Political, and Literary*, Eugene F. Miller, ed. (Indianapolis, 1985), 471.
161 "inexhaustible variety": Adams, *Thoughts on Government* (1776), in *FC*, I, 108.
162 "of a slow-fever . . . Attom in her Frame": *PGM*, I, 481–82.
163 "scarce able": *PGM*, I, 433.
"Man's Rank": *PGM*, I, 173.
"often wondered": *PGM*, III, 888.
"paced the rooms": quoted in Helen Hill Miller, *George Mason: Gentleman Revolutionary* (Chapel Hill, 1975), 100.
"all of my stock": *PGM*, I, 151.
164 "Every Gentlem[a]n . . . Brute Creation": *PGM*, I, 173.
165 "ancient usage": *PGM*, I, 178 n.12.
"every body's attention . . . glorious Commonwealth": *PGM*, I, 190.
166 "a premeditated Design": *PGM*, I, 204.
"seldom medled": *PGM*, I, 71.
"Content thyself": Joseph Addison, *Cato*, act IV, scene 4.
167 "a good deal pressed . . . universal NO": *PGM*, I, 241, 250.
"the Bablers": *PGM*, I, 255.
169 "first Bottle": *PGM*, I, 430.
"curious Problem . . . Government of a Colony": *LDC*, II, 347–48.
170 "the sneers . . . can contrive?": *FC*, I, 108, 110.
"poor Scrap . . . the Barons": *LDC*, V, 3; III, 431.
171 "in miniature" and "equal interest": *FC*, I, 108.
Congress adopted: *JCC*, IV, 342.
"a work of the most interesting nature": *PTJ*, I, 292.
172 "smart fit . . . their Hearts": *PGM*, I, 271.
"political Cooks": *PTJ*, I, 296.
173 "a dissolution of government": Locke quotations taken from Peter Laslett, ed., *Two Treatises of Government* (New York, 1988), 408–9, 411, 407.
176 "the balance of a great": *PTJ*, I, 134–35.
178 "no candid Man . . . of the People": *PGM*, I, 174 n.8.
"countries so sterile": *FC*, I, 672.
179 "Contemptible little Tract": *LDC*, III, 667.
"deputies, or sub-electors": *PGM*, I, 299–300.
"common interest": *PGM*, I, 300.
181 "confined to those" and "permanent intention": *PTJ*, I, 489–90, 504–5.
182 "Suspicious Characters": *To the Several Battalions of Military Associators in the Province of Pennsylvania* (Philadelphia, 1776).
183 "all Men" and "enormous Proportion": *PBF*, XXII, 530–33.
185 "Age of political Experiments": *LDC*, V, 302.
"not hold ourselves bound": *AFC*, I, 370.
"fruitfull a Source": *PJA*, IV, 211–12.
187 "very sensible Gentleman": *D&A*, II, 134.
188 "no free goverts. . . . how will they finish?": *DPDC*, II, 952–53.
"very bad govert. . . . in despotism": *DPDC*, II, 956–57.
"mean to do . . . science of Legislation": *DPDC*, II, 959.
189 "Surpass in Iniquity . . . Private Station": *DPDC*, II, 974–75.
190 "such great revolutions": *DPDC*, II, 976.
"a second Ireland": *DPDC*, II, 976.
"save a third": *DPDC*, II, 988.
191 "redeemed in time": *DPDC*, II, 1007–8.
192 "well timed delays": Livingston to William Duer, June 12, 1777, Robert R. Livingston Papers, New-York Historical Society.
"The same body": Oscar Handlin and Mary Handlin, eds., *The Popular Sources of Political Authority: Documents on the Massachusetts Constitution of 1780* (Cambridge, 1966), 153.

193 "laborious Piece": *PJA*, VIII, 140.
194 "innocently intended": *PJA*, VIII, 212.
"Constitution monger": *PJA*, VIII, 279.
"Men of Wealth": *PJA*, VIII, 276.
195 "the Moses": *LDC*, V, 161.
"fatal Experience" and "divide and distract" *PJA*, V, 221, 174.
"two thirds": Handlin, ed., *Popular Sources*, 432–33.

5. *Vain Liberators*

200 "wild inconsistent claims": *The Dying Negro* (London, 1775), ix.
201 "True sincere love of Country": *PHL*, IX, 354.
202 "all men are *by nature*": *PGM*, II, 287 (italics indicate additions to original wording).
"I think we Americans": Quoted in Massey, *John Laurens*, 63.
203 "augur misgovernment": *FC*, I, 5.
"noisome smell" and "putrefaction": *South Carolina Gazette*, June 1 and 8, 1769, quoted in Robert Olwell, *Masters, Slaves, and Subjects: The Culture of Power in the South Carolina Low Country* (Ithaca, 1998), 223.
"bare mention": *PHL*, VII, 544.
204 "four Sons . . . his Family": *PHL*, IX, 309.
"charging me": *PHL*, I, 179–80, 182–83, 188.
205 "Sheer Envy": *PHL*, IX, 336.
"directly from *Africa*": *PHL*, I, 241–42.
"very pretty order": *PHL*, I, 317.
"contrary to the opinion": *PHL*, I, 307–8, 318–19.
"Never put your Life": *PHL*, VI, 438.
206 roughly 60,000 slaves: Slave populations from Morgan, *Slave Counterpoint*, 61.
a typical profit: For discussion, see *PHL*, VI, 81–82, and 82 n.5; Kenneth Morgan, "Slave Sales in Colonial Charleston," *Economic History Review*, 113 (1998), 913–14, 924–26.
"poor Protestant Christians": *PHL*, VI, 149–50.
"poor pining creature": *PHL*, VII, 192.
207 "a piece of service": *PHL*, VII, 282–83.
"often where a Man" and "wishd that our oeconomy": *PHL*, III, 356, 373–74.
"insensibly drawn" and "plan of life": *PHL*, V, 668; VII, 170, 172.
208 "two careful midling hands": *PHL*, IX, 262.
"reflection is comfortable": *PHL*, V, 669.
"jealousy & disquiet": *PHL*, III, 248.
"old Domestics": *PHL*, X, 2–3.
"hot weather": *PHL*, VII, 468.
"Rice will bear": *PHL*, VIII, 9, 11.
209 "a Candle": *PHL*, VIII, 147.
"clumsy close Chaise . . . in Sculpture": *PHL*, VIII, 368; VII, 553–54.
"friendly attention": *PHL*, VIII, 376.
"ought not to abandon myself": *PHL*, VIII, 450–51.
"grief at heart" and "Satisfied": *PHL*, VIII, 463, 525.
210 "a Talkative Gentleman . . . last century": *PHL*, IX, 168.
211 "put to trial . . . we are advancing to": *PHL*, IX, 266–69.
"So wise": *PHL*, IX, 277.
"treating with propriety . . . Retraction on this Side": *PHL*, IX, 300–1.
"This resolution . . . forlorn Americans": *PHL*, IX, 401, 415.
212 "Boston is pitched" and "Power can support": *PHL*, IX, 387, 364.
"a good platform": *PHL*, IX, 366–67
"3000 Strokes . . . of Truth & Virtue": *PHL*, IX, 223–26.
213 "so stiff": *PHL*, IX, 478–79.
"wash your Mouth": *PHL*, IX, 633, 639, 619.
"like a Bird": *PHL*, X, 58–59.
"a Croud . . . Life more cheap": *PHL*, IX, 650, 647.
"Reserve your Life": *PHL*, X, 57.
"the pleasure of embracing you": *PHL*, X, 139–41.
214 "most unhappy of Sons": *PHL*, X, 451–54.
"I will obey you": *PHL*, X, 512.
"dishonour": quoted in Massey, *John Laurens*, 55.
"half a Soldier": *PHL*, XI, 13.
215 "Even at this Moment . . . before me": *PHL*, XI, 234.
"devising means . . . West Indies": *PHL*, XI, 223–24.
"not the man who enslaved them": *PHL*, XI, 224–25.

216 "long and comical Story" and "until we meet": *PHL*, VIII, 353, 435–36.
217 "Without Slaves": *PHL*, XI, 275–77.
"the more important Engagements": quoted in Massey, *John Laurens*, 68.
"Dear Daughter": *PHL*, XI, 348.
218 "Talents & his diligence": *PHL*, XI, 424.
"promised Support": *PHL*, XI, 459.
"a great Assembly": *PHL*, XI, 445–46.
"useless Service": *PHL*, XI, 487–89.
219 "loose Tongues": *PHL*, XI, 554–55.
"not his fault . . . Courage & temerity": *PHL*, XI, 546–49.
220 "If I may judge": *PHL*, XII, 344–45.
"my mind": *PHL*, XII, 274–75.
"the upper Road": *PHL*, XI, 454.
221 "running whole days": *PHL*, XII, 225.
"diabolical motto": *PHL*, XII, 118, 112.
"prompters & Actors": *PHL*, XII, 270–71.
"You are one": *PHL*, XII, 165.
"so much influence": *PHL*, XIII, 52.
"augment the Continental Forces": *PHL*, XII, 305.
222 "means of restoring": *PHL*, XIII, 52.
"not refuse this": *PHL*, XII, 328.
"reproach of Quixotism": *PHL*, XII, 368.
223 "that monster popular Prejudice . . . his own escape": *PHL*, XII, 390–92.
"long deplored . . . Public Good": *PHL*, XII, 392.
"set them at full liberty . . . whole Nations": *PHL*, XII, 412–13.
224 "this excentric Scheme": *PHL*, XII, 446–47.
"Have you consulted" and "He is convinced": *PHL*, XII, 368, 392.
"a Distinction": *PGW: RWS*, II, 125.
225 "discharge all the Negroes": *LDC*, II, 67.
"served faithfully": *JCC*, IV, 60.
"a Battalion of Negroes": *PGW: RWS*, XIII, 114, 125.
226 "rapidity and indecision": *PHL*, XIII, 532–37, 545–46.
"tempt the fates": *PHL*, XIV, 30, 80.
227 "expected plunder . . . that frontier": *PHL*, XIV, 361–63, 545.
228 "his brave conduct": Congressional resolutions in *JCC*, XII, 1105–7.
229 "Negroes sufficient": *LDC*, XI, 494–95.
230 "greatest source of Danger": *LDC*, XI, 538.
"my black project . . . enjoy it himself": *PHL*, XV, 59–60.
"spend as much time . . . anxious": *PHL*, XV, 64–65; John C. Fitzpatrick, ed., *The Writings of George Washington* (Washington, 1931–1944), XIV, 245–46.
231 "very proper soldiers": *PAH*, II, 17–18.
"unable to make any effectual efforts" and other excerpts from the various drafts of this report: *LDC*, XII, 242–44, 246–47; *JCC*, XIII, 386–89.
232 "lay a foundation": *LDC*, XII, 258.
233 "noble proposal": *LDC*, XII, 281.
"arms for 3000": *PHL*, XV, 66.
"policy of our arming Slaves . . . my thoughts": *Washington: Writings*, XIV, 267.
234 "pleasureable amusement": *PHL*, XV, 106.
235 "when solicited": *LDC*, XIII, 188–89.
"a Demosthenes": *PAH*, II, 102–3.
"long since foresaw . . . South Carolina": *PHL*, XV, 169–70.
236 "black Air Castle" and "animated and persuasive eloquence": *PHL*, XV, 177; *PAH*, II, 166–67.
237 "might have accepted": *PHL*, XV, 193–94.
"severest conflict": *PAH*, II, 230–31.
"determined to pursue": *PHL*, XV, 227.
"procure the levy": *PHL*, XV, 232–33.
238 "natural tardiness": *PHL*, XV, 300, 331, 333.
239 "a bounty of a Negro" and "instruments for enlisting": *PJM*, II, 182–83, 209.
240 "concerning Slaves": *PTJ*, II, 470–71.
"emancipate all slaves": *TJ: Writings*, 264–70.

6. The Diplomats

242 "silently" and "huzza'd": *AFC*, III, 224 n.1, 234.
243 "Mr. Jays name": *PJA*, VIII, 214.
"very private": *PHL*, XV, 338–39.
"a brisk gale . . . verdant, romantic": *JJ*, I, 680–86.
244 "It was impossible": *JJ*, I, 698–702.
"if You and all the family": *AFC*, III, 243.
245 "Eels, Sardines . . . mounted on stilts": *D&A*, II, 407, 413, 424.

247 "deliberating Executive assembly": *LDC*, XIV, 109.
249 "plentifull Abuse . . . true as well as false": *D&A*, IV, 72–74.
255 "Suppose all *their* Complaints": *PBF*, XX, 398.
"noble China Vase": *PBF*, XXII, 520.
"leaning on the Bar" and "*Hereditary* Legislators": *PBF*, XXI, 581–82.
256 "base Reflections": *PBF*, XXI, 598.
258 "a separate Peace": *PBF*, XXV, 305–7.
259 "so excentric": *PBF*, XXV, 436–39.
"your Nation is hiring": *PBF*, XXV, 650–51.
"chuckling, laughing": *PBF*, XXI, 112–13.
260 "too large"; "man of spirit"; "French blood": John J. Meng, ed., *Despatches and Instructions of Conrad Alexandre Gérard, 1778–1780* (Baltimore, 1939), 431, 424–25, 433–34.
261 "speak confidentially": Meng, ed., *Despatches*, 554.
dinner chez Gérard: *Despatches*, 779–83.
"as much Intrigue": *JJ*, I, 588.
"undertake to attend": *JJ*, I, 637.
262 "Why was I born": *AFC*, III, 233.
"Wisdom of Solomon" and "greatest Relief": *D&A*, II, 347, 353.
"his Name is so great": *PJA*, VIII, 213.
263 "a Passion for Reputation": *D&A*, II, 367.
"We live upon good Terms": *PBF*, XXXII, 186.
264 "young and virtuous . . . more obliged": *PBF*, XXXIII, 162–63.
"Air of Amsterdam": *AFC*, III, 424.
"difficult to discover": *PJA*, X, 435.
265 "old Goody . . . continued fit": *JJ*, I, 703, 709–10.
"What can I fear": *JJ*, I, 711.
"faithless and dangerous": *JJ*, I, 109.
267 "tie you up": *JCC*, XX, 651–52.
"base Jealousy": *PJA*, XIII, 189.
271 "singular Circumstance": *PBF*, XXXVI, 115.
"a Lillipution": *PJA*, XII, 134.
"will evaperate": *AFC*, IV, 267.
272 "full house": *PBF*, XXXVI, 621–22.
"dishonourable" and "Goodnight": Ian R. Christie, *The End of North's Ministry* (London, 1958), 364–65, 369.
273 "send Agents": *PJA*, XII, 305.
274 "Spain has taken": *PBF*, XXXVII, 198.
"want your Advice": *PBF*, XXXVII, 178.
"The American Cause": *AFC*, IV, 325.
"Signal Epocha": *PJA*, XII, 444–45.
275 "has been actuated": *PJA*, XIII, 189.
"hooping cough": *JJ*, II, 461, 464.
276 "Shelburne would not": *PJA*, XIII, 188–89.
"deep wounds": *JJ*, II, 286–87.
277 "Jay was a Lawyer": *JJ*, II, 301.
"hate us": *PJA*, XIII, 237–38.
"never to my Knowledge": *PJA*, XIII, 414.
278 "make friends . . . *our construction of it*": *RDC*, VI, 30–31.
"Facts and future events" and "prudence and self-possession": *RDC*, V, 740; Henry P. Johnston, ed., *Correspondence and Public Papers of John Jay* (New York, 1890–1893), III, 14.
"can it be wise": *RDC*, V, 373.
279 "Those Chains": *PJA*, XIII, 524–25.
"famous Altar Piece . . . I could not comprehend": *AFC*, V, 33; *D&A*, III, 32–33, 35–36.
"last comer": *D&A*, III, 40 n.
280 "remarkably attentive . . . la Negotiation": *D&A*, III, 47–50.
281 "The Dr. heard me": *D&A*, III, 82.
282 "no Notion": *D&A*, III, 43–44.
"making common cause . . . Ruin on the Nation": *D&A*, III, 48–49.
283 "the Tories": *D&A*, III, 51–52.
"brought on and encouraged . . . those People": *PBF*, XXXVIII, 350–56; *D&A*, III, 72, 75.
284 "Blossoms fall" and "our Fishermen": *D&A*, III, 73.
"recover their Estates . . . carry off no Negroes": *D&A*, III, 82.
"private letters": *PBF*, XXXVIII, 68–69.
285 "in some Respects . . . from the Distance": *PBF*, XXXVIII, 410–11.
"not a moral people": *D&A*, III, 46–47.
286 "extremely liberal . . . such allies to deal with": *PJM*, VI, 328–29.
"a tragedy to America": *PJM*, VI, 362, 358, 384.
288 "ask the King": *PBF*, XXXVIII, 461–62.
"more sensible than I am": *BF: Writings*, 1061.

"behind any of them": *BF: Writings*, 1079.

289 "He means well": *BF: Writings*, 1065.

7. *The Optimist Abroad*

293 "courageous Philosophers" and "Sheaves of Straw": *PBF*, XL (not yet published): from Franklin to Sir Joseph Banks, Nov. 21, 1783.
"anchored in the Air . . . from the Clouds": *Writings of Franklin*, IX, 83, 156.

294 "a common Carriage": *Writings of Franklin*, IX, 123.
"if the Balloon": *AFC*, V, 236.
"Don't you begin . . . pursuits of Commerce": *JJ*, II, 647, 649–51.
"run over their laces": *PTJ*, VI, 542.
"phaenomena . . . in a baloon": *PTJ*, VII, 136.
"2 tickets": James A. Bear Jr. and Lucia C. Stanton, eds., *Jefferson's Memorandum Books, 1767–1826* (Princeton, 1997), I, 548 and n.33.

295 "with certainty": *PTJ*, VII, 358.

296 "a Martyr" and "most violently": *PTJ*, VII, 538–39, 441.
"pour balm" and "suffered as much": *PTJ*, VII, 635; VII, 441–42.
"A Smile from you": *JJ*, II, 647.

297 "holes of different shapes": Thomas Jefferson Papers, Library of Congress.

298 "Negro problem": Gunnar Myrdal, *An American Dilemma: The Negro Problem and American Democracy* (New York, 1944, 1964), lxix–lxx.

299 "distinguished love . . . the American cause": *PTJ*, I, 522–24.
"Every letter": *PTJ*, I, 483.
"in the persuasion": *TJ: Writings*, 37.
"all men" and "preservation": *PTJ*, I, 423.

300 "ease, comfort, security": *FC*, I, 108.
"common reader . . . principal figure": *PTJ*, I, 74–80.
"unchequered happiness": *TJ: Writings*, 46.

301 "should still believe": *PTJ*, VII, 559.

302 "legislative drafting bureau": *PTJ*, VI, 306.
"foreigners" and "advantageous": *PTJ*, I, 558–59, II, 210 (for Mediterranean immigrants).

303 "those who entertain": *PTJ*, I, 548.

304 "heard little": *PTJ*, II, 230.
"far distant": *GW: Writings*, 332.
"Your Country": *PTJ*, II, 22.

305 "so Fruitfull": *FC*, I, 395.
"every person . . . an appropriation": *PTJ*, I, 362.
"permanent intention": *PTJ*, I, 504.

306 "encourage Marriage": *PTJ*, II, 140.
"publick happiness . . . accidental condition or circumstance": *PTJ*, II, 526–27.

307 "putting the Bible": *TJ: Writings*, 273.

308 *administrator: PTJ*, I, 360.
"enemy were ascending": *PTJ*, IV, 260–61.

309 "for the public good": Laslett, ed., *Two Treatises*, 375 (§160–61).
"unfortunate passages": *PTJ*, VI, 104–8.
"combining public service": *PTJ*, VI, 112.
"taken my final leave": *PTJ*, VI, 117–18.

310 "inflicted a wound": *PTJ*, VI, 185–86.
"All my plans": *PTJ*, VI, 198.
"wiped away": *PTJ*, VI, 203.
"change of scene": *TJ: Writings*, 46.
"His vanity": *PTJ*, VI, 241.

311 "Can any man": *PTJ*, VI, 248–49.
"the best tutors": *PTJ*, VI, 355.

312 "each individual": *PTJ*, I, 133.
"All the world . . . Western world": *PTJ*, VII, 26–27.

313 "thro' the Eastern states": *PTJ*, VII, 292.
"The Charters": *PTJ*, IV, 166–67.

314 "peruse it carefully": *PTJ*, VIII, 147–48.
"several judicious friends": *PTJ*, IX, 38.
"I fear the terms": *PTJ*, VIII, 229.
"formed when we were new . . . other legislatures": *TJ: Writings*, 243–50.

315 "pray for his death": *PTJ*, VII, 558.

316 "To emancipate all slaves": *TJ: Writings*, 264–70 for quotations in this and following paragraphs.

318 "*particular* customs" and other quotations in this paragraph: *TJ: Writings*, 288–89.
Don E. Fehrenbacher, ed., *Lincoln: Speeches and Writings, 1859–1865* (New York, 1989), 686–87.

319 "interesting spectacle": *PTJ*, VIII, 357.
"blotch of errors": *TJ: Writings*, 56.

320 "lay my bones there": *BF: Writings*, 1093.

"Amsterdam Fever": *AFC*, V, 124–26.
"whole tenor": *AFC*, V, 258.
"return to our Cottage": *AFC*, V, 237.
"Met my Wife": *D&A*, III, 170.

321 "such a phenomenon": *AFC*, VI, 171.
"Romain his companion": *TJP*, VIII, 245.
"The pleasure of the trip": *PTJ*, VIII, 233.

322 "polite, self-denying": *PTJ*, VIII, 239.
"I doubt whether": *PTJ*, VIII, 249.
"the most benevolent": *PTJ*, X, 244.
"best calculated for happiness . . . restlessness and torment": *PTJ*, VIII, 404, 568–69.
"careless jaunty": *AFC*, V, 437–38.

323 "contemplation of the hardness": *PTJ*, VIII, 404.
"finest soil": *PTJ*, X, 532.
"Surrounded by so many blessings": *PTJ*, X, 244–45.
"nineteen millions": *PTJ*, VIII, 404.
"If any body thinks": *PTJ*, X, 244.
"come occasionally" and other quotations in this paragraph: *PTJ*, VIII, 681–83.

324 "Here it seems": *PTJ*, VIII, 569.
"objects of an useful American education . . . promoted by them": *PTJ*, VIII, 636–37.
"poor compensation": *PTJ*, VIII, 409.
"Legislators cannot invent" and other quotations in this paragraph: *PTJ*, VIII, 682.

325 "by far the most important" *PTJ*, X, 244.
"in science": *PTJ*, VIII, 569.
"common sense": *PTJ*, X, 244.

326 "I am thoroughly cured": *PTJ*, VIII, 409.
"strongest of all the human passions": *PTJ*, VIII, 636–37.

327 "pyratical states" and "drop to a bucket": *PTJ*, IX, 468, 448.
"with derision": *PTJ*, IX, 446.
"enable me to estimate": *PTJ*, IX, 369.
"It will be long . . . come in Pilgrimage": *D&A*, III, 184–86.

328 "the article in which it surpasses": *PTJ*, IX, 445.
"the famous Boulton": *PTJ*, X, 609–10.
"an unlucky dislocation": *PTJ*, X, 602.

329 "lying Messengers": *PTJ*, X, 445.
"most effectual means" and other quotations in this paragraph: *PTJ*, X, 448–50.

330 "tedious sermon": *PTJ*, X, 452.
"promontory of noses . . . Italian singing": *PTJ*, XIII, 104.
"useless": *PTJ*, XI, 515.
"Architecture, painting": *PTJ*, XI, 515.

331 "Mascarponi, a kind of curd": *PTJ*, XI, 439.
"porters, carters": *PTJ*, XI, 446.
"gazing whole hours" and "filled with alarms": *PTJ*, XI, 226–28.
"market-women" and "how cheap a thing": *PTJ*, XI, 270–72.

332 "buying spree": Garry Wills, "The Aesthete," *New York Review of Books*, vol. 40, Aug. 12, 1993, 6–10.
"born to lose": *PTJ*, XI, 519–20.

333 "breakfast every day": *PTJ*, XI, 520.
"domestic cortege": *PTJ*, XII, 540.
"verbal jousting match": Joseph Ellis, *American Sphinx: The Character of Thomas Jefferson* (New York, 1997), 97. For the record, this seems to me a strangely titled book, given the voluminous and often quite revealing nature of Jefferson's massive correspondence.

335 "I confess": *PTJ*, XII, 350–51.
"very good": *PTJ*, XII, 356–57.

336 "piled upon" and "Above all things": *PTJ*, XII, 442.
"Compared with the authors": *PTJ*, XII, 557.
"Men, women, children" and "good sense": *PTJ*, XIII, 151.

337 "will elude": *PTJ*, XIV, 330.
"thorough reformation": *PTJ*, XIV, 330.
"most famous": Herbert Sloan, *Principle and Interest: Thomas Jefferson and the Problem of Debt* (New York, 1995), 50; "Whether one generation": *PTJ*, XV, 392.

338 "This principle": *PTJ*, XV, 396.

339 "my Congé . . . times and places": *PTJ*, XV, 305–6.
"Pray take me": *PTJ*, XV, 351.
"fruitless": *PTJ*, XV, 509.
"the first swallow": *PTJ*, XV, 521.

8. The Greatest Lawgiver of Modernity

341 "others of great genius": *JM: Writings*, 860.

342 "kill some time": Machiavelli's famous letter to Francesco Vettori, Dec. 19,

1513; for one of many translations, see *Machiavelli and His Friends: Their Personal Correspondence*, James B. Atkinson and David Sices, trans. and ed. (DeKalb, 1996), 262–65.

343 "the curiosity": *JM: Writings*, 840.

345 "sensations for many months": *PJM*, I, 75.

346 "Boston may conduct": *PJM*, I, 105.
"I have squabbled": *PJM*, I, 106.

347 "all men": *PJM*, I, 172–75, reprints the various drafts of the article on religion.
"personal solicitation": *PJM*, I, 193.

349 "a defect of adequate Statesmen": *PJM*, II, 6.

351 "by means perfectly constitutional": Archibald Stuart, as quoted in Ralph Ketcham, *James Madison: A Biography* (Charlottesville, 1971), 161.
"with great ease": *PJM*, VIII, 473.
"I flatter myself": *PJM*, VIII, 474.

352 "*wisdom* and steadiness": *PJM*, VIII, 350–52.
"A considerable itch": *PJM*, VIII, 477.

353 "the regulation of these varying": *Federalist* No. 10, in *JM: Writings*, 162.

354 "chose rather, to do nothing": *PJM*, VIII, 476–77.
"The expedient is no doubt liable": *PJM*, VIII, 483.

355 "temporizing or partial" and "present paroxysm": *PJM*, VIII, 505–6.

356 "partial experiment": *PJM*, IX, 55.
"Many gentlemen": *PJM*, IX, 96–97.

357 "to meet at Philadelphia": *AH: Writings*, 144.

358 "thought advisable": *PJM*, IX, 166.
"without giving offence": *PJM*, IX, 170–71.
"clogged with Ice": *PJM*, IX, 259.

359 "from the Measure": John Jay to John Adams, Feb. 21, 1787, Adams Family Papers, microfilm reel 368, Massachusetts Historical Society.
"arcana of futurity": *PJM*, IX, 294.
"postpone his actual attendance": *PJM*, IX, 378.

360 "found within the states": *PJM*, IX, 353–54.

361 "absolutely necessary": *PJM*, IX, 383.
"Contrary to the prevailing Theory": *PJM*, IX, 357.
"Whenever therefore an apparent interest": *PJM*, IX, 355–56.

363 "least possible encroachment": *PJM*, IX, 383.
"Want of sanction to the laws" and other quotations in the next four paragraphs: *PJM*, IX, 351–52.

365 "the ground work": *PJM*, IX, 383.
"their present populousness": *PJM*, IX, 383.

366 "a violent hd. ach" and "warmly and kindly pressed": Donald Jackson et al., eds., *The Diaries of George Washington* (Charlottesville, 1976–), V, 153–55.
"These delays": *RFC*, III, 22.
"two or three hours": *PGM*, III, 880.

367 "so corrected & enlarged" and "legislate in all cases": *RFC*, I, 20–21.
"delivered with much diffuseness": *RFC*, I, 445.

368 "profound & solemn conviction": *JM: Writings*, 841–42.
"such an attempt": *RFC*, I, 10–11.

369 "Whatever reason": *RFC*, I, 37.

370 "In point of manners": *RFC*, I, 447.
"You see the consequences": *RFC*, I, 242 n*.

371 "provide a Governmt.": *RFC*, I, 315–16.
"the great difficulty": *RFC*, I, 321.
"no particular abuses" and "danger of combinations": *RFC*, I, 484.
"every peculiar interest . . . Northern and Southern": *RFC*, I, 486.

373 "most exact transcript": *RFC*, I, 132.

375 "founded on the supposition . . . strangely misinterpreted": *RFC*, II, 19–20.

376 "time was wasted": *RFC*, II, 20.
"captivated by the compromise": *PTJ*, XII, 440.
"lesser evil": *Federalist* No. 62, in *JM: Writings*, 339.
"the most mild": *RFC*, II, 28.

377 "tedious and reiterated discussions": *PJM*, X, 208.

378 "be men": *RFC*, II, 100.
"drudgery": *PJM*, X, 105.

379 "to make war": *RFC*, II, 318–19.

380 "eventual choice": *RFC*, II, 527.

381 "may perhaps surprize you": *PJM*, X, 163–64 (italics indicate cipher).

382 "The State Conventions": *RFC*, II, 561.

383 "might be prepared": *RFC*, II, 587–88.

"retired to meditate": *Diaries of Washington*, V, 185.
"the best judges": *PJM*, X, 171.
384 "immoderate digression" and "evils issuing": *PJM*, X, 212–14.
"natural diversity": *PJM*, X, 208.
"The real wonder": *Federalist* No. 37, in *JM: Writings*, 200.
385 "newspapers here": *PJM*, X, 197.
386 "one wd. suppose": *PJM*, X, 199–200.
"rights of private judgment . . . designing men": *PJM*, X, 355–56.
387 "glory": *Federalist* No. 14, in *JM: Writings*, 172.
388 "inviolable . . . unavoidable inaccuracy": *Federalist* No. 37, in *JM: Writings:* 196–99.
389 "impetuous vortex": *Federalist* No. 48, in *JM: Writings*, 281.
"encroachments" and other quotations in this and next paragraph: *Federalist* No. 49, in *JM: Writings*, 286–90.
390 "involve me": *PJM*, X, 526–27.
"thrown himself fully" and "lame figure": *PJM*, XI, 77.
391 "bilious indisposition": *PJM*, XI, 134.
"parties continue": *PJM*, XI, 157.
392 "recommendatory alterations": *PJM*, XI, 181.
"There is no doubt": *PJM*, XI, 276.
393 "next in rank . . . premature advancement": *PJM*, XI, 296.
"never thought the omission": *PJM*, XI, 297.
"great reason to fear" and quotations in next two paragraphs: *PJM*, XI, 297–99.
395 "nauseous project": *PJM*, XII, 346–47.

9. *The State Builder*

396 "circumstances" and other quotations in this paragraph: *PAH*, IV, 275–77.
398 "upon mature reflection": William Loughton Smith to Edward Rutledge, June 21, 1789, *South Carolina Historical Magazine*, 69 (1968), 8.
"can there be a Govt.": *PAH*, III, 193.
"*heads or chief*": *PAH*, III, 208.
"Nations in general": *PAH*, III, 484–85.
"best in the world" and "unites public strength": *RFC*, I, 288.
399 "'Purge the constitution'": *TJ: Writings*, 670–71.
"Give to this influence": "Of the Independency of Parliament," in David Hume, *Essays Moral, Political, and Literary*, Eugene F. Miller, ed. (Indianapolis, 1985), 47.
"devoured": *TJ: Writings*, 1228.
401 "indecent and very suspicious": as quoted in Forrest McDonald, *Alexander Hamilton: A Biography* (New York, 1979), 6.
"grov'ling": *PAH*, I, 4.
402 "appointed Aide-De-Camp": *PAH*, I, 196.
403 "a stranger in this country . . . terrestreal Country": *PAH*, II, 254–55.
404 "She must neither love money": *PAH*, II, 37–38.
"confirmed to me": *PAH*, I, 176–77.
"private sentiments": *PAH*, I, 207–9.
405 "defects of our present system" and other quotations in this paragraph: *PAH*, II, 400–418.
"a little *nut brown maid*": *PAH*, II, 455.
406 "a conspicuous part": *PAH*, II, 509.
"'friendship and gratitude'": *PAH*, II, 517.
"'I must tell you'": *PAH*, II, 563–65, 569.
"for three years past . . . great importance to it": *PAH*, II, 566–67 (editing omitted).
"how you will" and "embarrassed": *PAH*, II, 601.
407 "my early entrance": *PAH*, II, 637.
"the extreme of vanity": *PAH*, II, 649.
"the business of finance": *PAH*, II, 673.
"endeavour by all means": *PAH*, II, 647.
"thought improper": *PAH*, II, 667.
408 "*a boy*": *PAH*, II, 678.
"honor obliged me": *PAH*, II, 682.
"Tomorrow Cornwallis": *PAH*, II, 683.
"the difficulties I met . . . public life": *PAH*, III, 67–68.
"my military situation": *PAH*, III, 89.
409 "The reason of allowing Congress": *PAH*, III, 105.
"whose ignorance": *PAH*, III, 139.
"has a method": *PAH*, III, 151.
410 "*to take the direction*" and "preserve the confidence": *PAH*, III, 253–55.
"becoming extremely unpopular" and "conductor of the army": *PAH*, III, 265–66.
"same steady line": *PAH*, III, 278.

411 "grown gray": Samuel Shaw to Rev. Eliot, April 1783, in John Rhodehamel, ed., *The American Revolution: Writings from the War of Independence* (New York, 2001), 788.
"within the bounds of duty" and "I confess": *PAH*, III, 306.
"I write to you unreservedly": *PAH*, III, 335–36.
"Continental views," "think continentally," and "local prejudices": *PAH*, III, 292, 321, 330.
412 "so little disposition": *PAH*, III, 376.
"occasionally uttering offensive words": *PJM*, VII, 176–77.
"insinuated": *PAH*, III, 408–9.
413 "too often compelled": *PAH*, III, 425.
"road to popularity": *PAH*, III, 417.
"decline public office": *PAH*, III, 482.
414 "acknowledged abilities": *PAH*, IV, 148.
"public sentiment" and other quotations in this paragraph: *PAH*, IV, 223–24.
415 "his dislike" and "No man's ideas": *PAH*, IV, 243, 253.
"well regulated": *Federalist* No. 29, in *AH: Writings*, 286–87.
416 "the wants of the states": *Federalist* No. 34, in *AH: Writings*, 311.
"a vigorous executive" and "decision, activity": *Federalist* No. 70, in *AH: Writings*, 374–75.
"the love of fame": *Federalist* No. 72, in *AH: Writings*, 389.
"the passions ought to be controlled": *Federalist* No. 49, in *JM: Writings*, 290.
"the prospect of being allowed to finish": *Federalist* No. 72, in *AH: Writings*, 389–90.
417 "I take it for granted": *PAH*, V, 201–2.
"the world and Posterity": *PAH*, V, 207.
"It would be inglorious" and other quotations in this paragraph: *PAH*, V, 220–22.
418 "those great revolutions" and other quotations in this paragraph: *PAH*, V, 348.
"the etiquette proper": *PAH*, V, 335–37.
419 "Hamilton is most talked of": *PJM*, XII, 271–72.
"arduous trust" and "always give me pleasure": *PAH*, V, 370–71.
"prepare a plan": *PAH*, VI, 66 n.99.
421 "to form a Commercial treaty": *PAH*, V, 482–83.
"I was much surprized": *PAH*, V, 488.
"the anxieties": *PAH*, VI, 66.
422 "plain and undeniable truths": *PAH*, VI, 67.
"a justifiable surmise": Charlene Bangs Bickford, ed., *Documentary History of the First Federal Congress* (Baltimore, 1972–), XII, 5–7.
424 "the voice of humanity": *PJM*, VI, 493.
425 "a long conversation": *PAH*, XI, 428.
"very considerable alienation": *PAH*, XI, 428.
"I go on the principle": *PJM*, XIII, 148.
"certainly be wrong" and "Novelty and difficulty": *PJM*, XIII, 147–48.
426 "dejected beyond description": *PTJ*, XVII, 205–7.
"backwards & forwards": *TJ: Writings*, 668.
"supporting one another": *PTJ*, XVII, 205–7.
"rally around" and "save the union": *TJ: Writings*, 668.
427 "determined to resign": *PTJ*, XVII, 206.
"a labyrinth": *PJM*, XIII, 252.
428 "such further provision": *PAH*, VII, 225 n.35.
"have my mind": *PAH*, VI, 557.
"As long as men": *PAH*, VII, 229.
"Public opinion being the ultimate arbiter": *PAH*, VII, 305–6.
430 "prudence of the President": *PTJ*, XIX, 263.
"chiefly for cases": *PTJ*, XIX, 280.
431 "uncommon solicitude": *PAH*, VIII, 97.
"every power vested": *PAH*, VIII, 98.
"the clerkship for foreign languages": *PTJ*, XIX, 351.
432 "ultimate arbiter": *PAH*, VII, 305.
"opinion of the public": *PTJ*, XIX, 263.
"Public opinion sets bounds": *PJM*, XIV, 170.
433 "pasionate courtship": *PAH*, VIII, 478.
"In Europe, charters of liberty": *PJM*, XIV, 191–92.
434 "partitions and internal checks": *PJM*, XIV, 218.
435 "real wish": *TJ: Writings*, 674.
"single object of revenue": *TJ: Writings*, 676.
"a single source of these discontents": *TJ: Writings*, 676–78.

436 "comes through so many channels": *PAH*, XI, 429–30.
"If I were disposed": *PAH*, XI, 444.
"corrupt squadron": *PTJ*, XXIII, 539.

437 "a decay of his hearing" and other quotations in this paragraph: *TJ: Writings*, 678–80.
"in strict confidence": *PAH*, XII, 129–34.
"introducing a monarchy": *PAH*, XII, 251–52.

438 "arraigns, the principal measures": *PAH*, XII, 190–91.
"wounding suspicions": *GW: Writings*, 817–20.
"greatest frankness": *PAH*, XII, 347–49.

439 "internal dissentions" and "deepest regret": *PTJ*, XXIV, 352.
"periodical change" and "not suffer my retirement": *PTJ*, XXIV, 357–58.

440 *"a womanish attachment"*: *PAH*, XI, 439.

441 "then to be liberated": *PTJ*, XXVII, 449.

442 "Those great revolutions": *PAH*, V, 348.

Sources and Further Reading

In the brief bibliography that follows, I list books that I have particularly relied upon in each of the chapters, and others that may interest general readers. The body of scholarly writings on the American Revolution is, of course, enormous, and the works listed here are only a small sample of a substantial literature.

Prologue: The World Beyond Worcester

Fred Anderson, *Crucible of War: The Seven Years' War and the Fate of Empire in British North America* (New York, 2000)

Bernard Bailyn, *The Ideological Origins of the American Revolution*, enlarged edition (Cambridge, 1992)

Ian Christie and Benjamin W. Labaree, *Empire or Independence: A British-American Dialogue on the Coming of the American Revolution* (New York, 1976)

Pauline Maier, *From Resistance to Revolution: Colonial Radicals and the Development of American Opposition to Britain, 1765–1776* (New York, 1972)

Edmund S. Morgan and Helen M. Morgan, *The Stamp Act Crisis: Prologue to Revolution*, rev. ed. (New York, 1963)

Chapter One: Advocates for the Cause

Bernard Bailyn, *The Ordeal of Thomas Hutchinson* (Cambridge, 1974)

Richard D. Brown, *Revolutionary Politics in Massachusetts: The Boston Committee of Correspondence and the Towns, 1772–1774* (Cambridge, 1970)

David Hackett Fischer, *Paul Revere's Ride* (New York, 1994)

Benjamin W. Labaree, *The Boston Tea Party* (New York, 1964)

Jack N. Rakove, *The Beginnings of National Politics: An Interpretive History of the Continental Congress* (New York, 1979)

Peter D. G. Thomas, *Tea Party to Independence: The Third Phase of the American Revolution, 1773–1776* (Oxford, U.K., 1991)

Chapter Two: The Revolt of the Moderates

Jane E. Calvert, *Quaker Constitutionalism and the Political Thought of John Dickinson* (New York, 2009)
Pauline Maier, *American Scripture: The Declaration of Independence* (New York, 1997)
Walter Stahr, *John Jay: Founding Father* (New York, 2005)

Chapter Three: The Character of a General

Wayne Bodle, *The Valley Forge Winter: Soldiers and Civilians in War* (University Park, Pa., 2002)
E. Wayne Carp, *To Starve the Army at Pleasure: Continental Army Administration and American Political Culture, 1775–1783* (Chapel Hill, 1984)
Caroline Cox, *A Proper Sense of Honor: Service and Sacrifice in George Washington's Army* (Chapel Hill, 2004)
David Hackett Fischer, *Washington's Crossing* (New York, 2004)
Ira Gruber, *The Howe Brothers and the American Revolution* (New York, 1972)
Don Higginbotham, *George Washington and the American Military Tradition* (Athens, Ga., 1985)
Edward G. Lengel, *General George Washington: A Military Life* (New York, 2005)
Paul Longmore, *The Invention of George Washington* (Berkeley and Los Angeles, 1988)
Piers Mackesy, *The War for America, 1775–1783* (Cambridge, 1964)
Stephen R. Taaffe, *The Philadelphia Campaign, 1776–1777* (Lawrence, Kans., 2003)

Chapter Four: The First Constitution Makers

Willi Paul Adams, *The First American Constitutions* (Chapel Hill, 1980)
Jeffrey Broadwater, *George Mason, Forgotten Founder* (Chapel Hill, 2006)
Pamela Copeland and Richard McMaster, *The Five George Masons: Patriots and Planters of Virginia and Maryland* (Charlottesville, 1975)
Ronald Hoffman, with Sally D. Mason, *Princes of Ireland, Planters of Maryland: A Carroll Saga, 1500–1782* (Chapel Hill, 2000)
Gordon S. Wood, *The Creation of the American Republic, 1776–1787* (Chapel Hill, 1969)

Chapter Five: Vain Liberators

Gregory D. Massey, *John Laurens and the American Revolution* (Columbia, S.C., 2000)
Daniel J. McDonough, *Christopher Gadsden and Henry Laurens: The Parallel Lives of Two American Patriots* (Cranbury, N.J., 2000)
Philip Morgan, *Slave Counterpoint: Black Culture in the Eighteenth-Century Chesapeake and Low Country* (Chapel Hill, 1998)
Henry Wiencek, *An Imperfect God: George Washington, His Slaves, and the Creation of America* (New York, 2003)

Chapter Six: The Diplomats

Edith B. Gelles, *Abigail and John: Portrait of a Marriage* (New York, 2009)
James C. Hutson, *John Adams and the Diplomacy of the American Revolution* (Lexington, Ky., 1980)
Richard Morris, *The Peacemakers: The Great Powers and American Independence* (New York, 1965)
Stacy Schiff, *A Great Improvisation: Franklin, France, and the Birth of America* (New York, 2005)
William C. Stinchcombe, *The American Revolution and the French Alliance* (Syracuse, 1969)
Gerald Stourzh, *Benjamin Franklin and American Foreign Policy* (Chicago, 1954)

Chapter Seven: The Optimist Abroad

William Howard Adams, *The Paris Years of Thomas Jefferson* (New Haven, 1997)
Annette Gordon-Reed, *The Hemingses of Monticello: An American Family* (New York, 2008)
Jon Kukla, *Mr. Jefferson's Women* (New York, 2007)
Peter S. Onuf, *Jefferson's Empire: The Language of American Nationhood* (Charlottesville, 2000)
Howard C. Rice Jr., *Thomas Jefferson's Paris* (Princeton, 1976)
Herbert Sloan, *Principle and Interest: Thomas Jefferson and the Problem of Debt* (New York, 1995)

Chapter Eight: The Greatest Lawgiver of Modernity

Lance Banning, *The Sacred Fire of Liberty: James Madison and the Founding of the American Republic* (Ithaca, 1995)
Ralph Ketcham, *James Madison: A Biography* (Charlottesville, 1990)
Stuart Leibiger, *Founding Friendship: George Washington, James Madison, and the Creation of the American Republic* (Charlottesville, 1999)
Drew McCoy, *The Last of the Fathers: James Madison and the Republican Legacy* (New York, 1989)
Jack N. Rakove, *James Madison and the Creation of the American Republic* (New York, 2006)
———, *Original Meanings: Politics and Ideas in the Making of the Constitution* (New York, 1996)
Colleen Sheehan, *James Madison and the Spirit of Republican Self-Government* (New York, 2008)

Chapter Nine: The State Builder

Max Edling, *A Revolution in Favor of Government: Origins of the U.S. Constitution and the Making of the American State* (New York, 2003)
John Lamberton Harper, *American Machiavelli: Alexander Hamilton and the Origins of U.S. Foreign Policy* (New York, 2004)
Forrest McDonald, *Alexander Hamilton: A Biography* (New York, 1979)
Gerald Stourzh, *Alexander Hamilton and the Idea of Republican Government* (Stanford, 1970)

Index